Applied Time Series and Box–Jenkins Models

Walter Vandaele

ACADEMIC PRESS, INC.

(Harcourt Brace Jovanovich, Publishers)

New York London Paris San Diego San Francisco

São Paulo Sydney Tokyo Toronto

To my three girls, A. A., and L. P.

Academic Press, Inc.
111 Fifth Avenue
New York, N.Y. 10003

United Kingdom Edition Published by
Academic Press, Inc. (*London*) Ltd.
24/28 Oval Road, London NW1

ISBN: 0-12-712650-3
Library of Congress Catalog Card Number:
82-73937

Printed in the United States of America

TABLE OF CONTENTS

CHAPTER 7   UNITED STATES RESIDENTIAL CONSTRUCTION   161

CHAPTER 8   UNITED STATES UNEMPLOYMENT   221

CHAPTER 9   THE CLOROX COMPANY CASE   245

CHAPTER 10   TRANSFER FUNCTION MODELS   257

CHAPTER 11   IDENTIFICATION OF TRANSFER FUNCTION MODELS   267

PREFACE

In the Preface to the 1973 edition of Kendall's (1976) *Time Series*, Kendall wrote that "[i]n the last thirty years the theory of time-series has been transformed into a new subject." An even greater transformation has taken place since Box and Jenkins published their *Time Series Analysis: Forecasting and Control* (First edition 1970). Since this publication the terms *Box–Jenkins models* and *ARIMA (AutoRegressive Integrated Moving Average) models* have become synonymous. Indeed Box and Jenkins have really revolutionized this subject in developing a practical approach for constructing and evaluating ARIMA time series models.

Although a number of books have been published since the Box and Jenkins book appeared, none has been written at a level accessible to students and practitioners with very little statistical training and none has been written presenting the Box–Jenkins models in a truly applied way. When I developed, in 1977, a Business Forecasting course for MBA students at the Harvard University Graduate School of Business Administration, the need for such a book became very clear. This textbook is an outgrowth of one section of this course. The students or practitioners who work through this book will acquire confidence that they can effectively construct and use ARIMA forecasting models as well as transfer function and intervention models. Similarly, the practitioner will be able to recognize situations where these models can be successfully used, and will be able to translate the results obtained from these models in decision actions.

An added feature of this book are several chapters (Chapters 10–14) devoted to transfer function models and intervention analysis. Although these models are certainly more elaborate to construct and evaluate, the reader will be able to work through these chapters without any additional difficulties. These chapters clearly build on the material presented in earlier chapters.

METHODOLOGICAL APPROACH

In order to achieve the above objectives, I tried to build up the intuition of the time series user. Therefore, where possible, I not only derive the important theoretical results, but immediately confront the reader with generated data that reproduced these results. For example, in Chapter 4 I use data generated according to known ARIMA processes to reinforce the results derived in Chapter 3 and to build a bridge between theoretical and applied results, showing, in particular, the need for measures of uncertainty. In the transfer function chapters (Chapter 10 and following) because the methodology is inherently more complex, I first show the important building blocks, relying on generated data, before deriving the theoretical generalizations.

Another important methodological tool is the computer. Currently there are quite a number of commercially available Box–Jenkins computer programs. I have developed a collection of highly interactive computer time series programs under the name TS. A Primer to this computer package is available as Appendix C of this text. I strongly recommend that the readers have access to a time series computer package and work through the analysis of the many time series presented in the book. With this purpose in mind, I have listed in Appendix A all the data for the time series used in this book. Nobody should believe that it is possible to successfully construct time series models without touching real data.

Because I am convinced that a hands-on experience is a most important didactic tool I have not included the traditional exercises that can be found in a number of other textbooks. I find many of these exercises of very little real interest to the applied-oriented audiences addressed by this book. In order to encourage this hands-on approach, I have devoted three chapters to the analysis of real data. Chapter 7 is written in a semiprogrammed text format. I ask the readers to analyze the data on U.S. residential construction jointly with me. This chapter should acquaint readers with the type of judgment required to formulate Box–Jenkins models. In Chapter 8 I summarize my analysis of the U.S. unemployment, again asking readers to use the computer to evaluate their own models. Finally, Chapter 9 is a case study presented for the purpose of solving a real decision problem. In this chapter readers must construct time series models and combine these with a linear programming problem to formulate a managerial solution to the underlying decision problem.

OVERVIEW OF THE BOOK

The first 9 chapters are devoted to the construction of univariate time series models. Chapters 10 to 13 cover the transfer function models, and in the final chapter the intervention model is presented.

Chapter 1 contains an overview of the whole book and introduces the univariate and transfer function models. It also contains an overview of the many fields in which these time series models have been successfully used.

In Chapter 2, the notion of stationarity is introduced, the importance of working with stationary data is discussed, and how data can be made stationary is shown. I rely extensively on a quarterly U.S. gross national product series to illustrate intuitively the important concepts. In Chapter 3, a taxonomy of the most important ARIMA models and their properties is presented. The next chapter is devoted to the first prong of the Box–Jenkins time series methodology, the identification phase. After identifying some possible models that could have generated a particular data series, the reader is then presented in Chapter 5 with the estimation and diagnostic checking methodology. Finally in Chapter 6, how to construct optimal time series forecasts is discussed.

Chapters 7, 8, and 9 are devoted to the analysis of real time series. These chapters should serve the important function of reaffirming that the reader can, after having worked through the earlier chapters, effectively construct ARIMA models.

Chapters 10 through 13 present the transfer function models. In many forecasting situations, other variables (events) will systematically influence the series to be forecasted and therefore there is a need to go beyond a univariate time series model. Chapter 10 introduces the transfer function model as a model that incorporates more than one time series and takes explicitly into account the dynamic characteristics of the system. As with the univariate model building process, the transfer function model construction involves three stages that are similarly labeled identification, estimation and diagnostic checking, and forecasting. Chapter 11 presents identification and shows that the cross correlations form an important new tool. Chapter 12 is devoted to estimation and diagnostic checking and demonstrates that most of the tools are identical to those presented in Chapter 5. Finally in Chapter 13, I discuss how to forecast with a transfer function model. These last chapters again illustrate the transfer function methodology with the use of a sales-advertising example.

The final chapter of the book shows that the intervention model is nothing but a structured dummy variable model of which the underlying model is either a univariate ARIMA model or a transfer function model. The methodology is illustrated with data on directory assistance calls before and after the introduction of a service charge.

Appendix A contains all the data explicitly used in this book. In Appendix B a basic explanation of the expectations operator is presented. Finally, Appendix C is the Primer to the Time Series package of interactive computer programs that I developed.

INTENDED AUDIENCE

The book is intended for students, researchers, and practitioners who have had little statistical training but either want to study these new important time series methods or are confronted with decisions based on these methods. Of course, those with more rigorous statistical training but unfamiliar with the Box–Jenkins time series models will certainly benefit from studying this book. I have primarily written this book with an applied-oriented audience in mind. I refer to the above section on the Methodological Approach for suggestions on how to derive the greatest benefit from the text.

This book can be used in an advanced undergraduate course or beginning graduate course on time series forecasting. Such a course could be part of a business school, department of economics, or engineering curriculum. Social science departments (education, psychology, public health, medicine) have also started to introduce such a course in their curriculum and researchers in these fields are using the time series methodology covered in this book in their applied work.

No matter who makes use of this book, it is important to remember that its orientation is an applied one. Therefore, I again strongly encourage readers to develop a hands-on experience using real data encountered in their respective fields of study or work.

Walter Vandaele
Washington, D.C.
Lincoln's Birthday, 1983

ACKNOWLEDGMENTS

This book would not have been written if, in the Spring of 1977, I had not chosen to teach a course on Business Forecasting for MBA students at the Harvard University Graduate School of Business Administration. In September 1979 I continued to work on this book while I was associated with the Center for Computational Research in Economics and Management Science (CCREMS), Sloan School of Management, MIT and with the Department of Economics, Harvard University. The manuscript was completed during the late and wee hours of the days (more late than wee) while I was Economic Advisor to the Director, Bureau of Competition, Federal Trade Commission.

I also used the manuscript of the book at a series of time series seminars that I conducted for audiences of business and government managers as part of the Educational Program of Data Resources, Inc. The transfer function chapters were initially written for these seminars. I benefited enormously from this interaction with the business community. Certainly these seminars formed a fruitful laboratory for my ideas on how to make these methods understandable to an applied-oriented audience.

I am indebted to a great many colleagues at the Harvard Business School, CCREMS at MIT, and the Department of Economics, Harvard University. In particular I want to thank John Pratt, Robert Schlaifer, and Art Schleifer of the Harvard Business School, and Mark Watson of the Department of Economics at Harvard for reading and commenting on various drafts of the manuscript. I also want to acknowledge the help received from John Schmitz and Dan O'Reilly of Data Resources, Inc. They participated in the DRI Time Series seminars as instructors and they helped me clarify my thinking in innumerable ways. In addition E. W. Swift, Georgia State University and H. H. Stokes, University of Illinois at Chicago Circle as well as several referees made many important suggestions. Only I can be blamed for not following all these suggestions. Furthermore, I want in particular to thank Robert Schlaifer for discussing with me how to write an efficient interactive computer package and for subsequently answering the many questions that I had about the programming of my interactive Time Series package. Without having access to the very proficient program library written by Robert Schlaifer, the writing of this interactive package would not have been possible.

From my former students, I want to specially recognize Sergio Koreisha, now at the University of Oregon. We had numerous discussions about the most effective way to present this material. He developed the idea of writing Chapter 7 in a programmed text format. Also under my guidance he wrote Chapters 8 and 9. I want to thank him for his contribution to this book.

My wife Annette, to whom this book is dedicated, owns my thanks for her love. She is always able in a few words to bring matters down to earth.

Several secretaries have repeatedly typed different versions of the chapters and in doing so they brought the manuscript to its final form. Nancy Hayes, Cheryl Levin, Martha Laisne, Kate Doyle, and in particular Karen Glennon are due many thanks. Michelle DuPree read the whole manuscript and provided invaluable editorial help. I also greatly appreciate the effort and guidance of Susan Elliott Loring, Senior Editor, and Georgia Lee Hadler, Project Editor at Academic Press.

Permission from Holden-Day to adapt from Box and Jenkins (1976) *Time Series Analysis: Forecasting and Control,* Figures 5.9(b) and 8.3(a) and Appendix A9.1, and from Nelson (1973) *Applied Time Series Analysis for Managerial Forecasting,* Figures 5.1 and 5.2, is gratefully acknowledged.

Of course, I assume the usual responsibility for those errors that undoubtedly still lurk somewhere in the book.

W. V.

CHAPTER 1

INTRODUCTION

1.1 THE IMPORTANCE OF A GOOD FORECASTING SYSTEM

The managerial need for accurate and reliable forecasts can certainly not be denied. Every day inventory, production, scheduling, financial, and marketing decisions are made which depend on projections of future sales. Inaccurate forecasts can have serious disruptive effects on business operations. For example, if demand for a product in a certain region of the country is underestimated, then production schedules of the plant supplying this territory and possibly those of nearby plants will have to be modified to accommodate the unanticipated demand, or otherwise sales will be lost. Additional overtime and interplant shipping costs may be incurred. Special and more costly orders may have to be placed to assure continuation of production as raw material supplies become insufficient to meet demand. Furthermore, permanent losses in sales may result if the lead times involved in placing new orders and receiving shipments from the other plants are too long. Overestimation of demand can also be costly. Investment opportunities may be missed because unnecessary capital is tied up in inventory and warehousing costs.

Aside from the need for accurate and reliable forecasts, it is also essential for a company to operate with a uniform set of projection figures. Complications are bound to arise if the marketing department bases its promotion and advertising plans on forecasts which are different from the ones used by the production department to set manufacturing schedules.

Uniformity in forecasts can be achieved by arranging meetings with the department heads responsible for formulating projections in order to obtain a consensus on the set of figures to be used throughout the company. It should be noted, however, that forecasts based solely on executive opinions may be influenced by political factors such as the necessity to compromise in order to be in line with upper management's disposition toward various interrelated matters. Yet forecasts generated using only mathematical models cannot generally assimilate all the managerial information, such as possible changes in production and delivery schedules caused by strikes or equipment failure. A good forecasting system is one in which systematically derived figures can be altered, if necessary, using available managerial expertise. It is also important to realize that the implementation of any forecasting system will create conflicts in many company settings. A forecasting system, if not properly introduced, will be foreign to the existing decision-making process.

1.2 THE SYSTEMATIC APPROACH

There exist many methods and approaches for formulating forecasting models, but in this book we will deal exclusively with the time series forecasting model. In particular, we will discuss the Autoregressive Integrated Moving

Average (ARIMA) models described by George Box and the late Gwilym Jenkins in their 1970 book entitled *Times Series Analysis: Forecasting and Control.*[1] Although time series models have been studied for many years,[2] Box and Jenkins popularized their use, demonstrated how to extend their application to seasonal data, and made the methodology operational. Their contribution to the field of time series analysis cannot be overestimated.

The Box–Jenkins approach possesses many appealing features. It allows the manager who only has data on past years' sales to forecast future sales without having to search for other time series data such as consumer's income, prices, etc. However, the Box–Jenkins approach also allows for the use of several time series to explain the behavior of another series if these other time series data are available.

In the next section we will discuss what a time series is, introduce some of the objectives of time series analysis, and present the distinction between univariate and multiple time series models. Throughout the chapter we will use examples to illustrate the concepts introduced.

1.3 WHAT IS A TIME SERIES?

A *time series* is a collection of observations generated sequentially through time. The special features of a time series are that the data are ordered with respect to time, and that successive observations are usually expected to be dependent. Indeed, it is this dependence from one time period to another which will be exploited in making reliable forecasts. The order of an observation is denoted by a subscript t. Therefore, we denote by z_t the tth observation of a time series. The preceding observation is denoted by z_{t-1}, and the next observation as z_{t+1}.

It also will be useful to distinguish between a time series process and a time series realization. The observed time series is an actual realization of an underlying time series process. By a *realization* we mean a sequence of observed data points, and not just a single observation. The objective of time series analysis is to describe succinctly this theoretical process in the form of an observable model that has similar properties to those of the process itself.

In this book we will analyze measurements or readings made at predetermined and equally or almost equally spaced time intervals[3] to generate hourly,

---

[1] A revised edition was published in 1976.

[2] See, e.g., Wold (1954) and Quenouille (1957) for a discussion of earlier work on time series models.

[3] Little theoretical work has been done on time series observed at unequal intervals. We can refer the reader to Quenouille (1958) on autoregressive series, and to Granger (1963) and Cleveland and Devlin (1980, 1982) on the effect of varying month length on economic series.

daily, monthly, or quarterly data. The calendar month is an example where the interval between observations is not quite constant. Such time series are called *discrete* time series, in contrast with continuous time series, which exist at every point in time. An example of a continuous time series is the temperature in a given place.

Discrete time series can arise in several ways. Given a continuous time series, it is possible to construct a discrete time series by taking measurements at equally spaced intervals of time. Granger and Newbold (1977) define such series as *instantaneously recorded.* Examples of this type of series are daily temperature readings at 3:00 P.M. and the Dow Jones index at closing time on successive days. Note again that because the New York Stock Exchange is closed on weekends, the time interval between successive daily observations of the Dow Jones index is not quite constant. Alternatively, discrete time series can arise by accumulating or aggregating a realization for a predetermined time interval. These are *accumulated series,* as described by Granger and Newbold (1977). Examples are monthly rainfall, quarterly industrial production, daily miles traveled by an airline, and monthly traffic fatalities. Notice that monthly traffic fatalities are actually an aggregation of discrete events. Therefore, although we do not formally analyze continuous time series, sometimes continuous data can be transformed into discrete data, which can then be analyzed by the methods presented in this book.

In many cases the data will exhibit a number of nuisance effects. The difference in the lengths of months is one such case; the fact that a month may include either four or five weekends will influence the observed data points. Movable feasts and holidays contribute their share of confusion, especially Easter, which may fall either in the first or the second quarter of the year. (We refer to footnote 3 for references dealing more specifically with these issues.) Strikes also introduce aberrations into the series.

In many situations the "raw" data may have to be "cleaned up" before we can start a formal analysis. The following are a few ways to accomplish this task:

1. A certain measure of comparability may be obtained for calendar month production data by correcting the figures to correspond to a standard month of 30 days, e.g., by multiplying the production in February by 30/28.
2. Short-term effects may sometimes be eliminated by aggregation. Variations due to a movable Easter can be eliminated by working with six-month rather than with three-month periods.
3. Data relating to nominal values may be divided by some index measuring changes in money values in order to create constant-value series (i.e., series in real terms).

No hard-and-fixed rules exist for when and how the data should be cleaned, or preprocessed, or for when to use the raw data. The analyst has to consider the purpose of the forecast. If there is a need for quarterly fore-casts, it makes no sense to use semi-annual data in order to eliminate the Easter problem. It is also unnecessary to clean up all the series we want to analyze. We may want to leave the monthly sales data unadjusted for the different lengths of the calendar months but allow the seasonal index to include the effect of changes in length from one month to the other.

In other cases, the user may be confronted with possible outliers in the data that would play havoc in analyzing such a series. Sometimes the user may be able to identify such outliers because their cause is known, such as a strike or holiday effect. In other cases the user may feel uneasy in identifying such outliers. In both cases the methods to deal with these problems are beyond the scope of this book. For more details the reader is referred to Cleveland et al. (1978), Kleiner et al. (1979), and Martin et al. (1983).

1.4 EXAMPLES

Time series occur in a variety of fields. Some examples are:

- *Business* Successive weekly and/or monthly sales figures. Figure 1.1, taken from Chatfield and Prothero (1973), presents the monthly sales figures of an engineering product from 1965 to May 1971.[4] The seasonality of this monthly series is very typical. November sales are high each year, and April and May sales are low. In addition, the series shows evidence of a trend; although the sales mean over the whole period equals 298.4 (see the dotted line in Figure 1.1), we would certainly predict a higher sales level for 1972 than for 1968. Other business examples include the number of tourists visiting Hawaii (Geurts and Ibrahim, 1975), the hourly electricity demand (Brubacher and Wilson, 1976), and the monthly total airline miles flown in the United Kingdom (Montgomery and Johnson, 1976).
- *Economics* Agricultural commodity pricing (Leuthold et al., 1970), the mar-ket yield on U.S. government three-month Treasury Bills (see Figure 1.2), monthly export totals, monthly employment and unemployment statistics, the Federal Reserve Board Index of Industrial Production, and money supply statistics are but a few of many other examples.
- *Sociology* Crime statistics (Boston armed robberies—see Deutsch and Alt, 1977), suicide rates, birth rates, and divorce rates are typical sociological examples.
- *Physics and Engineering* Numerous time series occur in the branches of the natural sciences: meteorology (Ledolter, 1976), marine science, and geo-physics. Examples are wind velocity, rainfall statistics, degree-days by month,

[4] All series are enumerated in Appendix A. The dotted line in the main body of Figures 1.1–1.4 indicates the position of the sample mean of the data.

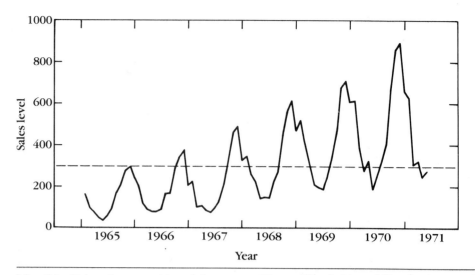

FIGURE 1.1   Monthly Sales, January 1965–May 1971. (From Chatfield and Prothero, 1973.)

and solar activity (see Morris, 1977, for an analysis of the sunspot data). An example of engineering data is the yield data from a batch chemical process (Jenkins and Watts, 1968). This yield data is represented in Figure 1.3. In contrast with the Chatfield and Prothero sales data in Figure 1.1, this series does not show any signs of a trend. The mean of the series, therefore, could be used as an initial forecast of the level of the series (see the dotted line in Figure 1.3).

FIGURE 1.2   Market Yield on U.S. Government Three-Month Treasury Bills, January 1956– December 1968. (From Federal Reserve Board.)

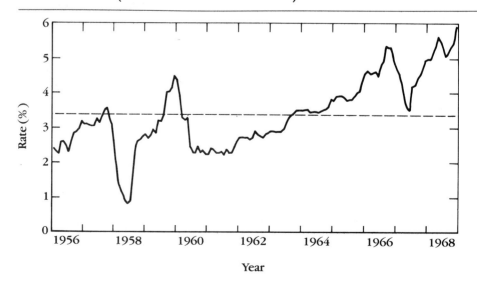

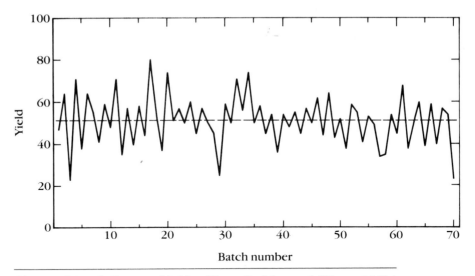

FIGURE 1.3   Chemical Batch Process Yields. (From Jenkins and Watts, 1968.)

■ *Medicine and Public Health*   Epidemiological statistics (see Figure 1.4. for the reported cases of rubella in the United States[5]), immunization statistics, immunogenetics, electrocardiograms, and electroencephalograms are some of the more familiar examples in the fields of medicine and public health.

FIGURE 1.4   Reported Cases of Rubella, 1966–1968, biweekly. (From Montgomery and Johnson, 1976.)

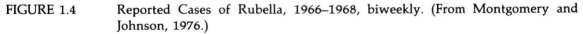
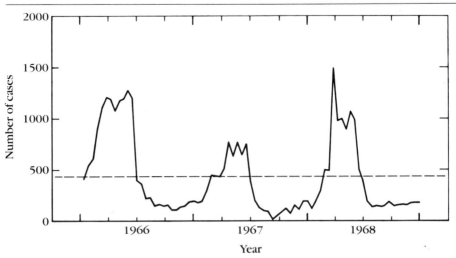

[5] The reported cases of rubella by two-week periods in Ohio, Indiana, Illinois, Michigan, and Wisconsin have been collected by the Center for Disease Control, Atlanta, Georgia, and are reported in Montgomery and Johnson (1976).

1.5 OBJECTIVES OF TIME SERIES ANALYSIS

Since the reasons for studying time series often determine the choice of methods to use, it is helpful to be given an overview of some study objectives:

1. to obtain a concise *description* of the features of a particular time series process;
2. to construct a model to *explain* the time series behavior in terms of other variables and to relate the observations to some structural rules of behavior;
3. based on the results of (1) or (2), to use the analysis to *forecast* the behavior of the series in the future based upon a knowledge of the past. From (1) we assume that there is sufficient momentum in the system to ensure that past and future behavior will be the same. From (B) we have more insight into the underlying forces of the time series process and can exploit these to obtain more accurate forecasts; and
4. to *control* the process generating the series by examining what might happen when we alter some of the parameters of the model, or by establishing policies that intervene only when the process deviates from a target by more than a prescribed amount.

In this book we primarily will concentrate on the objectives specified in (1), (2), and (3). However, in Chapter 14 we also will discuss the intervention model, which will allow us to evaluate the effect of interventions.

1.6 UNIVARIATE AND MULTIPLE TIME SERIES MODELS

Aside from the distinction between discrete and continuous time series models, it is also important to classify time series models according to the number of variables included in the model. A time series model consisting of just one variable is appropriately called a *univariate* time series model. A univariate time series model will use only current and past data on one variable. For example, if we forecast the unemployment rate next month or two months from now using a univariate model, we could only use current and past unemployment data. Implicit in the formulation of such a model is the assumption that factors which influence unemployment have not changed or are not expected to change sufficiently to warrant introducing these explicitly into the model.

A time series model which makes explicit use of other variables to describe the behavior of the desired series is called a *multiple* time series model. The model expressing the dynamic relationship between these variables is called a *transfer function model.* The terms transfer function model and multiple time series model are used interchangeably. A *transfer function model* is related to the standard regression model in that both have a dependent variable and one or more explanatory variables. But, as will be shown in Chapter 10, a transfer function model can allow for a richer dynamic structure in the relationship between the dependent variable and each explanatory variable, and between the error term.

The usefulness of such a transfer function can be better appreciated if we consider an example. Suppose that aside from unemployment data, we also have data on the money supply. Then, by constructing a transfer function model, we could exploit the dynamic relationship between unemployment and the money supply. For instance, if we discover that a change in the money supply this month will trigger a response in the unemployment situation two months from now, we may then be in a much better position than if we had just used a univariate model to predict future unemployment.

Finally, a special form of transfer function model is the *intervention model.* The special characteristic of such a model is not the number of variables in the model, but that one of the explanatory variables captures the effect of an intervention, a policy change, or a new law.

The next eight chapters will be devoted to univariate time series models. The univariate ARIMA model will also be the cornerstone of the transfer function analysis. Chapters 10 through 13 will discuss the transfer function model. The intervention model will be treated in Chapter 14.

1.7 THE UNIVARIATE BOX–JENKINS TIME SERIES METHODOLOGY

In the next eight chapters we will be discussing the univariate ARIMA analysis, commonly called the *Box–Jenkins approach.* This Box–Jenkins approach for analyzing time series data consists of extracting the predictable movements from the observed data. The time series is decomposed into several components, sometimes called filters. The Box–Jenkins approach primarily makes use of three linear filters: the autoregressive, the integration, and the moving average filter.

If we think of these filters as being special types of sieves, then the Box–Jenkins method can be viewed as an approach by which time series data are sifted through a series of progressively finer sieves. As the data pass through each sieve, *some* characteristic component of the series is filtered out. This process will terminate when what continues to go through the sieves is judged to be so fine that no additional information can be filtered out of it.

Figure 1.5 contains a very schematic representation of how the observed data, z_t, is transformed by the ARIMA filters. All the terms used in Figure 1.5 will be carefully explained in the next chapters. After applying the integration filter to the observed data, we obtain a filtered series, w_t. Next, the

FIGURE 1.5   The ARIMA Model.

autoregressive filter produces an intermediate series, e_t, and finally the moving average filter generates random noise, a_t. The objective in applying these filters is to end up with random noise which is unpredictable.

The natures of the different sieves and the grid sizes of the sieves are all the information we need to describe the behavior of the time series. Indeed, finding the number and nature of the filters is equivalent to finding the structure, identifying the form, and constructing the model for the series. The Box–Jenkins method provides a unified approach for *identifying* which filters are most appropriate for the series being analyzed; for *estimating* the parameters describing the filters (i.e., for estimating the grid sizes of the sieves), for *diagnosing* the accuracy and reliability of the models that have been estimated, and, finally, for *forecasting*.

In the next chapters we shall describe the details of each of these filters and the process of formulating such Box–Jenkins models. In Chapter 2 we will discuss, using an example, the integration filter which is closely related to the concept of trend. Chapter 3 will give an overview of the different ARIMA models and their characteristics. In Chapter 4 we will present the identification strategy. Chapter 5 is devoted to the estimation of the ARIMA models and the important model diagnostic checking. Chapter 6 presents the forecasting. Chapter 7 is an overview chapter allowing the reader to verify the understanding of the material presented in earlier chapters. Chapter 7 is written in a programmed text format. The effectiveness of the Box–Jenkins technique is examined in Chapter 8. Finally, Chapter 9 concludes the univariate part of the book with a case study.

CHAPTER 2

STATIONARITY

In this chapter we analyze the integration filter presented in Figure 1.5. To focus the discussion, a modified section of this figure is represented in Figure 2.1.

In discussing the integration filter we will define a related concept called *stationarity,* and indicate how to transform nonstationary data into stationary data. We shall begin by presenting some motivating issues for explaining this stationarity concept.

2.1 CALCULATING THE MEAN

How would you calculate the mean or average of a time series of a specified length? Calculating the mean of a sequence of observations might appear to be a trivial problem, as we would just add all observations and divide this total by the number of observations. However, if the series is steadily increasing over time (i.e., shows evidence of a *trend*[1]) and we make decisions based on this mean, we would certainly not want to select the same value for the start as for the end of the series. We would be hard pressed to claim that the dotted line representing the mean of the monthly sales data from Chatfield and Prothero (see Figure 1.1), or of the market yield on U.S. government three-month Treasury Bills (see Figure 1.2), would be a good forecast of the future level of each series. Indeed, if we regard the observed series as one realization of all possible series that could be generated by the same mechanism for that same time interval, we have only a sample of size one. We are therefore faced with the difficult task of estimating a mean for *each* time period based on *one* observation for that time period, and, clearly, with the even harder task of estimating variances and autocorrelations.[2]

The observed value of the series at a particular time period should be viewed as a random value; that is, if a new realization could be obtained under similar conditions, we would not obtain the identical numerical value. Let us measure at equal intervals the thickness of a steel wire made on a continuous extraction machine.[3] Such a list of measurements can be interpreted as a *realization* of wire thicknesses. If we were to repeatedly stop the process, service the machine, and restart the process to obtain new wires under *similar* machine conditions, we would be able to obtain *new* realizations from the *same stochastic process.* These realizations could be used to calculate the mean thickness of the wire after one minute, two minutes, etc. The term

---

[1] We define a *trend* as any systematic change in the level of a time series. In Section 2.5 we will present a more detailed discussion of trend and trend removal methods.

[2] The autocorrelations (to be defined below) are crucial elements in deciding which process is generating the data.

[3] This example is taken from Granger and Newbold (1977).

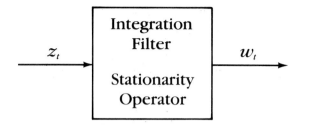

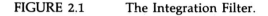

FIGURE 2.1   The Integration Filter.

stochastic simply means random, and the term *process* should be interpreted as the mechanism generating the data.

Let us denote the realizations by z_{jt}, $j = 1, 2, \ldots, J$; $t = 1, 2, \ldots, n$; with J being the number of realizations, and n the total time length of each realization. Therefore, we really have J time series of the same process. Then a possible estimator of the mean of the wire thickness after t time units, denoted as $\hat{\mu}_t$, is defined as

$$\hat{\mu}_t = \frac{1}{J} \sum_{j=1}^{J} z_{jt}. \tag{2.1}$$

However, in most situations we can only obtain one realization.[4] For example, we cannot stop the economy, go back to some arbitrary point in time, and then restart the economic process to observe a new realization. With a single realization, we cannot estimate with any precision the mean at each time period t, and it is impossible to estimate the variance and autocorrelations. Therefore, to estimate the mean, variance, and autocorrelations parameters of a stochastic process based on a single realization, the time series analyst must impose restrictions on how the data can be generated.

2.2 STATIONARITY

In the previous section we indicated that it is not advisable to estimate the mean for *each time period* based on just one realization of a general stochastic process. But if there is no trend in the series, we might be willing to *assume* that the mean is constant for each time period and that the observed value at each time period is representative of that mean.

We could then estimate the mean of the series by averaging over all the data as expressed by the following standard formula:

$$\hat{\mu} = \frac{1}{n} \sum_{t=1}^{n} z_t. \tag{2.2}$$

[4] For work on repeated measurements on the same process, see Anderson (1978).

Notice that we dropped the subscript j from z_{jt} because we have only one realization. Thus, to assume that the observed value at each time period is representative of the mean value, we must restrict the mean of the series to be constant. For the chemical batch process yield data represented in Figure 1.3, such an assumption could be quite plausible. This assumption is just one of the conditions for stationarity. In Section 2.5 we will discuss what can be done if this assumption is violated.

A second condition for stationarity is that the variance of the process be constant.[5] The variance of the series expresses the degree of variation around the assumed constant mean level and as such gives a measure of uncertainty around this mean. If the variance is not constant but, say, increases as time goes on, it would be incorrect to believe that we can express the *uncertainty* around a forecasted mean level with a variance calculated based on *all* the data. If we were to do so, we would really have a kind of average variance which would deflate the uncertainty around the forecasted values. We will therefore impose that the series has a constant variance.[6] In Section 2.4 we will present some transformations that in certain situations will make it more plausible to assume a constant variance for a transform series.

Finally, we must also impose a condition on the nature of the correlation between data at different time periods. The autocorrelation measures the correlation[7] between an observation at time t, z_t, and an observation at time s, z_s. Given that this is a correlation between observations of the *same* series at different time periods, it is appropriately called *auto*correlation. Given the values z_1, z_2, z_3, . . . , z_n, the autocorrelation between z_t and z_{t+1} measures the correlation between the pairs (z_1, z_2), (z_2, z_3), . . . , (z_{n-1}, z_n) and is denoted by ρ_1. Likewise, the autocorrelation between z_t and z_{t+2} equals the correlation between the $(n-2)$ pairs (z_1, z_3), (z_2, z_4), . . . , (z_{n-2}, z_n), and is similarly denoted by ρ_2. In general, ρ_k measures the correlation between pairs of observations k periods apart.

If, for example, we have for ρ_1 a value that is strongly positive, close to $+1$, and we currently have observed a value for z_n above the mean value of the series, we would again expect the next value to be above the mean. Similarly, if ρ_1 is strongly negative, close to -1, we would expect the next value to be below the mean if the current value is above the mean. It should

---

[5] For a definition and some properties of a *variance*, see Appendix B.

[6] This assumption is similar to the assumption of homoscedasticity of the disturbance terms in regression analysis.

[7] It will suffice for now to know that if two series are strongly positive correlated, then when one series increases (decreases) we expect to see that the other series also increases (decreases); that if two series are strongly negative correlated, then when one series increases (decreases) we expect to see that the other series decreases (increases); and that if two series are not correlated, there will be no relationship between one series increasing (decreasing) and the other. A precise definition of autocorrelation is given in Appendix B. In Chapter 3, we will explicitly calculate the autocorrelations of several ARIMA time series processes.

be intuitively clear that autocorrelations will play an important role in forecasting time series.

Let us now divide the series in two halves and calculate the autocorrelations for the first half as well as for the second half of the series. If we observe really different values, say for ρ_1, based on the first half and based on the second half, we then should not use the autocorrelations for the first half in making predictions for the future but we should rather rely on the second half only. And, similarly, the ρ_1 based on the whole sample, which can be viewed as a kind of average autocorrelation over the whole series, would be misleading. We therefore will impose the condition that an autocorrelation should not depend on which segment of the data is used to calculate the correlation. That is, we assume that the autocorrelations between z_t and z_s are independent of the t and s and are only determined by the lag between t and s.

Finally, we remark that if the autocorrelations depend only on the time interval between two points, then $\rho_{t-s} = \rho_{s-t}$, and in particular, $\rho_{-k} = \rho_k$. In other words, the autocorrelation ρ_k is the same whether the series is lagged forward or backward. Because the autocorrelations are symmetrical about lag zero, only the autocorrelations for positive values of k need to be examined.

Let us now summarize these three conditions. A process z_t, $t = 1, 2,$ ..., n is stationary[8] if

$$\text{mean[9] of } z_t = Ez_t = \mu \tag{2.3}$$

$$\text{variance of } z_t = E[(z_t - \mu)^2] = \sigma^2 \tag{2.4}$$

and

$$\text{autocorrelation } (z_t, z_s) = E[(z_t - \mu)(z_s - \mu)]/\sigma^2 = \rho_{t-s}.$$

That is, a time series process is stationary if the mean and the variance are constant over time (and both are finite) and if the autocorrelation between values of the process at two time periods, say t and s, depends only on the distance between these time points and not on the time period itself. (We arbitrarily assume that $t > s$.)

In this section we have tried to make it intuitively clear that if the process is stationary, we can meaningfully estimate the mean, variance, and autocorrelations from just one realization. In fact, we can then estimate the mean, variance, and autocorrelations of the process using methods as if a realization

---

[8] Strictly speaking, we define weak or covariance stationarity, also sometimes called *stationarity in mean and variance*. A stronger form of stationarity refers to the joint distribution of a set of random variables. If $F(z_{t+1}, z_{t+2}, \ldots, z_{t+m})$ is the joint distribution of any set of m consecutive z's, then the series is strictly stationary if F is independent of t for all $m > 0$. The distribution of any set of m consecutive observations is the same wherever in the series we choose those m observations.

[9] See Appendix B for an explanation and properties of the expectation operator E.

of length n constituted n samples of the same process. For example, the formula (2.2) constitutes a valid estimator of the mean.

Since most economic and business series are not stationary (trends are frequently present in such series), our next task is to show how nonstationary series can be transformed into stationary series. In explaining in greater detail nonstationarity and the possible transformation, we will explicitly rely on an economic time series example.

2.3 NONSTATIONARY DATA: A TIME SERIES PLOT

The first step in any time series analysis should be to plot the available observations against time. This is often a very valuable part of any data analysis since qualitative features such as trend, seasonality, discontinuities and outliers will usually be visible if present in the data.[10] Although the desirability of making such a plot at the outset is self-evident, we still come across studies in which an incorrect conclusion was reached because the data had not been plotted.

In this and subsequent sections we will make explicit use of a quarterly gross national product (GNP) data series. Figure 2.2 contains a plot of the GNP of the United States series from the first quarter of 1947 through the fourth quarter of 1970. The data,[11] seasonally unadjusted, are in billions of current dollars and, contrary to the convention, are expressed in quarterly rates rather than in annual rates (quarterly rates multiplied by four).

Examining this plot, we observe that for the entire 23-year period the GNP has been steadily growing, and because of this trend this series is therefore nonstationary. Changes in population, industrial and agricultural productivity, as well as changes in the standard of living are just some of the factors which have contributed to this long-term trend.

Also apparent from the graph is a recurring pattern within each year. The GNP data for the fourth quarter of each year is always higher than the data for the previous three quarters. The occurrence of such a seasonal pattern is generally attributed to a pickup in sales during the Christmas season. Seasonal variations, however, are also related to climate and customs. For example, unemployment among building and catering workers tends to be highest in the winter, while sales of gasoline in a summer resort area tend to be higher during the summer months.

Seasonal variations in general recur with a high degree of regularity

[10] Outliers are somewhat harder to visualize in time series data because of the dependence of one observation on another. Indeed time series outliers could be located in the 'middle' of the series. See Martin et al. (1983).

[11] This example is from Roberts (1974). The data are listed in Appendix A, Table A.5.

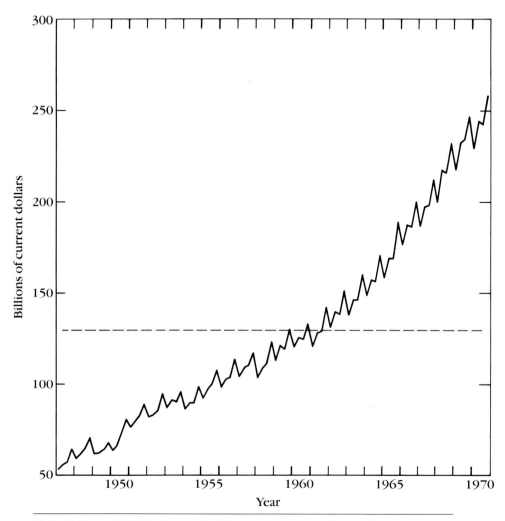

FIGURE 2.2   United States' GNP, Quarterly Rates 1947–1970. (From Roberts, 1974.)

every year; that is, they have an *annual* periodicity or cycle. However, we can easily extend the concept to include periodicities over other time intervals such as daily, hourly, weekly, etc. Retail sales, for example, tend to be higher on Fridays and Saturdays than on Mondays or Tuesdays.

Finally, although we might be pushing the issue too far, from Figure 2.2 we can also see that the GNP data seems slightly more volatile as time goes on—the variance among the observations increases as the level of the GNP series increases.

We can now summarize the findings based on an examination of a plot of the GNP data:

1. the variance of the series is not constant over time;
2. there is a trend in the data; and
3. there is a seasonal pattern.

All three observations indicate that the GNP data is nonstationary. Finding 1 violates the constant variance assumption (2.4), and Findings 2 and 3 are indications that the mean in the series is not constant over time.

In the following sections we shall examine each of these findings in more detail and suggest ways to remove these stationarity violations. First we shall discuss how to eliminate problems associated with nonconstant variances, and then we shall discuss how to remedy the problems associated with trend and seasonality.

2.4 STABILIZATION OF THE VARIANCE

The analysis of a time series requires that the series be stationary, and in particular that the variance, the volatility of the series, be constant over time. Although point estimates of the parameters and subsequent forecasts are not necessarily incorrect, inferential statements will certainly be hampered when the variance of the series is not constant over time. In Chapter 3 we will introduce the concept of time series parameters and will discuss statistical inferences.

There are several possible data transformations that could be used to induce a constant variance. The basic idea is to transform the data such that an originally curved plot is straighter,[12] and at the same time make the variance constant over the whole series. Two of the frequently used transformations are the logarithmic[13] transformations and the square root transformation.[14]

The *logarithmic transformation* can be employed effectively when:

1. the variance of the series is proportional to the mean level of the series;
2. the mean level of the series increases/decreases at a constant percentage rate.

---

[12] A good discussion on straightening curves can be found in Chapters 4 and 5 of Mosteller and Tukey (1977).

[13] Unless stated to the contrary, all logarithms in this book are natural logarithms, although the logarithm to any base could be used.

[14] In general, most business and economic data will all be positive. If this was not the case, we could add an arbitrary constant to all the observations such that all are now positive. Such a constant would obviously shift the level of the series and would have to be readjusted in making forecasts.

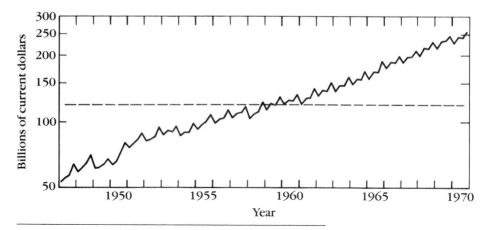

FIGURE 2.3   United States' GNP (Logarithmic Scale), 1947–1970.

Figure 2.3 contains the plot of the logarithm of the GNP series. As we can see, this transformation has tended to overcorrect the data. We now observe greater variability in the earlier years than in the later years. We must therefore find an intermediate transformation which will strike a balance between the original data and its logs. A useful transformation is the *square root transformation* of the data $\sqrt{z_t}$.

The time series plot of the square root transformation of the quarterly GNP data is presented in Figure 2.4. Although the trend and the seasonal

FIGURE 2.4   United States' GNP (Square Root Scale), 1947–1970.

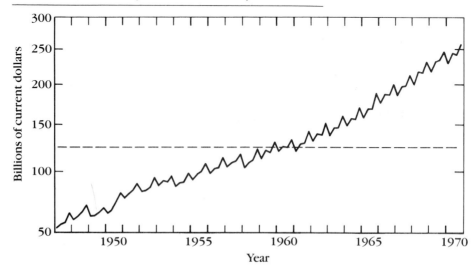

components are still present, the variation among the observations is relatively constant over time. The first step towards achieving stationarity has apparently been successful; the variance has been stabilized.

As mentioned in the previous section, we are using the GNP series to illustrate several features of stationarity, although this series may not entirely suit all these features. In particular, the variance transformations proposed are not and should not make a large difference for forecasting *this* particular series. We do hope that the reader has been able to understand what the logarithmic and square root transformations can accomplish.

Other transformations could also be evaluated to induce stationarity. A general class of such transformations,[15] called *power transformations*, can be specified by $z^{(\beta)}$, where z represents the untransformed data and (β) is a parameter taking on positive or negative values. For $\beta = 0$ we define $z^{(\beta)}$ as log z. The square root has a value for (β) equal to $1/2$. Similar transformations are discussed in Mosteller and Tukey (1977, Chapter 4). It has been found in practice that we can restrict the search for an appropriate transformation[16] to the logarithmic and the square root transformations.

2.5 REMOVAL OF THE TREND

Most economic time series, whether macro or micro, are characterized by movements along a trend line. This is particularly true for the series expressed in levels such as sales and unemployment, rather than in rate of change or in return, such as unemployment rate and stock market return. Although there is a general understanding of what a trend is, it is difficult to give a more precise definition of the term *trend* than "any systematic change in the level of a time series." While this definition lacks mathematical rigor, it is still the best definition. The difficulty in defining a trend stems from the

---

[15] Box and Cox (1964) suggested the following transformation:

$$z^{(\beta)} = \begin{cases} \dfrac{z^\beta - 1}{\beta} & \beta \neq 0 \\ \log z & \beta = 0 \end{cases}.$$

[16] Some time series programs allow for this general class of power transformations. The TS programs allow only for the logarithmic transformation and therefore if another transformation is more appropriate, the user should transform the data before running any of the TS programs.

The Time Series, TS, collection of programs is described in Appendix C. In this book we will make exclusive use of results generated by this collection of programs. However it is in no way necessary for the reader to have access to this particular package. We strongly recommend however that users have access to some time series package so that they can run examples of time series analyses and become proficient in the application of these methods. Currently there are a number of commercial packages available. For the availability of this TS package, the reader should contact the author.

REMOVAL OF THE TREND **21**

fact that what looks like a change in the level in a short series of observations may turn out not to be a trend when a longer series becomes available but be part of a cyclical movement.

Several methods for removing the trend have been suggested in the literature. One common method of detrending a time series is to use a regression model. Such a regression model can be written as

$$z_t = \beta_0 + \beta_1 t + a_t \qquad (2.5)$$

if the trend is assumed to be linear, or as

$$z_t = \beta_0 + \beta_1 t + \beta_2 t^2 + a_t \qquad (2.6)$$

if the trend is presumed to be a quadratic polynomial. Of course, a higher order polynomial could equally be fitted.[17] In all these models it is assumed that the trend is fixed and deterministic.

Intuitively, we may not like to assume that the trend is deterministic, but we may want to assume that the trend is stochastic. As mentioned in Box and Jenkins (1976, p. 300), we should make a clear distinction between fitting a series and modeling a series. If the objective is purely fitting, we could estimate a higher order polynomial that fits the data. However, we may be confronted with a series that does not really follow a trend but is slowly drifting upwards and downwards (i.e., there is a stochastic behavior that is creating a trend). The real problem is that it is extremely difficult to tell whether a change in the level of the series is due to a deterministic or to a stochastic trend.

For these and other reasons,[18] Box and Jenkins have advocated the use of an alternative method called *differencing*. We do not claim that this method will be the most effective method for removing the trend present in all types of series. Experience, however, does indicate that differencing is an extremely useful method for modeling a stochastic trend component in a large number of business and economic series.

The method of *differencing* a time series consists of subtracting the values of the observations from one another in some prescribed time-dependent order. For example, a *first* (order) difference transformation is defined as the difference between the values of two adjacent observations; *second* (order) differencing consists of taking differences of the *differenced* series; and so on.

---

[17] Some other approaches for removing the trend include the classical decomposition (see Vatter et al. (1978)); curve fitting using a Gompertz curve or a logistic curve (see Gregg et al. (1964); Harrison and Pearce (1972); and Levenbach and Reuter (1976)); filtering, such as calculating moving averages, exponential smoothing (see Kendall (1976); and Wheelwright and Makridakis (1980). Kendall is critical of the filtering of a single functional form over the whole course of the observations: "[A] polynomial might fit the segment of the series given but would clearly be unsafe to project far into the future," p. 29.).

[18] For a comparison between polynomial curve fitting and differencing, see, e.g., Chan et al. (1977) and Nelson and Kang (1981).

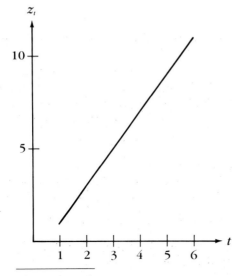

FIGURE 2.5   Linear Trend.

FIGURE 2.6   Nonlinear Trend (a) and Its First Differences (b).

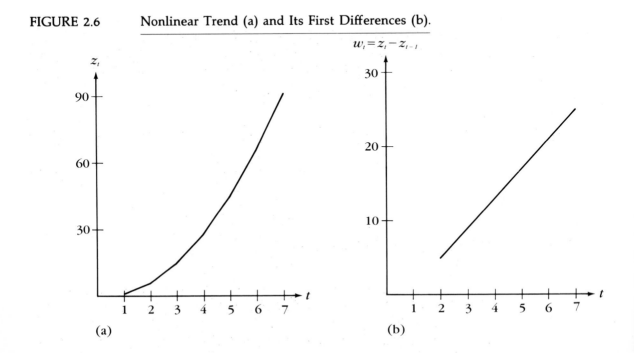

(a)   (b)

To illustrate the method, consider the following series with values 1, 3, 5, 7, 9 and 11, as plotted in Figure 2.5. This series exhibits a constant increase of two units from one observation to the next:

$$3 - 1 = 2$$
$$5 - 3 = 2$$
$$7 - 5 = 2$$
$$9 - 7 = 2$$
$$11 - 9 = 2 \qquad (2.7)$$

By taking the first difference of the series with a linear trend, the trend disappears. Now let us apply this to the series: 1, 6, 15, 28, 45, 66, and 91, as plotted in Figure 2.6(a). Note that this series exhibits a nonlinear trend. Calculating first differences yields the series: 5, 9, 13, 17, 21, 25. As the graph in Figure 2.6(b) shows, this differenced series exhibits a linear trend with a constant increase of 4. Therefore, by taking differences of the differences (i.e., *second* differences), we would obtain a trend-free series. This example should not be considered as a proof that by constructing second differences, we will always be able to remove a nonlinear trend. Quite to the contrary, only a special form, a quadratic, trend will be removed in this way.

Note that a second-order difference operation is *not* equivalent to subtracting the values of observations in time period t and $t - 2$. Let

z_t = original series

y_t = first-order differenced series, i.e., $y_t = z_t - z_{t-1}$

w_t = second-order differenced series.

Then we see that

$$w_t = y_t - y_{t-1}$$
$$= (z_t - z_{t-1}) - (z_{t-1} - z_{t-2})$$
$$= z_t - 2z_{t-1} + z_{t-2}. \qquad (2.8)$$

In Section 3.5 we will introduce the difference operator ∇ and backshift operator B. These operators will allow us to simplify the notation of higher order differencing.

Observe that each time we difference a series, we lose an observation. In our first example, the original series consisted of six observations. After first-order differencing, the transformed series contained only five points. Similarly, the second example consisted originally of seven points, but after we made the second-order difference transformation, only five observations were left.

Due to random fluctuations in the data, such neat results cannot always be obtained. However, for many economic time series, first- or second-order differencing will be sufficient to remove the trend component so that further analyses can proceed. In Figure 2.7 we have plotted the first differences of

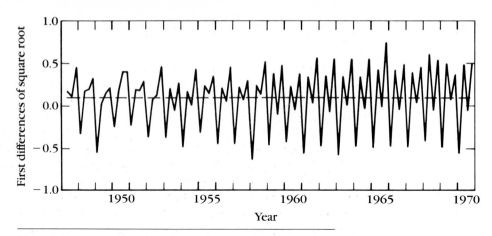

FIGURE 2.7   United States' GNP, First Differences of Square Root.

the square root of GNP. As can be seen there is apparently no longer a trend present.

One can show that once the trend is removed, further differencing will continue to produce a series without trend. However, each additional differencing results in one additional data point being lost. As will be discussed in Section 4.3, this *overdifferencing* will needlessly complicate the structure of the model, and should therefore be avoided. In subsequent chapters we will present additional tools to assess if a series can be considered to be stationary or if additional differencing is recommended.

Although, as mentioned above, there are fundamental differences between removing the trend by a polynomial and removing it by using differences, there are also connections between these two approaches. Also, as will be shown below, we can use the differencing approach to evaluate the existence of a deterministic trend.

Let us represent the series given in (2.7) as in equation (2.5):

$$z_t = \beta_0 + \beta_1 t, \qquad t = 1, 2, \ldots, 6 \tag{2.9}$$

with $\beta_0 = -1$ and $\beta_1 = 2$. Alternatively,[19] we could difference equation (2.9) to obtain

$$z_t - z_{t-1} = \beta_1 \tag{2.10}$$

or, solving for z_t,

$$z_t = z_{t-1} + \beta_1.$$

[19] It can easily be shown that the unique solution of the first-order difference equation (2.10) is given by (2.9) with $\beta_0 = 0$. For a reference on difference equations and their application to economics, see Chiang (1967).

In equations (2.9) and (2.10) we could easily add an error term as was done in equation (2.5). As a matter of course, we would not expect to be able to represent a time series exactly as a function of time alone. Adding an error term will not modify any of the basic conclusions.

While β_1 in equations (2.9) and (2.10) describes the same deterministic trend, it should be clear that β_1 in (2.10) is easily estimated as the mean of the differenced series $z_t - z_{t-1}$, whereas in equation (2.9) we would have to estimate the slope, say by least squares, a more cumbersome procedure.

In addition, equation (2.10) suggests a straightforward way to evaluate whether there is a deterministic trend by testing the significance of the mean of $z_t - z_{t-1}$. If this mean is statistically different from zero, we can conclude that there is a deterministic trend in the original series. This point will be discussed in more detail later on in this book.

The same results hold for higher-order polynomials. A second-order polynomial can be evaluated by testing the mean value of second-order differenced data. If the trend is of an exponential form, then a logarithmic transformation, as discussed in Section 2.4, can be used to straighten out the plot. Then, first differencing could be applied to this transformed series.

As a final remark, we remind the reader that the basic idea of removing the trend is not to forget that there is a trend in the series, but to obtain a new series that then can be analyzed and forecasted more effectively with the methods explained in subsequent chapters in this book. The original series will then be forecasted by reintroducing the trend. If, for example, we have

$$y_t = z_t - z_{t-1}$$

and the forecast for period $t + 1$ of the differenced data is y_f, we can then undo the difference filter using

$$z_t = y_t + z_{t-1}$$

and forecast the original series for period $t + 1$ as

$$z_f = y_f + z_t,$$

where z_t is the actual observed value of the series at period t. Later on in this book (Chapter 6 in particular) we will return to this issue and indeed show that this is an optimal forecasting method.

2.6 SEASONAL FLUCTUATIONS

A close look at the first differences of the square root GNP data in Figure 2.7 reveals a pattern of peaks and valleys spaced at *four-period* intervals. This should not be too surprising considering that we are analyzing *quarterly* data.

If we were analyzing monthly data, some kind of seasonal pattern would be likely to appear at 12-month intervals. Now, if we were to connect all the points of the first quarter changes in Figure 2.7, and then do the same for the second, third, and fourth quarters separately, we would observe that these four lines hardly overlap. We could therefore propose to separate the data into four distinct samples and analyze each one of them individually. However, this would not be prudent because the size of each sample would be only one fourth of the original size.

Many monthly or quarterly economic time series will exhibit similar effects which have a high degree of regularity. We define these effects as seasonal fluctuations which recur every year, that is, which have an annual cycle or periodicity. An example of this is the monthly sales volume of Sears, which peaks every December as a result of Christmas shopping. Everybody would normally forecast the December sales volume higher than the just-observed November figure.

For analyzing seasonality, it is impossible to determine the seasonal effects without paying attention to trend. Suppose that the monthly series can be represented exactly by a linear trend. In any year January would be the lowest month, whereas December would be the highest. But these are not seasonal effects; as a matter of fact, there are no seasonal effects. Therefore, it is generally recommended that the trend first be removed from a series.

As there are several methods to remove a trend, there are also several methods to deal with seasonality. Most of these methods are "autoadjustment" procedures, based only on information contained within the series being adjusted.[20] Other proposed methods focus on building a causal model for the seasonal and nonseasonal components based on the interrelationship between this series and many other series. Ideally, one should formulate a complete econometric model in which the causes of seasonality are incorporated directly into the equations. These model methods are, however, beyond the scope of this book. For a survey of recent developments in this new and interesting model building approach and its comparison with autoadjustment procedures, see Pierce (1980).

Here we will concentrate on the autoadjustment procedures. The transformed GNP data presented in Figure 2.7 could be represented with the following model:

$$z_I = s_I + y_I$$
$$z_{II} = s_{II} + y_{II}$$

---

[20] Some of these autoadjustment procedures for measuring or eliminating seasonal fluctuations are classical decomposition, moving averages, regression with dummy variables, and Census X-11 seasonal adjustment (see Shiskin et al. (1967); Cleveland and Tiao, (1976); and Dagum (1978)). Researchers at the AT&T Bell Laboratories have recently developed SABL (Seasonal Adjustment–Bell Laboratories), a new seasonal adjustment procedure which relies more heavily on graphic displays as diagnostic tools (see Cleveland et al. (1978), and Cleveland and Devlin (1980)).

$$z_{III} = s_{III} + y_{III}$$
$$z_{IV} = s_{IV} + y_{IV},$$

where the subscript $_I$ refers to the data in the first quarter, and similarly for $_{II}$, $_{III}$, and $_{IV}$. The variable z stands for the data, y is the seasonal adjusted series, and s stands for the seasonal component assumed constant across all first quarters (s_I), second quarters (s_{II}), etc.

Under this model we could propose the following adjustment procedure, called *seasonal differencing,* in contrast with *consecutive* differencing which was discussed earlier:

$$z_I - z_{I-4} = y_I - y_{I-4},$$

and similarly for the second, third, and fourth observations. This four-quarter seasonal differencing would totally remove the fixed seasonal component. For this model, several other autoadjustment procedures, such as regression with dummy variables to estimate the seasonal component, could be used equally well. However, the same remarks about detrending with a polynomial also apply here. If we want to allow for stochastic seasonal components, rather than deterministic, so that we can model, rather than fit a slowly evolving seasonal pattern through time, seasonal differencing is to be preferred. Also, we must remind the reader that this seasonal differencing is not intended to remove all seasonal components but just to induce stationarity so that all the observations can be modeled as one series, possibly including seasonal coefficients in the model. The proposed model to allow for seasonal coefficients will be one of the topics to be developed in subsequent chapters.

For now, let us expound on the seasonal differencing using the data in Figure 2.7 as a reference point. What do you suppose would happen if we calculated differences among observations spaced at four-period intervals, that is, calculated the differences between the first quarter value of each successive year and similarly the differences between the second, third, and fourth quarters?

A possible answer to this question would be that the plot of the quarterly differenced series would show that these observed seasonal characteristics would be substantially reduced, if not eliminated altogether. The presumption is that differences between the same quarters in different years are stable. (The fluctuations within quarters as seen in Figure 2.7—after drawing in the interconnecting lines—are small.) In fact, Figure 2.8 contains a plot of consecutive and seasonally differenced series which does not exhibit any recognizable pattern.

The seasonal transformation described above is called *seasonal differencing* of *order 1* with a *span* of 4 periods. Seasonal differencing of *order 1* indicates that we are taking the *first* differences among the same quarters (or months, if we were working with monthly data) in different years. A *span* of 4 implies that a lag of 4 periods is used in the seasonal differencing operation. To clarify this point, let us calculate first- and second-order quarterly differences. Let

$$z_t = \text{original series,}$$
$$y_t = \text{first-order quarterly differencing, and}$$
$$w_t = \text{second-order quarterly differencing;}$$

then

$$y_t = z_t - z_{t-4},$$

and

$$w_t = y_t - y_{t-4}, \tag{2.11}$$

or

$$w_t = (z_t - z_{t-4}) - (z_{t-4} - z_{t-8})$$
$$= z_t - 2z_{t-4} + z_{t-8}. \tag{2.12}$$

Just as we lost observations when we took consecutive differences, we lose observations when taking seasonal differences. For each seasonal differencing we lose as many observations as the length of the span.

Notice that the transformed data represented in Figure 2.8 is a series that has first been transformed with a square root transformation, then consecutively differenced, and finally seasonally differenced. That is, let

FIGURE 2.8   United States' GNP, Seasonal Adjusted and Detrended Square Root.

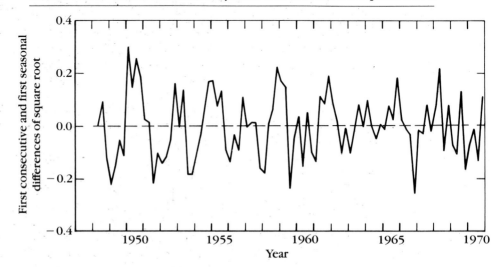

z_t = square root of the original series, and

$y_t = z_t - z_{t-1}$: first-order consecutive differences,

then the data in Figure 2.8 represent first-order seasonal differences with a span of 4; that is,

$$w_t = y_t - y_{t-4}.$$

Alternatively, we can re-express w_t as

$$w_t = (z_t - z_{t-1}) - (z_{t-4} - z_{t-5})$$
$$= (z_t - z_{t-4}) - (z_{t-1} - z_{t-5}). \qquad (2.13)$$

From (2.13) it follows that the order of differencing is arbitrary. We can as well calculate seasonal differences of consecutively differenced data or calculate consecutive differences of seasonally differenced data. As mentioned above, in Section 3.5 we will elaborate on consecutive and seasonal differencing, and at the same time introduce some simplifying notations.

2.7 SUMMARY

We showed that by means of three simple transformations, the nonstationary GNP data can be transformed into a stationary series which can then be used to estimate the parameters of the models. These transformations are:

- taking the square root of the data (variance stabilization);
- taking the first differences (elimination of trend); and
- taking first-order quarterly differences (elimination of seasonal component).

Referring to our analogy presented in Chapter 1, we have successfully sifted the data through the first sieve, the stationarity sieve. It has been found expedient to first induce a variance stabilization transformation before differencing. One of the major reasons for this is that differenced data will, by its very nature, include many negative values of which, without introducing needless technicalities, we cannot take a logarithm or a square root transformation.

You should keep in mind, however, that the number and type of transformations needed to induce stationarity clearly depend on the data being studied. There are examples throughout this book which will demonstrate the necessity of using different sets of differencing and variance stabilization operations.

Finally, a quick way to tell whether a series is stationary is to divide the data into two or three non-overlapping segments. If the series is stationary, then each "snapshot" of the data should exhibit a similar type of behavior and should have the same mean and variance. If one were to take snapshots of Figure 2.8, and compare the set of observations 5–20, with the set 40–

55, and the set 70–85, one would find that each snapshot shows a similar time series behavior.

In the following chapters we will show how to construct models which can exploit certain characteristics of stationary series. For example, you will observe that the data in Figure 2.8 appears to have some "stickiness"—that is, once the series is below the mean it remains there for several periods, and, similarly, once an observation is above the mean, the next observation will most likely also be above the mean. Box–Jenkins models can deal effectively with such forms of data behavior. In generating accurate forecasts, it will be important to rely on these characteristics.

CHAPTER 3

AUTOREGRESSIVE INTEGRATED MOVING AVERAGE MODELS

Figure 1.5 in Chapter 1 contains an overview of how the three Box–Jenkins filters—the integration, the autoregressive, and the moving average filters—are used to formulate a forecasting model. In Chapter 2 we analyzed the integration filter and showed how to transform nonstationary series into stationary series. In this chapter we shall discuss the properties of the autoregressive and the moving average filters. In presenting these filters we shall concentrate on the interpretation, rather than on the mathematical details. However, to make the discussion more lucid, it will be necessary from time to time to introduce some new mathematical concepts.

In Sections 3.1 and 3.2 we present the autoregressive and moving average models, respectively, followed in Section 3.3 by the mixed autoregressive moving average model. The models presented in these three sections allow us to model many stationary time series. However, applied researchers frequently have to forecast nonstationary series. We therefore present in Section 3.4 the autoregressive integrated moving average model that can be used for many such series. Then, in Section 3.5, we introduce some simplifying notation.

Finally, Section 3.6 is devoted to an introduction of the seasonal models. Again the importance of seasonal model need not be stressed, as many business and economic time series are of a quarterly or monthly frequency, exhibiting seasonal characteristics.

We recommend that the reader should at first concentrate on the material presented in Sections 3.1, 3.2, 3.3, and 3.5. Leave the content of Sections 3.4 and 3.6 for a second reading after having worked through Sections 4.1 and 4.2 in Chapter 4. In these two sections, using generated time series data we will show that the characteristics of the different time series models presented in this chapter indeed allow us to discriminate among these different time series models.

3.1 AUTOREGRESSIVE MODELS

3.1.1 First-Order Autoregressive Models

A time series[1] is said to be governed by a first-order autoregressive process if the current value of the time series[2], z_t, can be expressed as a linear function of the previous value of the series and a random shock a_t. If we denote the previous value of the series by z_{t-1}, this process can then be written as

$$z_t = \phi_1 z_{t-1} + a_t, \qquad (3.1)$$

---

[1] We assume that the time series is stationary or has been transformed into a stationary series.

[2] In this chapter the symbols z_t, z_{t-1}, z_{t-2}, . . . , denote the values of a process measured at equally spaced time periods t, $t-1$, $t-2$, . . . , and unless otherwise stated, are expressed as deviations from the mean of the series.

where ϕ_1 is the autoregressive parameter which describes the effect of a unit change in z_{t-1} on z_t, and which needs to be estimated. The random shocks a_t, also known as errors or white noise series, are assumed to be normally[3] and independently distributed with mean zero, constant variance σ_a^2, and independent of z_{t-1}; that is,

$$E(a_t) = 0 \tag{3.2}$$

$$E(a_t a_s) = \begin{cases} \sigma_a^2 & \text{if } t = s \\ 0 & \text{if } t \neq s \end{cases} \tag{3.3}$$

$$E(a_t, z_{t-1}) = 0. \tag{3.4}$$

Because of the normality assumption, some authors would define a_t as Gaussian white noise. Once again keep in mind that the z_t's are deviations from the mean, μ. Alternatively, one could also present (3.1) as

$$(y_t - \mu) = \phi_1(y_{t-1} - \mu) + a_t \tag{3.5}$$

or

$$y_t = (1 - \phi_1)\mu + \phi_1 y_{t-1} + a_t, \tag{3.6}$$

where μ is the mean of y_t and y_t now represents the actual data. The only formal difference between (3.1) and (3.6) is the inclusion of the intercept $(1 - \phi_1)\mu$. For convenience, unless otherwise stated, we shall work with deviations from the mean and interchangeably use z_t to denote the original data, y_t, and the deviation from the mean. We feel that this will not create any confusion as it will be clear from the context whether we are dealing with deviation or with the actual data.

If we view (3.1) as a linear regression model in which the dependent variable z_t is explained by and regressed on its previous values, z_{t-1}, then we can see why such a process is called *autoregressive*. The process described by equation (3.1) is called an autoregressive process of *order 1*, AR(1). The order of the process corresponds to the number of parameters that need to be estimated. In the case of an autoregressive process, that number corresponds to the number of lagged z's included in the model.

Figure 3.1 contains a schematic representation of an autoregressive process. If you compare this representation to the corresponding part of Figure 1.5, you will observe that the arrows are pointing in opposite directions (z_{t-1} and a_t in Figure 1.5 have been grouped together into e_t). The reason for this is that when we are trying to *construct a model* the only information we have available is z_t and *not* the shocks. Thus, to identify the model which governs the behavior of the series we must filter out the principal components of the series and then use these components, these filters, to deduce an

---

[3] For most of the result we do not really need the normality assumption.

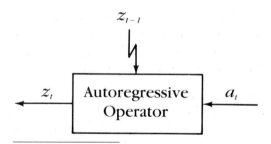

z_{t-1}

z_t | Autoregressive Operator | a_t

FIGURE 3.1   The AR(1) Model.

appropriate model. However, when we are trying to *explain the working of a process,* we have to describe how the principal components fit together.

3.1.2 Restrictions on the Autoregressive Parameters: Stationarity Condition

In Chapter 2 we introduced stationarity and indicated that if the series is not stationary it is not meaningful to estimate the mean, variance, and autocorrelations of such a series.

   If this stationarity condition has such a powerful effect on the analysis of the series, it is natural to inquire if this condition also introduces restrictions on the parameter values (ϕ's) of the autoregressive process. To answer this, let us estimate the mean and variance of an AR(1) model.

   Since ϕ_1 does not affect the calculation of the mean of the series represented by equation (3.1) (i.e., the mean can be calculated whether ϕ_1 is positive or negative, large or small), the constant mean condition (3.2) does not suggest that any restrictions should be imposed on ϕ_1. However, for the variance to be nonnegative (by definition *all* variances have to be nonnegative because the expectation of the square of a certain quantity cannot be negative), it is necessary that $|\phi_1|$ be less than or equal to 1. To see this, let us square both sides of equation (3.1) and take expectations:

$$E(z_t^2) = E[(\phi_1 z_{t-1} + a_t)^2]. \tag{3.7}$$

Since by assumption z_t is a deviation of the observed value from the mean, the left-hand side of (3.7) equals the variance of z_t. Evaluating (3.7) we get

$$\text{Var}(z_t) = E[(\phi_1 z_{t-1})^2 + 2\phi_1 z_{t-1} a_t + a_t^2]. \tag{3.8}$$

Working out the expectations of the right-hand side of (3.8) term by term, we obtain

$$\begin{aligned}
\text{Var}(z_t) &= E[(\phi_1 z_{t-1})^2] + E(2\phi_1 z_{t-1} a_t) + E(a_t^2) \\
&= \phi_1^2 E(z_{t-1}^2) + 2\phi_1 E(z_{t-1} a_t) + E(a_t^2) \\
&= \phi_1^2 \text{Var}(z_{t-1}) + 0 + \sigma_a^2.
\end{aligned} \tag{3.9}$$

In (3.9) we used the assumption (equation (3.4)) that z_{t-1} and a_t are independent, and the definition $E(a_t^2) = \sigma_a^2$.

Now, if the variance of the series is stationary, then it does not matter if we change the period over which we calculate the variance. Therefore,

$$\mathrm{Var}(z_t) = \mathrm{Var}(z_{t-1}).$$

Substituting this result into (3.9), we obtain

$$\mathrm{Var}(z_t) = \phi_1^2 \mathrm{Var}(z_t) + \sigma_a^2$$
$$(1 - \phi_1^2)\mathrm{Var}(z_t) = \sigma_a^2$$
$$\mathrm{Var}(z_t) = \frac{\sigma_a^2}{1 - \phi_1^2}. \tag{3.10}$$

Thus, for (3.10) to be nonnegative, $1 - \phi_1^2$ must be nonnegative, which implies that

$$\phi_1^2 \le 1 \quad \text{or} \quad |\phi_1| \le 1.$$

Note that for $|\phi_1| = 1$, the $\mathrm{Var}(z_t)$ is infinite, and then a series is also considered nonstationary. Therefore, for the variance to be both nonnegative and finite, the restriction imposed on the ϕ_1 value is

$$|\phi_1| < 1. \tag{3.11}$$

3.1.3 Special Characteristics of an AR Process

In the same way that a person's fingerprint can be used to identify an individual, there are certain *properties* which can be used to distinguish an AR process from any other process. In this section we shall discuss such properties.

AUTOCOVARIANCES AND AUTOCORRELATIONS

The covariance between two random variables is defined as

$$\mathrm{Cov}(x, y) = E[(x - Ex)(y - Ey)] \tag{3.12}$$

where Ex and Ey are the means of random variables x and y, respectively. The *auto*covariance of z_t at lag 1, denoted as λ_1, is a covariance between z_t and z_{t-1}, and is therefore defined as

$$\lambda_1 \equiv \mathrm{Cov}(z_t, z_{t-1}) = E(z_t z_{t-1}). \tag{3.13}$$

Again the z_t and z_{t-1} variables express deviations from the mean of the series. Note that because of stationarity assumptions, the autocovariance solely depends on the lag between z_t and z_{t-1}. We can also define the variance of z_t as the covariance between z_t and z_{t-1}, and we can therefore denote the variance of z_t as λ_0.

Substituting (3.1) for z_t in (3.13) and using the assumption $Ea_t = 0$, we find that

$$\lambda_1 = \phi_1 \text{Var}(z_{t-1}) + \text{Cov}(a_t, z_{t-1}). \tag{3.14}$$

Since a_t is independent of z_{t-1}, the last term of the right-hand side of (3.14) is zero, and therefore

$$\lambda_1 = \phi_1 \text{Var}(z_{t-1}). \tag{3.15}$$

Since the stationarity condition implies that $\lambda_0 \equiv \text{Var}(z_t) = \text{Var}(z_{t-1})$, we can rewrite (3.15) as

$$\lambda_1 = \phi_1 \lambda_0. \tag{3.16}$$

The autocovariance at lag 2, λ_2, is defined as the covariance between z_t and z_{t-2}. Again, by substituting (3.1) for z_t, it can be shown that λ_2 can be expressed as

$$\lambda_2 = \phi_1 \text{Cov}(z_{t-1}, z_{t-2}) + \text{Cov}(a_t, z_{t-2}). \tag{3.17}$$

Since the lag between z_{t-1} and z_{t-2} is one period, the $\text{Cov}(z_{t-1}, z_{t-2})$ is simply λ_1. Furthermore, since a_t is independent of z_{t-2}, (3.17) reduces to

$$\begin{aligned} \lambda_2 &= \phi_1 \lambda_1 \\ &= \phi_1^2 \lambda_0. \end{aligned} \tag{3.18}$$

Proceeding in the same way, it can be shown that for $k > 0$,

$$\lambda_k = \phi_1 \lambda_{k-1} \tag{3.19}$$

or

$$\lambda_k = \phi_1^k \lambda_0. \tag{3.20}$$

A useful parameter, as we shall see in Chapter 4, is the autocorrelation parameter. The autocorrelation at lag k, ρ_k, is defined as the ratio of the autocovariance at lag k to the autocovariance at lag zero ($\text{Var}(z_t)$) and is therefore a scaled autocovariance. For the autoregressive process of order 1, the autocorrelations are obtained as

$$\rho_k = \frac{\lambda_k}{\lambda_0} = \phi_1^k, \qquad k > 0. \tag{3.21}$$

MEMORY FUNCTION

Another distinguishing feature of the autoregressive process is its long "memory." Suppose that we write equation (3.1) in terms of past errors by successively eliminating the lagged z_t's. That is, substitute

$$z_{t-1} = \phi_1 z_{t-2} + a_{t-1} \tag{3.22}$$

into equation (3.1) to get

$$z_t = \phi_1^2 z_{t-2} + a_t + \phi_1 a_{t-1}. \tag{3.23}$$

Then, substitute

$$z_{t-2} = \phi_1 z_{t-3} + a_{t-2}$$

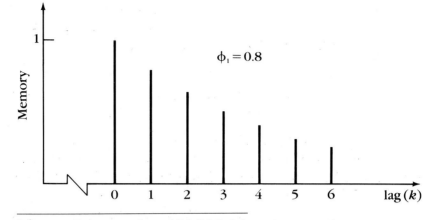

FIGURE 3.2   Memory Function of an AR(1) Model.

into equation (3.23) to obtain

$$z_t = \phi_1^3 z_{t-3} + a_t + \phi_1 a_{t-1} + \phi_1^2 a_{t-2}, \qquad (3.24)$$

and so on, until we get the *error-shock* form,

$$z_t = a_t + \phi_1 a_{t-1} + \phi_1^2 a_{t-2} + \phi_1^3 a_{t-3} + \cdots . \qquad (3.25)$$

Thus the AR(1) model is rewritten as a sum of current error and an infinite number of past error terms. Therefore, current observation, z_t, is still influenced by shocks, a_t's, that occurred in the distant past. We claim that an AR(1) process has an infinite memory. If the process is stationary, $|\phi_1| < 1$, the effect of shock will gradually dissipate, whereas this is not the case if the process is nonstationary.

A *memory coefficient* at lag one is defined as the coefficient of the a_{t-1} error. It also represents the effect of a shock at period t on the observation z_{t+1}. Alternatively, the memory coefficient at lag one indicates the effect on current observation z_t of a shock that occurred one period ago. The memory coefficients at any lag k, $k > 0$, are similarly defined and have a similar interpretation. The plot of the memory coefficients as a function of the lag k, $k \geq 0$, is called the *memory function* of the process. Figure 3.2 contains the memory function of a stationary AR(1) model.

3.1.4 An Intuitive Look at the Stationarity Condition

In order to gain a better appreciation for the stationarity condition expressed by equation (3.11), let us calculate a few autocorrelations using equation

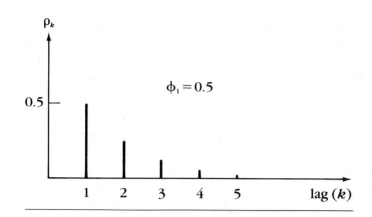

FIGURE 3.3   Autocorrelation Function of a Stationary AR(1) Model.

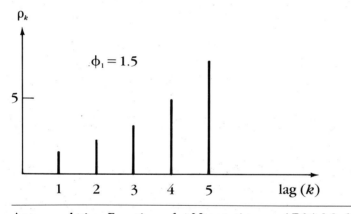

FIGURE 3.4   Autocorrelation Function of a Nonstationary AR(1) Model.

(3.21), using the value $\phi_1 = 0.5$ and $\phi_1 = 1.5$, respectively for the autoregressive parameter. Figures 3.3 and 3.4 represent the graph of the first five autocorrelations. Such a graph is defined as the autocorrelation function (see Section 4.2). Note how the autocorrelations seem to exponentially decay to zero when the autoregressive parameter $\phi_1 = 0.5$ and how they grow[4] when $\phi_1 = 1.5$. The reader should be aware that we used a different vertical scale in these two figures. Let us see what these two types of autocorrelation behavior imply about the data. If the observation at time $t - k$, z_{t-k}, has some effect on the observation at time t, z_t, then we would expect that ρ_k would be nonzero. In most situations we would expect that recent past values would have some effect on the current values, but that the effect would become smaller and smaller the further back we go in time. We have seen that if

[4] Technically, the autocorrelations are not defined for $|\phi_1| > 1$. However, a finite sample estimate can always be calculated using the formulas specified in Chapter 4. We also refer to Section 4.2.2 for the use of autocorrelations in deciding if a series is nonstationary.

$\phi_1 = 1.5$, ρ_k becomes larger and larger. This indicates that an event ten years ago has a much stronger effect on the current value of the data than an event that happened just last year. Furthermore, such autocorrelation behavior also implies that an event 100 years ago (if the data were available) would have even a more dramatic effect on today's observation than an event that occurred ten years ago. This, of course, for most data series is not very reasonable.

Another way to see why $|\phi_1|$ must be less than one is to look at the error-shock form of the process, equation (3.25). We see that the effect of a shock at time t, a_t, persists over a long period of time. Now it seems reasonable to assume that a recent error will have an impact on an event occurring today, and that as time goes by the effect of the error will gradually disappear. For this assumption to hold, $|\phi_1|$ must be less than one.[5] Hence, a reasonable memory function of an AR(1) process should look something like Figure 3.2.

3.1.5 Higher-Order Autoregressive Models

Equation (3.1) can be broadened to include more lagged variables. For example, if events two periods ago also had an effect on what is happening today, we could extend (3.1) to include z_{t-2}; that is, we could express z_t as

$$z_t = \phi_1 z_{t-1} + \phi_2 z_{t-2} + a_t, \tag{3.26}$$

where ϕ_1 and ϕ_2 are autoregressive parameters to be estimated. If we reexpress (3.26) in terms of actual data as opposed to deviations from the mean, we would primarily have to add a constant to the equation and change all z's to y's (see equation 3.6). For an AR(2) process this constant would be

$$\delta = (1 - \phi_1 - \phi_2)\mu, \tag{3.27}$$

with μ the mean of the original data y_t.

The model (3.26) is an autoregressive process of order 2, or AR(2). In general, a pth order autoregressive model, AR(p), is written as

$$z_t = \phi_1 z_{t-1} + \phi_2 z_{t-2} + \cdots + \phi_p z_{t-p} + a_t. \tag{3.28}$$

Stationarity conditions, as well as autocovariances, autocorrelations, and memory function characteristics can be derived for an autoregressive process

---

[5] In a later chapter (see Section 5.3.1) we will indicate that for an autoregressive process to be stationary, the size of the roots should be larger than one. Alternatively stated, the roots should lie outside the unit circle.

For an AR(1) process it is easy to show the connection with this requirement and the above condition that $|\phi_1| < 1$.

The first-order equation or polynomial that we have to solve for x in order to obtain the root is

$$1 - \phi_1 x = 0.$$

The solution or root is $x = 1/\phi_1$. If $|\phi_1| < 1$, we have $|x| > 1$. The size of the root is in this case defined as the value x^2.

of any order. The results, however, will not be as simple as for an AR(1) process. In general, these conditions and parameters will be complicated functions of the ϕ_i's parameters. Fortunately, for most business and economic data (omitting seasonality for now)[6] we mostly encounter AR processes of order not higher than two. In Chapter 4, we discuss how to identify potential models and shall explicitly analyze the characteristics of an AR(2) process.

3.2 MOVING AVERAGE MODELS

A simple extension of the AR(1) model would be to include past errors to see if they can improve on the time series representation of the data. Specifically, we could modify the AR(1) model to obtain

$$z_t = \phi_1 z_{t-1} + a_t - \theta_1 a_{t-1}, \tag{3.29}$$

where a_{t-1} is the error at period $t-1$, and θ_1 is called the *moving average parameter*, which describes the effect of the past error on z_t, and which needs to be estimated. For reasons that will become clear, it is also customary to write a negative sign in front of the parameter θ_1. Model (3.29) has the form of a multiple regression model with two independent variables z_{t-1} and a_{t-1}, but this analogy is imperfect because the lagged error term is not observed and, as such, cannot be used as a regressor.

3.2.1 First-Order Moving Average Models

A special model is obtained from (3.29) by omitting the lagged variable z_{t-1}. This model is called a *moving average* model of order 1 and expresses the current value of the series z_t as a linear function of the current and previous errors or shocks, a_t and a_{t-1}. Mathematically, we express a first-order moving average model,[7] MA(1), as

$$z_t = a_t - \theta_1 a_{t-1}, \tag{3.30}$$

where θ_1 is the moving average parameter. As with an autoregressive process, the random shocks in a moving average process are assumed to be normally and independently distributed with mean zero and constant variance σ_a^2; that is, they satisfy equations (3.2) and (3.3).

Figure 3.5 contains a schematic representation of a moving average process. Note that for the purpose of explaining the process the arrows once again point in the opposite direction from those in Figure 1.5. The moving

---

[6] We shall deal with these seasonal effects in Section 3.6.

[7] The order of the process, as in the autoregressive case, corresponds to the number of parameters that need to be estimated. In the case of a moving average process this number corresponds to the number of lagged a_t's included in the representation of the process.

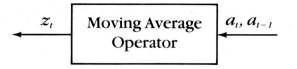

FIGURE 3.5   The MA(1) Model.

average filter, or operator, in the box indicates that we have to multiply the past residual a_{t-1} with the moving average parameter, θ_1, and subtract this from the current residual, a_t, to obtain z_t as output.

The MA(1) for the actual data, as opposed to deviations from the mean, would be written as

$$y_t - \mu = a_t - \theta_1 a_{t-1}$$

or

$$y_t = \mu + a_t - \theta_1 a_{t-1}, \tag{3.31}$$

where μ is the mean of the series. Unlike the AR(1) process, the intercept in the case of an MA(1) process is actually the mean of the series.

3.2.2 Parsimony

An important reason for relying on a moving average process is that the number of parameters to be estimated can be drastically reduced. And, since the parameters have to be estimated with a finite number of data points, it is important to represent a time series process with as few parameters as possible. That is, we want a parsimonious representation of the model.

To demonstrate how an MA process reduces the number of parameters to be estimated, we will show the equivalence between an MA(1) process and an AR(∞) process (i.e., an autoregressive model with an infinite number of parameters).

Let us start by solving (3.30) in terms of a_t; that is,

$$a_t = z_t + \theta_1 a_{t-1}. \tag{3.32}$$

Similarly, a_{t-1} can be expressed as

$$a_{t-1} = z_{t-1} + \theta_1 a_{t-2}. \tag{3.33}$$

Next, substituting (3.33) into (3.30), we obtain

$$z_t = -\theta_1 z_{t-1} - \theta_1^2 a_{t-2} + a_t. \tag{3.34}$$

Continuing the substitution for $a_{t-2}, a_{t-3}, \ldots$, the MA(1) process can be expressed as

$$z_t = -\theta_1 z_{t-1} - \theta_1^2 z_{t-2} - \theta_1^3 z_{t-3} - \theta_1^4 z_{t-4} - \cdots + a_t. \tag{3.35}$$

Equation (3.35), called the *inverted form* of a moving average process, represents a process with an *infinite* number of autoregressive terms and no lagged error terms. Although the coefficients of the lagged z_t's would approach zero if $|\theta_1| < 1$, we would still have to include many parameters if we rely solely on the autoregressive representation of an MA(1) process. The objection could be raised that the many parameters are just powers of the same basic coefficient θ_1. However, it remains the case that the number of observations actually available to estimate these parameters would be greatly reduced and determined by the highest lag term included.

Therefore, by introducing moving average models, we have discovered a *parsimonious* way of representing complicated models; that is, we have found a way to *reduce* the number of parameters necessary to adequately depict the process. And, as we will see in Chapter 5, methods exist for estimating the parameters of the MA(1) process (3.30) efficiently.

3.2.3 Restrictions on the Moving Average Parameters: Invertibility Condition

In Section 3.1.2 we evaluated the mean, variance, and autocorrelations of an AR(1) process to see what conditions could be imposed on the autoregressive parameter as a result of the stationarity requirement of the model. With the same purpose in mind we propose to carry out the same calculations for an MA(1) process.

It is easy to show that the mean and the variance of an MA model are constant over time without imposing any restrictions on the values of the moving average parameters. Using the white noise conditions specified in (3.2), (3.3), and (3.4), we see that for an MA(1) process represented by (3.30), the mean is zero and the variance is constant,

$$Ez_t = E(a_t - \theta_1 a_{t-1}) = Ea_t - \theta_1 Ea_{t-1} = 0 \tag{3.36}$$

$$\begin{aligned} \mathrm{Var}(z_t) &= E[(a_t - \theta_1 a_{t-1})^2] \\ &= E(a_t^2 - 2\theta_1 a_{t-1}a_t + \theta_1^2 a_{t-1}^2) \\ &= (1 + \theta_1^2)\sigma_a^2. \end{aligned} \tag{3.37}$$

The term $E(-2\theta_1 a_{t-1}a_t) = 0$ because of (3.3). Similarly, we see that the covariance between z_t and z_{t-1} is a constant:

$$\begin{aligned} \mathrm{Cov}(z_t, z_{t-1}) &= E[(a_t - \theta_1 a_{t-1})(a_{t-1} - \theta_1 a_{t-2})] \\ &= E(-\theta_1 a_{t-1}^2) \\ &= -\theta_1 \sigma_a^2. \end{aligned} \tag{3.38}$$

The second line in (3.38) follows from (3.3). Consequently, the autocorrelation is also constant.

Thus, the stationarity condition does not impose restrictions on the value of θ_1. However, if we were to allow $|\theta_1| > 1$, then we would have to accept a very unrealistic interpretation of equation (3.35), namely, that the influence of past observations (lagged z_t's) increases the further back we go in time. It is very unrealistic to think that an event that occurred many years ago has more influence on today's situation than an event that occurred recently. Therefore, we must restrict θ_1 to satisfy $|\theta_1| < 1$. This condition is called the *invertibility condition* of an MA(1) model.[8] In Chapter 4 we will justify this name and give another justification why $|\theta_1|$ must be less than 1.

3.2.4 Special Characteristics of an MA Process

The moving average process can also be distinguished from other processes by its autocovariances, autocorrelations, and memory function. In this section we shall examine the characteristics of these properties for an MA(1) process.

AUTOCOVARIANCES AND AUTOCORRELATIONS

In section 3.2.3 we calculated the autocovariances at lags zero and one. We found that

$$\lambda_0 = \mathrm{Var}(z_t) = (1 + \theta_1^2)\sigma_a^2 \tag{3.39}$$

and

$$\lambda_1 = \mathrm{Cov}(z_t, z_{t-1}) = -\theta_1\sigma_a^2. \tag{3.40}$$

Similarly, the autocovariance at lag 2, λ_2, is obtained as

$$\lambda_2 = \mathrm{Cov}(z_t, z_{t-2}) = E[(a_t - \theta_1 a_{t-1})(a_{t-2} - \theta_1 a_{t-3})]$$

$$= 0. \tag{3.41}$$

Equation (3.41) is obtained by repeated use of equation (3.3), $E(a_t a_s) = 0$ for $t \neq s$. Similarly, it can be shown that $\lambda_k = 0$ for $k \geq 2$. Therefore, the autocorrelations, defined as $\rho_k = \lambda_k/\lambda_0$, are

$$\rho_1 = -\theta_1/(1 + \theta_1^2);$$
$$\rho_k = 0, \quad k \geq 2. \tag{3.42}$$

In comparing these autocorrelations with the autocorrelations of the AR(1) process as specified in (3.21), we observe that whereas the autocorrelations for the AR(1) process die out gradually, for an MA(1) process they die out abruptly, only ρ_1 is not zero. (By definition, the autocorrelation ρ_0 always equals 1, regardless of the process being analyzed.)

MEMORY FUNCTION

The memory function, as you will recall, is a plot of the memory coefficients, the coefficients of the error terms if we represent the current value of the series, z_t, in terms of past errors only. Therefore, the memory function for

---

[8] Again, this invertibility condition can be translated in a condition about the root of the equation. As in footnote 5, we can again show that for a first-order equation $|\theta_1| < 1$ corresponds to the size of the root being larger than one.

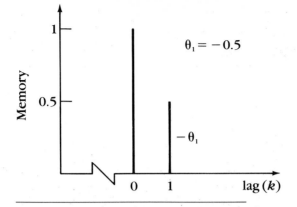

FIGURE 3.6 Memory Function of an MA(1) Model.

an MA process can be obtained directly from the definition of the process.

From equation (3.30) it follows directly that for an MA process, a shock at time t, a_t, will influence the observations at time t and $t + 1$, but will have no effect beyond period $t + 1$. At time t the system will feel the full impact of the shock, but at time $t + 1$ its effect will be proportional to θ_1. Therefore, the memory of an MA(1) process only lasts for one period. In Figure 3.6 we have drawn the memory function for an MA(1) with $-1 < \theta_1 < 0$. Specifically, in Figure 3.6 we have $\theta_1 = -0.5$. Note that we represent the coefficient in front of a_{t-1} with it's negative sign; that is, $-\theta_1$ and not θ_1.

3.2.5 Higher-Order Moving Average Models

The MA(1) model in (3.30) can easily be extended to include additional lagged residual terms. A qth order MA process, MA(q), for instance, can be expressed as

$$z_t = a_t - \theta_1 a_{t-1} - \theta_2 a_{t-2} - \cdots - \theta_q a_{t-q}. \qquad (3.43)$$

The memory function for an MA(q) model is a plot of the values (1, $-\theta_1, \ldots, -\theta_q$). The effect of a shock, a_t, will persist only for q periods. As we shall see in Chapter 6, the memory function of this process will play an important role in generating forecasts.

The invertibility condition for the general MA(q) process is not as simple to derive as for the MA(1) process, as it involves nonlinear functions of all the q parameters in the model. Again, it is fortunate that many social science series can be represented with process of order 1 or 2. In Chapter 4 we will discuss the MA(2) process in more detail and state the invertibility condition that needs to be imposed on θ_1 and θ_2.

3.3 MIXED AUTOREGRESSIVE MOVING AVERAGE MODELS

3.3.1 ARMA(1,1) Models

In equation (3.29) we represented a process with an autoregressive and a moving average term. Such a model can also be obtained as follows.
Rewrite an AR(1) model as

$$z_t - \phi_1 z_t = e_t, \tag{3.44}$$

where e_t, instead of being white noise, represents an MA(1) process,

$$e_t = a_t - \theta_1 a_{t-1}. \tag{3.45}$$

Then, combining (3.44) with (3.45), we obtain

$$z_t - \phi_1 z_{t-1} = a_t - \theta_1 a_{t-1}$$

or

$$z_t = \phi_1 z_{t-1} + a_t - \theta_1 a_{t-1}. \tag{3.46}$$

Again, the z_t's represent deviations from the mean of the series. Otherwise, a constant δ, equal to

$$\delta = (1 - \phi_1)\mu, \tag{3.47}$$

would have to be included in an equation as (3.46) but with z_t's replaced by y_t's.

Models of the form (3.46) are called (mixed) *autoregressive moving average models* and are denoted as ARMA(p,q). The p refers to the number of autoregressive parameters, and the q to the number of moving average parameters. Model (3.46) is an ARMA(1,1) process and Figure 3.7 contains a schematic representation of the model. The operator on the right represents the moving average part (i.e., equation (3.45)), and the one on the left represents the autoregressive part (i.e., equation (3.44)).

It is important to recognize that the introduction of mixed autoregressive moving average models achieves great parsimony in model specification. It can readily be shown that (3.46) can be rewritten as an infinite order MA process by successively eliminating z_{t-1}, z_{t-2}, etc. Similarly, (3.46) can be

FIGURE 3.7 The ARMA Model.

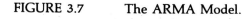

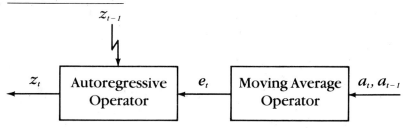

rewritten as an infinite order AR process by successively eliminating a_{t-1}, a_{t-2}, etc. But, as we remarked in Section 3.2.2, a pure AR or a pure MA model would only be obtained at the cost of a very large number of parameters to be estimated, and of very inefficient use of the data.

3.3.2 Restrictions on the Parameters of the Mixed Model

Again, we calculate the mean and covariances of the process represented by (3.46) to see what restrictions the stationarity and invertibility conditions impose on the parameters ϕ_1 and θ_1. It is easy to see that $Ez_t = 0$ irrespective of the values of ϕ_1 and θ_1. (Note that (3.46) is written in deviations from the mean.) The variance can be calculated by squaring the right-hand side of (3.46) and taking expectations:

$$\lambda_0 \equiv \text{Var}(z_t) = E(z_t^2) = E[(\phi_1 z_{t-1} + a_t - \theta_1 a_{t-1})^2]$$
$$= E(\phi_1^2 z_{t-1}^2 + a_t^2 + \theta_1^2 a_{t-1}^2 + 2\phi_1 z_{t-1} a_t - 2\theta_1 \phi_1 z_{t-1} a_{t-1} - 2\theta_1 a_t a_{t-1})$$
$$= \phi_1^2 (E z_{t-1}^2) + \sigma_a^2 + \theta_1^2 \sigma_a^2 - 2\theta_1 \phi_1 E(z_{t-1} a_{t-1}). \tag{3.48}$$

The last line was obtained using conditions (3.3) and (3.4). In calculating $E(z_{t-1} a_{t-1})$ we recognize that

$$z_{t-1} = \phi_1 z_{t-2} + a_{t-1} - \theta_1 a_{t-2} \tag{3.49}$$

and that, therefore, using conditions (3.3) and (3.4),

$$E(z_{t-1} a_{t-1}) = \sigma_a^2. \tag{3.50}$$

Substituting (3.50) into (3.48) and using the fact that because of stationarity,

$$\text{Var}(z_t) = E(z_t^2) = E(z_{t-1}^2),$$

we have that

$$\lambda_0 = \text{Var}(z_t) = \frac{1 + \theta_1^2 - 2\theta_1 \phi_1}{1 - \phi_1^2} \sigma_a^2. \tag{3.51}$$

The autocovariance at lag 1 can similarly be calculated as

$$\lambda_1 = E(z_t z_{t-1}) = E(\phi_1 z_{t-1}^2 + a_t z_{t-1} - \theta_1 a_{t-1} z_{t-1})$$
$$= \phi_1 \lambda_0 - \theta_1 \sigma_a^2,$$

since $E(a_t z_{t-1}) = 0$ and by using (3.50). Substituting (3.51) for λ_0 and rearranging terms, we obtain

$$\lambda_1 = \frac{(1 - \phi_1 \theta_1)(\phi_1 - \theta_1)}{1 - \phi_1^2} \sigma_a^2. \tag{3.52}$$

Finally, the autocovariance at lag 2 is

$$\lambda_2 = E(z_t z_{t-2}) = E(\phi_1 z_{t-1} z_{t-2} + a_t z_{t-2} - \theta_1 a_{t-1} z_{t-2}) = \phi_1 \lambda_1. \quad (3.53)$$

Note in the derivation of (3.53) that $E(a_t z_{t-2}) = 0$ and $E(a_{t-1} z_{t-2}) = 0$, using (3.4), and that, because of stationarity, $E(z_{t-1} z_{t-2}) = E(z_t z_{t-1})$. Similarly, it can be shown that for all $k \geq 2$

$$\lambda_k = \phi_1 \lambda_{k-1}$$
$$= \phi_1^{k-1} \lambda_1. \quad (3.54)$$

By considering what (3.54) is saying about the correlations between past observations, and by examining what would happen if past shocks a_{t-1}, a_{t-2}, . . . were successively eliminated from (3.46) and replaced by lagged values of z_t, we can discern the restrictions that the stationarity and invertibility conditions impose on the parameters ϕ_1 and θ_1 in order for those equations to make intuitive sense. In order for the correlation between observations to decrease the greater the lag between the two series, it is necessary to restrict $|\phi_1|$ to be less than one. Otherwise, as (3.54) implies, the present would be more strongly correlated with the distant past than with more recent events. Similarly, in order to understand the generating process of the ARMA(1,1) model, we have to restrict $|\theta_1|$ to be less than one. This will be explained in the next section where we will derive the memory function of this process by successively replacing lagged z_t's with lagged a_t's. We therefore conclude that for the ARMA(1,1) model stationarity requires $|\phi_1| < 1$ and invertibility requires $|\theta_1| < 1$.

3.3.3 Special Characteristics of an ARMA(1,1) Process

We will now show how the autocovariances, autocorrelations, and memory function of an ARMA(1,1) process are distinct from those of an AR(1) or an MA(1) process.

AUTOCOVARIANCES AND AUTOCORRELATIONS

In section 3.3.2. we derived the autocovariances for the ARMA(1,1) process. For convenience we summarize these results below:

$$\lambda_0 = \frac{1 + \theta_1^2 - 2\theta_1 \phi_1}{1 - \phi_1^2} \sigma_a^2,$$

$$\lambda_1 = \frac{(1 - \phi_1 \theta_1)(\phi_1 - \theta_1)}{1 - \phi_1^2} \sigma_a^2,$$

$$\lambda_k = \phi_1 \lambda_{k-1}, \quad k \geq 2. \quad (3.55)$$

The autocorrelations, defined as $\rho_k = \lambda_k / \lambda_0$, therefore, are

$$\rho_1 = \frac{(1 - \phi_1 \theta_1)(\phi_1 - \theta_1)}{1 + \theta_1^2 - 2\theta_1 \phi_1}$$

$$\rho_k = \phi_1 \rho_{k-1}, \quad k \geq 2. \quad (3.56)$$

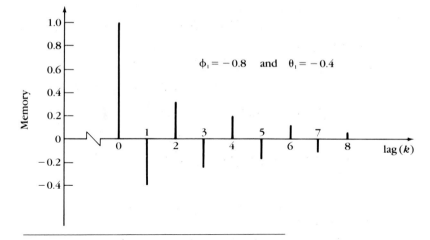

FIGURE 3.8   Memory Function of an ARMA(1,1) Model.

As for a stationary AR(1) process, the autocorrelations of a stationary ARMA(1,1) die out gradually. The rate of decline is determined by the AR parameter ϕ_1. However, the first autocorrelation, ρ_1, is not equal to ϕ_1, but is influenced by both the AR and the MA parameters. Based on these properties, it could be difficult in practice to distinguish between an ARMA(1,1) process and an AR(1) process. However, in the next chapter we will present additional statistics which will facilitate the discrimination.

MEMORY FUNCTION

We can evaluate the memory function by successively eliminating the lagged observations z_{t-1}, z_{t-2}, . . . from (3.46). For the ARMA(1,1) process we obtain

$$z_t = a_t + (\phi_1 - \theta_1)a_{t-1} + \phi_1(\phi_1 - \theta_1)a_{t-2} + \phi_1^2(\phi_1 - \theta_1)a_{t-3}$$
$$+ \phi_1^3(\phi_1 - \theta_1)a_{t-4} + \cdots. \qquad (3.57)$$

Figure 3.8 contains the memory function for this ARMA(1,1) process with $\phi_1 = -0.8$ and $\theta_1 = -0.4$.

3.3.4 Higher-Order ARMA Models

The ARMA(1,1) model (3.46) can also be extended to include more autoregressive and moving average parameters. In general, an ARMA(p,q) model is represented by

$$z_t = \phi_1 z_{t-1} + \phi_2 z_{t-2} + \cdots + \phi_p z_{t-p} + a_t - \theta_1 a_{t-1} - \cdots - \theta_q a_{t-q}. \quad (3.58)$$

It is customary to precede the values of the θ_i parameters with a negative sign and the ϕ_i parameters with a positive sign. It is important to be familiar with this convention because in most computer programs the estimates of the ϕ_i's and the θ_i's are printed assuming that the sign convention is the one given in (3.58). In section 3.5.3 we will introduce some simplifying notation for these models, at which time it will become clear that it is quite consistent to have positive signs for the ϕ_i parameters and negative signs for the θ_i parameters.

At this point we have discussed the AR, MA, and ARMA models which can be used to model a wide variety of stationary time series. However, in practice we frequently encounter time series that are nonstationary. Fortunately, several of such series may be made stationary by simple transformations. In Section 3.4 we will present and analyze a more general class capable of modeling many such nonstationary series. Another class includes the seasonal model. (These seasonal models form the topic discussed in Section 3.6.) Section 3.5 is devoted to some simplifying notation. Again, an understanding of seasonal modeling is extremely important for applied time series analysis because many series are quarterly/monthly series in which there are important seasonal characteristics. As mentioned in the beginning of this chapter, the reader might first read Section 3.5 and then proceed to Chapter 4, returning to Sections 3.4 and 3.6 after studying the material presented in Sections 4.1 and 4.2. In these two later sections we will show, using generated data, that the empirical characteristics of the different models allow us to discriminate between the different theoretical models.

3.4 NONSTATIONARY MODELS

If the series is not stationary, then we cannot use as such any of the models discussed so far. Often transformations of the data will be necessary to induce stationarity and the general nature of these transformations has already been discussed in Chapter 2. In this section we will complete the discussion of the Box–Jenkins filters and indicate how all these filters fit together to form the Autoregressive Integrated Moving Average (ARIMA) model. The ARIMA model is introduced with a discussion of the random walk model.

3.4.1 Random Walk Model

We have shown that for an AR(1) model to be stationary, $|\phi_1|$ must be less than 1. A special model is obtained if $\phi_1 = 1$. In this case, the AR(1) model can be written as

$$z_t = z_{t-1} + a_t \tag{3.59}$$

or as

$$z_t - z_{t-1} = a_t. \tag{3.60}$$

Model (3.59) is known as a *random walk model,* a model in which *changes* are brought about by a white noise series. If the a_t represent steps taken forwards or backwards at time t, then z_t will represent the position of the walker at time t, and the decision where to go during the next period is not influenced by where the walker is now. Figure 3.9 shows the realization of a random walk process. The series makes wide swings. *It drifts.* If we had only a short realization of the process, we might conclude that the process follows a trend. Note how this figure resembles the stock market history seen daily in *The Wall Street Journal.* Indeed, studies have shown (e.g., Fama, 1970), that the stock market prices behavior conforms well to a random walk model, as successive price *changes* are essentially independent.

MEMORY FUNCTION

By successively eliminating the lagged $z_{t-1}, z_{t-2}, \ldots$ from (3.59), we obtain

$$z_t = a_t + a_{t-1} + a_{t-2} + \cdots. \qquad (3.61)$$

FIGURE 3.9   A Random Walk Series.

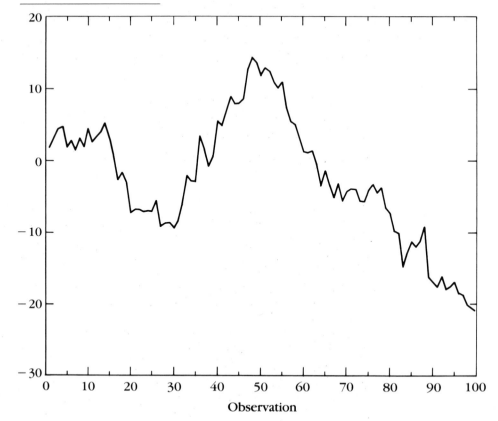

Observation

Equation (3.61) clearly demonstrates that a random walk process is a stochastic process wherein successive random shocks accumulate over time. From (3.61) we see that the memory function of the random walk model is constant and takes on the value 1 for all lags. This means that the effect of a shock incurred at time t will never die out. A memory function that does not die out is a general characteristic of a nonstationary series.

3.4.2 Autoregressive Integrated Moving Average Models

Note that although z_t in the random walk model (3.59) is nonstationary, equation (3.60) shows that there is a rather simple way to model the random walk. The first differences of the series, $z_t - z_{t-1}$, constitute a stationary series. Indeed, by differencing the random walk process, we obtain a time series whose observations are the random shocks $a_1, a_2, \ldots, a_t$. In other words, differencing transforms the random walk into a white noise process.

As seen in Chapter 2, nonstationary series can often be transformed into stationary series by means of differencing. By working with changes in the realizations, one can often induce stationarity so that some of the AR, MA, or ARMA filters discussed above can then be used to model the differenced series.

Now, suppose that the first differences of a series are stationary. Then, by defining the difference between consecutive values of z_t as

$$w_t = z_t - z_{t-1}, \tag{3.62}$$

we could replace z_t and z_{t-1} in the mixed autoregressive moving average model, the ARMA(1,1) model (see (3.46)) with w_t and w_{t-1} to obtain

$$w_t = \phi_1 w_{t-1} + a_t - \theta_1 a_{t-1}. \tag{3.63}$$

The process for z_t expressed by (3.62) and (3.63) is called an *autoregressive integrated moving average* (ARIMA) model. In particular, this model is labeled an ARIMA(1,1,1) model. The numbers inside the parentheses refer to the order of the autoregressive process, the degree of differencing required to induce stationarity, and the order of the moving average process, respectively. In general, autoregressive integrated moving average models are denoted as ARIMA(p,d,q).

The term *integrated* is possibly misleading. A more appropriate term would be "summed together." From the first difference transformation, (3.62), it can be shown that the actual realizations, z_t, can be written as an infinite *sum* of past and present differences:

$$z_t = w_t + w_{t-1} + w_{t-2} + \cdots. \tag{3.64}$$

We are now in a position to summarize the general structure of an ARIMA process. Figure 3.10 pictures the three operators discussed in this chapter. The moving average operator transforms white noise into an intermediate

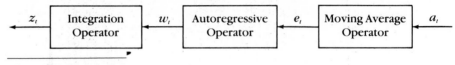

FIGURE 3.10   The ARIMA Model.

series, e_t. For a first-order moving average filter, e_t would be represented as

$$e_t = a_t - \theta_1 a_{t-1}. \tag{3.65}$$

The autoregressive operator transforms e_t into another intermediate series, w_t. Again, for a first-order autoregressive filter, w_t would be expressed as

$$w_t = \phi_1 w_{t-1} + e_t. \tag{3.66}$$

Finally, the integration operator transforms w_t to z_t.

The models described in Sections 3.1, 3.2, and 3.3 are also ARIMA models. An AR(p) model can always be denoted as an ARIMA($p,0,0$), an MA(q) model as an ARIMA($0,0,q$), and an ARMA(p,q) as an ARIMA($p,0,q$).

3.5 NOTATION

In this section we shall introduce some new notation which will simplify the writing of ARIMA models.

3.5.1 Difference Operator

If z_t represents the "raw" data or the series obtained after making a transformation to stabilize the variance, such as a logarithmic or a square root transformation, then we can write the first differences as

$$w_t = z_t - z_{t-1}. \tag{3.67}$$

The symbol w_t for first differences is not universally accepted. Sometimes the symbol ∇ is used to denote the difference operator; that is, the first differences can be represented by ∇z_t, defined as

$$\nabla z_t = z_t - z_{t-1}. \tag{3.68}$$

Second-order consecutive differencing can now be defined as

$$\begin{aligned}
\nabla^2 z_t &= \nabla(\nabla z_t) \\
&= \nabla(z_t - z_{t-1}) \\
&= (z_t - z_{t-1}) - (z_{t-1} - z_{t-2}) \\
&= z_t - 2z_{t-1} + z_{t-2}.
\end{aligned} \tag{3.70}$$

In general, the dth-order consecutive difference is represented by $\nabla^d z_t$, and is calculated by consecutively taking differences of the differences.

In the next section we shall be discussing seasonal models. In Section 2.6 we introduced seasonal differencing. Again we can use the ∇ symbol for seasonal differencing. We define the first-order *seasonal* difference with a *span* of s periods as

$$\nabla_s z_t = z_t - z_{t-s}. \tag{3.71}$$

If we were dealing with quarterly data, the span would be equal to four. First-order seasonal differencing with a span of four is denoted as

$$\nabla_4 z_t = z_t - z_{t-4}. \tag{3.72}$$

Second-order seasonal differences with a span of 12 are denoted by $\nabla_{12}^2 z_t$, and are equal to

$$\begin{aligned} \nabla_{12}^2 z_t &= \nabla_{12}(z_t - z_{t-12}) \\ &= (z_t - z_{t-12}) - (z_{t-12} - z_{t-24}) \\ &= z_t - 2z_{t-12} + z_{t-24}. \end{aligned} \tag{3.73}$$

Sometimes we will take seasonal differences of consecutive differences. The first seasonal differences, with span equal to four, of the first consecutive differences are defined as

$$\begin{aligned} \nabla_4 \nabla z_t &= \nabla_4(z_t - z_{t-1}) \\ &= (z_t - z_{t-1}) - (z_{t-4} - z_{t-5}) \\ &= (z_t - z_{t-4}) - (z_{t-1} - z_{t-5}) \\ &= \nabla \nabla_4 z_t. \end{aligned} \tag{3.74}$$

Note, from (3.74), that the order in which we take the differences does not matter. We can as well first take consecutive and then seasonal differencing or reverse the order of the differencing.

In general, combinations of seasonal and consecutive differencing are written as

$$\nabla_s^D \nabla^d z_t, \tag{3.75}$$

where D represents the order of the seasonal difference operator, s the span, (i.e., the length of the seasonal cycle), and d the order of the consecutive difference operator.

3.5.2 Backward Shift Operator

Another useful notation is the *backward shift operator*, or backshift operator. The backward shift operator, B, is defined as

$$Bz_t = z_{t-1}. \tag{3.76}$$

Hence, $B^2 z_t = B(Bz_t) = Bz_{t-1} = z_{t-2}$, and in general

$$B^k z_t = z_{t-k}. \tag{3.77}$$

Note that there exists a relationship between the backward shift operator and the difference operator. From the definition of ∇ and using the backshift operator we can express ∇z_t as

$$\nabla z_t = z_t - z_{t-1} = z_t - Bz_t = (1 - B)z_t. \tag{3.78}$$

Therefore, the relationship between the difference operator and the backshift operator is

$$\nabla = 1 - B. \tag{3.79}$$

Box and Jenkins (1970) almost exclusively rely on the B operator. However, we feel that the ∇ operator allows us to further simplify the notation.

An nth-order consecutive difference operator expressed in terms of the backward shift operator is obtained by expanding the polynomial $(1 - B)^n$. For example, for a second-order differencing, we have

$$\nabla^2 z_t = (1 - B)^2 z_t = (1 - 2B + B^2)z_t = z_t - 2z_{t-1} + z_{t-2}. \tag{3.80}$$

3.5.3 ARIMA Models

Using the ∇ and B notations, we can represent any ARIMA(p,d,q) model by the general formula

$$\phi(B)w_t = \theta(B)a_t, \tag{3.81}$$

where the polynomials in B are defined as

$$\phi(B) = 1 - \phi_1 B - \cdots - \phi_p B^p \tag{3.82}$$

$$\theta(B) = 1 - \theta_1 B - \cdots - \theta_q B^q, \tag{3.83}$$

and

$$w_t = \begin{cases} \nabla^d z_t & d > 0 \\ z_t & d = 0. \end{cases} \tag{3.84}$$

Note in (3.82) and (3.83) that all coefficients now have a minus sign in front of each parameter. This is the consistent notation used in this book. Of course, if you want to express current data as a function of lagged data and in doing so move the lagged terms to the right of the equality sign in (3.81), we will then have a positive sign in front of each ϕ_i parameter.

CONSTANT TERM

Sometimes a better fit of the data can be obtained by including a constant term in (3.81). In that case we simply add the constant δ to the right-hand side of (3.81); that is,

$$\phi(B)w_t = \delta + \theta(B)a_t. \tag{3.85}$$

This constant δ should not be interpreted as the mean of the w_t data. Since stationarity implies that the mean of $B^k w_t = w_{t-k}$ equals the mean of w_t, it follows that

$$E(1 - \phi_1 B - \cdots - \phi_p B^p)w_t = (1 - \phi_1 - \cdots - \phi_p)Ew_t. \tag{3.86}$$

Furthermore, since $Ea_t = 0$,

$$E[\theta(B)a_t] = 0, \tag{3.87}$$

and, therefore, we have that the mean of the series w_t is related to δ by

$$Ew_t = \delta/(1 - \phi_1 - \cdots - \phi_p). \tag{3.88}$$

It is only when the model is a pure MA model that $Ew_t = \delta$.

3.6 SEASONAL MODELS

In time series observed at quarterly or monthly intervals we often encounter seasonal patterns. Peaks and valleys have a tendency to occur around the same quarters or months of successive years. Unemployment, for example, has a tendency to rise during the summer months, usually because there are more teenagers looking for jobs, and to gradually level off around late fall and winter as manufacturers start gearing their production of the Christmas rush. Champagne sales increase around December because of the holidays, and generally tail off afterwards, probably because there are fewer subsequent celebrations traditionally associated with drinking champagne. In Figure 1.1 we showed the peak and valley behavior of the Chatfield and Prothero sales data of an engineering product. Such a behavior showing a sequence of peaks and valleys is also generally associated with many economic time series.

It seems logical, therefore, to try to exploit the correlation between the same quarters or months in successive years, as well as the correlation between successive quarters or months, whenever we are trying to model such time series. The ARIMA models discussed above can be extended to analyze seasonal variations.

3.6.1 Seasonal Autoregressive Models

A time series observed, say, at quarterly intervals is governed by a first-order seasonal autoregressive process if the current value of the series, z_t, can be expressed as a linear function of the value of the series attained one year ago, z_{t-s} ($s = 4$), and a random shock, a_t. That is,

$$z_t - \Phi_1 z_{t-s} = a_t;$$

which, using the backshift operator can be rewritten as

$$(1 - \Phi_1 B^s)z_t = a_t, \tag{3.89}$$

where Φ_1 is the seasonal autoregressive parameter. This model could be denoted as an AR(1)$_s$ model. However, we will denote this as an SAR(1) model and let the span value be inferred from the context.

In the same way that we extended the AR models to include more parameters, we can extend (3.89) to include more seasonal autoregressive parameters. In general, a seasonal autoregressive model of order P can be written as

$$\Phi(B^s)w_t = a_t, \tag{3.90}$$

where

$$\Phi(B^s) = 1 - \Phi_1 B^s - \Phi_2 B^{2s} - \cdots - \Phi_P B^{Ps} \tag{3.91}$$

$$w_t = \nabla_s^D \nabla^d z_t, \tag{3.92}$$

and ∇_s^D and ∇^d are respectively the seasonal and consecutive difference operators used to induce stationarity in the series z_t.

The autocorrelation function of a seasonal autoregressive model is similar in general characteristics to the one for the regular autoregressive model except that the values of the autocorrelations appear at multiples of the span. For example, for a first-order seasonal autoregressive model with a positive value for the parameter Φ_1 and with a span of four, the autocorrelation values appear at multiples of four and gradually decay to zero. The autocorrelations of the AR(1) model, given in (3.22), carry over to the seasonal AR(1) model, but the nonzero values will only occur at lags that are multiples of the span. That is,

$$\rho_{sk} = \Phi_1^k \qquad s = \text{span}, \qquad k > 0$$

Figure 3.11 contains the autocorrelation function for an autoregressive model with $\Phi_1 = 0.5$. In Chapter 4 we shall discuss autocorrelation functions for higher-order seasonal models.

FIGURE 3.11   Autocorrelation Function of a Seasonal AR(1) Model.

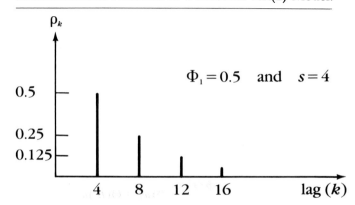

3.6.2 Seasonal Moving Average Models

A stationary time series is said to be governed by a first-order seasonal moving average process if the current value of the series z_t can be represented by a current shock, a_t, and a shock occurring exactly s observations earlier, a_{t-s}, where s equals the span of the seasonal model. Such a model is written as

$$z_t = a_t - \Theta_1 a_{t-s},$$

or equivalently as

$$z_t = (1 - \Theta_1 B^s) a_t, \tag{3.93}$$

where Θ_1 is the seasonal moving average parameter. This model could be denoted as an MA(1)$_s$ model. We prefer the notation SMA(1).

If more seasonal moving average parameters are necessary to model a series, then we can expand (3.93) to include these parameters. In general, a seasonal moving average model of order Q can be expressed as

$$w_t = \Theta(B^s) a_t, \tag{3.94}$$

where

$$\Theta(B^s) = 1 - \Theta_1 B^s - \Theta_2 B^{2s} - \cdots - \Theta_Q B^{Qs} \tag{3.95}$$

$$w_t = \nabla_s^D \nabla^d z_t, \tag{3.96}$$

and ∇_s^D and ∇^d, as in the case of the seasonal autoregressive process, are the difference operators representing the degree of seasonal and consecutive differencing, respectively, needed to make the series z_t stationary.

Again the autocorrelation function of seasonal moving average models behave similarly to the autocorrelation function of the regular moving average models except that the values of the autocorrelations appear at lags which are multiples of the span. For a seasonal moving average model of order 1, the autocorrelation function will have only *one* nonzero value and that value will appear at the lag corresponding to the span. Therefore, the equation (3.42) becomes $\rho_s = -\Theta_1/(1 + \Theta_1^2)$ and all other autocorrelations are zero. Higher-order seasonal moving average models will have the same number of nonzero autocorrelations as the order of the process, and these autocorrelations will also occur at multiples of the span. Figure 3.12 contains the autocorrelations of a first-order seasonal moving average model, with $\Theta_1 = -0.8$ and $s = 4$.

3.6.3 Mixed Seasonal Models

We can also combine the seasonal autoregressive and seasonal moving average processes into a single class of models. In fact, a series which is governed by a *mixed* seasonal process can be expressed as

$$\Phi(B^s) w_t = \Theta(B^s) a_t, \tag{3.97}$$

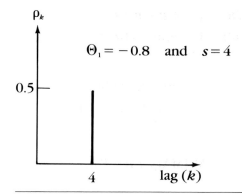

$\Theta_1 = -0.8$ and $s = 4$

FIGURE 3.12   Autocorrelation Function of a Seasonal MA(1) Model.

where again

$$\Phi(B^s) = 1 - \Phi_1 B^s - \Phi_2 B^{2s} - \cdots - \Phi_P B^{Ps} \tag{3.98}$$

$$\Theta(B^s) = 1 - \Theta_1 B^s - \Theta_2 B^{2s} - \cdots - \Theta_Q B^{Qs} \tag{3.99}$$

$$w_t = \nabla_s^D \nabla^d z_t \tag{3.100}$$

and the ∇_s^D and ∇^d are the difference operators used to induce stationarity.

The only dissimilarity between the autocorrelations of a seasonal mixed process and those of a regular mixed process are that the nonzero autocorrelations appear at lags which are multiples of the span.

Using the same convention as for the regular autoregressive integrated moving average models (see Sections 3.4.2 and 3.5.3), we can denote the general mixed seasonal models as ARIMA$(P,D,Q)_s$ where

P = order of the seasonal autoregressive process,

D = number of seasonal differencing,

Q = order of the seasonal moving average process, and

s = the span of the seasonality.

3.6.4 General Multiplicative Seasonal Models

Finally, we can now combine all the models discussed so far into a single general class of time series models which, for many time series, when properly analyzed, can yield excellent fits, and generates accurate forecasts. This broad class of models is called *multiplicative ARIMA models,* and is expressed as

$$\phi(B)\Phi(B^s) w_t = \theta(B)\Theta(B^s) a_t, \tag{3.101}$$

where

$$\phi(B) = 1 - \phi_1 B - \phi_2 B^2 - \cdots - \phi_p B^p \tag{3.102}$$

$$\Phi(B^s) = 1 - \Phi_1 B^s - \Phi_2 B^{2s} - \cdots - \Phi_P B^{Ps} \tag{3.103}$$

$$\theta(B) = 1 - \theta_1 B - \theta_2 B^2 - \cdots - \theta_q B^q \qquad (3.104)$$

$$\Theta(B^s) = 1 - \Theta_1 B^s - \Theta_2 B^{2s} - \cdots - \Theta_Q B^{Qs} \qquad (3.105)$$

and

$$w_t = \nabla_s^D \nabla^d z_t. \qquad (3.106)$$

Note that equation (3.106) indicates that seasonal and consecutive differencing may be required to induce stationarity.

Alternatively, we can summarize (3.101) as

$$\text{ARIMA}(p,d,q) \times (P,D,Q)_s.$$

The meaning of the letters inside and outside the parentheses should, by now, be obvious.

A special case of (3.101) is the first-order consecutive and seasonal autoregressive model ARIMA$(1,0,0) \times (1,0,0)_4$. Suppose that quarterly data z_t can be represented by a first-order seasonal AR model or SAR(1) model:

$$z_t - \Phi_1 z_{t-4} = e_t, \qquad (3.107)$$

where Φ_1 is the seasonal first-order AR parameter or, alternatively can be rewritten using the backshift operator, as

$$(1 - \Phi_1 B^4)z_t = e_t. \qquad (3.108)$$

It is assumed that the error term e_t is not white noise but that it can be represented by a first-order autoregressive process in the consecutive quarters, AR(1),

$$e_t = \phi_1 e_{t-1} + a_t, \qquad (3.109)$$

or as

$$(1 - \phi_1 B)e_t = a_t, \qquad (3.110)$$

with a_t a white noise series. Specifically, (3.110) can be thought of as representing influences of successive quarters. Then, on multiplying (3.108) by $(1 - \phi_1 B)$, we obtain the first-order consecutive and seasonal autoregressive model, ARIMA$(1,0,0) \times (1,0,0)_4$:

$$(1 - \phi_1 B)(1 - \Phi_1 B^4)z_t = a_t. \qquad (3.111)$$

The stationarity conditions for this model are

$$|\phi_1| < 1 \quad \text{and} \quad |\Phi_1| < 1.$$

Another special model is the first-order consecutive and seasonal moving average model[9], ARIMA$(0,1,1) \times (0,1,1)_4$,

[9] The model has been very frequently encountered in the analysis of seasonal data. See, e.g., Thompson and Tiao (1971) for an early case study using this model.

$$\nabla_4\nabla z_t = (1 - \theta_1 B)(1 - \Theta_1 B^4)a_t. \tag{3.112}$$

This model can be thought of as arising from a quarterly first-order moving average model, SMA(1) in the quarterly differenced series,

$$\nabla_4 z_t = (1 - \Theta_1 B^4)e_t, \tag{3.113}$$

with the consecutive differences of the error term e_t governed by a first-order moving average process,

$$\nabla e_t = (1 - \theta_1 B)a_t. \tag{3.114}$$

Combining (3.114) with (3.113), we obtain (3.112). The invertibility conditions for this model require that $|\theta_1| < 1$ and $|\Theta_1| < 1$.

3.7 SUMMARY

In this chapter we have presented the different types of time series models that will be analyzed further in this book. Sections 3.1, 3.2, and 3.3 covered the different forms of mixed autoregressive moving average models, ARMA (p,q). In Section 3.4 we completed this taxonomy with the introduction of differencing, leading to the autoregressive integrated moving average models, ARIMA(p,d,q). In these four sections we not only presented the models but also derived the autocorrelation function and memory function and evaluated the stationarity and invertibility conditions. After a section devoted to notation, we introduced the reader in Section 3.6 to the very relevant seasonal models.

The content of this chapter was also primarily theoretical in that theoretical results were derived. If the time series model is an AR(1) process, then its autocorrelations should be as presented in equation (3.21). In Chapter 4 we will reverse the roles. There we will use generated data to show that the many characteristics discussed in Chapter 3 allow us to effectively discriminate between the different theoretical models.

CHAPTER 4

IDENTIFICATION

We now present the Box–Jenkins iterative approach for constructing linear time series models. This approach basically consists of four steps:

1. *Identification* of the preliminary specifications of the model;
2. *estimation* of the parameters of the model;
3. *diagnostic checking* of model adequacy; and
4. *forecasting* future realizations.

Figure 4.1 contains a flow diagram of this approach. Box and Jenkins in their (1970) book have to be credited for proposing the above four steps and making these steps operational. Although the methodology underlying several of these steps was known in the literature long before 1970, Box and Jenkins first unified these steps into the Box–Jenkins approach.

In the *identification* stage, one chooses a particular model from the general class of ARIMA models specified in Chapter 3 [see equation (3.101)]. That is, one selects the order of consecutive and seasonal differencing required to make the series stationary, as well as specifies the order of the regular and seasonal autoregressive and moving average polynomials necessary to adequately represent the time series model. In the identification phase, we use the autocorrelations and other properties introduced in Chapter 3.

The procedures employed in the identification stage are inexact and require a great deal of judgment. However, one is not irrevocably committed to a chosen model, as an initially inappropriately chosen model can always be altered and improved upon by subsequent analysis.

After a tentative model has been identified, the parameters of that model are *estimated.* Then, by applying various *diagnostic checks,* one can determine whether or not the model adequately represents the data. If any inadequacies are detected, a new model must be identified and the cycle of identification, estimation, and diagnostic checking repeated. Later in Chapter 7 we specifically show that by studying the residual patterns of an inappropriate model, one can make logical modifications to arrive at a formulation which more adequately depicts the behavior of the series.

If it were not for the rather simple nature of many economic and business time series, or social science time series in general, the identification would be a horrendous task. Fortunately, there are a number of features which simplify the task. First of all, the user generally knows the value of the span s. For monthly data, $s = 12$; for quarterly data, $s = 4$. Next, for many social science time series, the orders of the process are generally low, mostly 0 or 1, sometimes 2. Time series from other sciences may involve a more complicated ARIMA structure and as a result may be harder to identify. In this book, where we are primarily, although not exclusively, concerned with economic and business time series, it seems logical to concentrate most attention on these simpler ARIMA structures, the ARIMA models with first- or second-order polynomials.

Finally, the model which passes all the checks is used to generate *forecasts.* It should be noted that models which adequately depict the behavior of the data can at times generate forecasts which are not at all acceptable. Whenever this occurs, one may have to go back to the identification stage and restart the process.

4.1 IDENTIFICATION STRATEGY

Since the general class of ARIMA models given in equation (3.101) is very extensive, it is desirable to seek simple screening procedures to tentatively identify a small subclass of models for further detailed study; that is to deter-

FIGURE 4.1   Functional Diagram of the Box-Jenkins Approach.

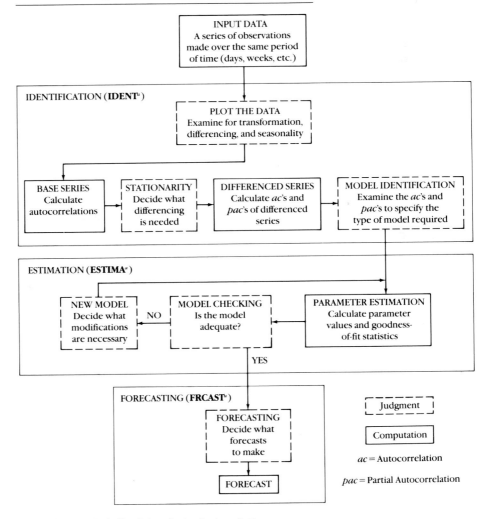

[a]Name of programs in the Time Series collection. See Appendix C.

mine the values of p, d, and q (and P, D, and Q, if seasonal parameters must also be included). Identification is the key to time series model building. The two most useful tools for time series model identification are the autocorrelation function (*acf*) and the partial autocorrelation function (*pacf*). In Section 4.2 we discuss the characteristics of the autocorrelation function and show how the autocorrelations are used in formulating simple ARIMA models. Similarly, in Section 4.3 we define partial autocorrelations and then analyze its properties and show how the partial autocorrelations are used to either verify the formulations based on the properties of the autocorrelations or to identify other potential models. In Section 4.4 we discuss the characteristics of the autocorrelation and partial autocorrelation functions of seasonal models.

The balance of this chapter is devoted to synthesizing the results obtained in the previous sections, to show how to estimate the autocorrelation and partial autocorrelation functions based on a time series realization in order to obtain sample autocorrelations and sample partial autocorrelations, and to discuss the differences between the theoretical and the sample autocorrelations.

As pointed out earlier, the identification stage requires a great deal of judgment. The sample autocorrelations and sample partial autocorrelations never match with the theoretical autocorrelations. Furthermore, there is no exact deterministic approach for identifying an ARIMA model. Experience with the application of the general principles described in this chapter will have to serve as the surrogate for an exact procedure.

Before starting the formal analysis, the user should always carefully inspect a plot of the raw time series. Such a plot would quickly reveal if there are gross problems with the data, such as incorrect data, keypunch errors, gross outliers, etc. Some of these problems should, of course, be corrected before further analysis. Indeed, an examination of the *acf* or *pacf* of such a series might not be able to allude to these underlying gross errors in the data. In addition, the plot should also be examined for sources of nonstationarity. As will be discussed below, some sources of nonstationarity, such as systematic trend, will show up in the *acf* of the series. Other sources of nonstationarity, such as variance nonstationarity, will in general be visible in the time series plot, but will not show up in the *acf*.

As discussed in Section 2.4, if the variance is nonstationary, the series should first be transformed by applying a logarithmic or square root transformation, or any other appropriate variance stabilizing transformation.

4.2 AUTOCORRELATION FUNCTIONS OF ARIMA MODELS

In Chapter 2 we defined the autocovariances and autocorrelations and applied these concepts in Chapter 3 to some simple time series processes. We will now apply these concepts to some generated time series data.

4.2.1 The Autocorrelation Function

A useful device in interpreting a set of autocorrelations is a graph of the autocorrelations plotted against the lag. This graph is called the *autocorrelation function (acf)* or *correlogram.* Since, as discussed in Section 2.2,

$$\rho_{-k} = \rho_k, \tag{4.1}$$

the *acf* is symmetrical about the lag zero and only the positive half of the *acf* need to be examined.

Figure 4.2 contains a typical autocorrelation function. This function shows that the autocorrelations ρ_1, ρ_2, and ρ_{12} are large, the first two positive and the last one negative.

In practice, the autocorrelations of the underlying stochastic process, the *population* autocorrelations, are not known. Therefore, one must rely on *estimates* of the population autocorrelations based on realizations of a given time series. These estimates are called *sample autocorrelations* and are denoted by r_k. In identifying an appropriate model from the class of ARIMA models, we will therefore have to use sample autocorrelations. We initially will present and discuss the population autocorrelation, but at several places we will use autocorrelations calculated from simulated series—series generated according to a known process—to familiarize the reader with the different time series models and to demonstrate some of the effects resulting from the use of estimates rather than true population values. In Section 4.5 we present a more detailed discussion of the sample autocorrelations.

FIGURE 4.2   Autocorrelation Function (*acf*).

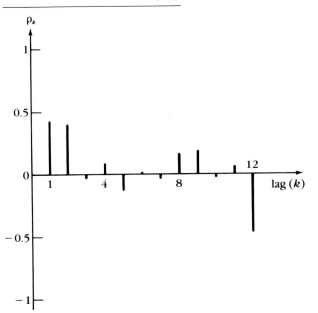

4.2.2 Deciding If a Series Is Stationary

After having made the necessary transformations of the data to insure stationarity, and in particular to insure that the variance is constant, one should check the autocorrelation function to see if the autocorrelations, ρ_k, appear to die out quickly as the lag k increases. If the *acf* fails to die out quickly, we may have to apply further differencing.[1] Although, in general, for nonstationary data the low lag (i.e., for small values of k) autocorrelations will be very large, this in itself is not sufficient to claim that the series is nonstationary. The important characteristic of an *acf* suggesting further differencing is that the *acf* dies out slowly, and therefore that autocorrelations remain nonzero even for high lags.

In order to see why large autocorrelations at high lags imply that the series is not stationary, consider the interpretation of the autocorrelations for a series which exhibits an upward trend. Since successive values of such a series are highly correlated, even for observations spaced quite far apart, the autocorrelation at high lags will still be large. There is a persistent tendency for observations which are above the mean of the series to be followed again by observations above the mean.

A note of caution is appropriate. In theory, ρ_k's for a nonstationary model, as discussed in Sections 2.2 and 3.1, are not defined. However, given a finite realization of a nonstationary time series, it is always possible to estimate autocorrelations using the formulas given in Section 4.5. Although the interpretation of the r_k's as specified in the preceding paragraph is clear, for a nonstationary series these values are not good estimates of ρ_k and certainly could not be used to characterize the model. Therefore, if the data are nonstationary, the text in the preceding paragraphs really refers to sample autocorrelations and not population autocorrelations. The persistence of high autocorrelation between data points spaced further and further apart, as measured by r_k, should be looked upon as a possible indication that the data is nonstationary.

Finally, let us look at the autocorrelation function of a specific process. In Section 3.1.2 we derived the autocorrelation of an AR(1) process as

$$\rho_k = \phi_1^k. \tag{4.2}$$

This equation clearly shows that if the AR process is stationary (i.e., for $|\phi_1| < 1$), the autocorrelations will die out for large values of k. In our discussion

---

[1] Roy (1977) showed that for integrated moving average processes the autocovariances increase as the number of observations increases, although he conjectures that the autocorrelations will converge to one. He also generated ARIMA processes including both autoregressive and moving average terms and found that using sample autocovariances and sample autocorrelations (see Section 4.5) these conclusions still hold. However for an AR(1) process the sample autocorrelations failed to converge to one. The conclusion therefore is that in evaluating stationarity both autocovariances and autocorrelations could be valuable, and that (sample) autocorrelations that fail to converge to zero should be looked upon as possible evidence of nonstationarity.

of the random walk model, we indicated that the random walk model is a nonstationary AR(1) model with $\phi_1 = 1$, and that it can be made into a stationary model by taking the first differences.

Figures 4.3(a), 4.3(b), and 4.3(c) contrast the (sample) *acf* of a simulated random walk model with the (sample) *acf* of the first and second differences. As expected, the *acf* of the random walk model does not die out rapidly, whereas the *acf* of the first difference does. Therefore, if we observe that the autocorrelation function does not die out quickly, it is good practice to evaluate additional differencing. Also, as expected, second differences of a random walk [see Figure 4.3(c)] behave as an MA(1) process with $\theta_1 = 1$.

In the process of trying to make the series stationary by taking differences of the series, one may at times go too far. That is, one may *overdifference* the series. Differencing a stationary series produces another stationary series. However, the model generated by the overdifferenced series will be more complicated than a model generated by a stationary series obtained with the minimum amount of differencing. Fortunately, in many situations the *acf* will clearly show that one has overdifferenced the data.

As an example, consider the pure white noise model:

$$z_t = a_t. \tag{4.3}$$

Remember that all the autocorrelations of this model are zero. Now if we difference (4.3), as

$$z_t - z_{t-1} = a_t - a_{t-1}$$

and rewrite this equation as

$$w_t = a_t - a_{t-1}, \tag{4.4}$$

with $w_t = z_t - z_{t-1}$, we obtain an MA(1) model with $\theta_1 = 1$. The autocorrelation function of (4.3) is given in Figure 4.3(b), and the *acf* of (4.4) is presented in Figure 4.3(c) for the simulated data. Clearly this model is more complicated than the original white noise model. The *acf* of this model has a spike at lag 1 with $\rho_1 = -0.5$, with ρ_1 defined in (3.42) as $\rho_1 = -\theta_1/(1 + \theta_1^2)$, and is 0 thereafter. Consequently, an estimate close to -0.5 for the first autocorrelation, together with small values for the other autocorrelations, should be an indication of overdifferencing.[2]

What kind of autocorrelation pattern should we expect for seasonally nonstationary data? When spikes in the autocorrelation function at lags of multiples of the span of the seasonality do not die out quickly, they are again evidence that the series is not stationary. This would indicate that seasonal differencing may be required. Figure 4.4 contains an example of the *acf* of a seasonally nonstationary series. If, in analyzing monthly data, for example, we observe spikes in the *acf* at lags 12, 24, 36, 48, etc., which do not die

[2] For a more detailed discussion of the pitfalls of overdifferencing, see Plosser and Schwert (1971).

FIGURE 4.3(a)   **Autocorrelation Function of a Random Walk Model.**

```
SERIES WITH D = 0 DS = 0 S = 0
MEAN =    4.116      SD =    10.14    (NOBS = 200)

AUTOCORRELATIONS
  LAGS ROW SE
  1-12  .07   .96  .93  .90  .87  .83  .80  .76  .72  .68  .64  .59  .55
  13-24 .28   .51  .47  .44  .40  .37  .34  .31  .28  .26  .24  .22  .21
  25-36 .30   .20  .19  .18  .17  .17  .16  .16  .15  .15  .14  .13  .13

CHI-SQUARE TEST          P-VALUE
Q(12) =  .152E+04  12 D.F.    .000
Q(24) =  .184E+04  24 D.F.    .000
Q(36) =  .192E+04  36 D.F.    .000

                    AUTOCORRELATION FUNCTION
                    WITH  D = 0 DS = 0 S = 0

              -0.8  -0.4   0.0   0.4   0.8
              * . . . * . . . * . . . * . . . * . . . *
              .                 ( . )           R    1 .
              .               (     .  )         R    2 .
              .               (     .  )         R    3 .
              .              (      .   )        R    4 .
              .             (       .    )      R    5 .
              .             (       .    )      R    6 .
              .            (        .    )   R       7 .
              .            (        .    )   R       8 .
              .            (        .   )  R        9 .
              .            (        .    ) R       10 .
              .            (        .   )R       11 .
              .            (        .    R       12 .
              .            (        .    R       13 .
              .            (        .   R )     14 .
              .            (        .   R )     15 .
              .            (        .  R   )    16 .
              .            (        .  R   )    17 .
              .            (        . R    )    18 .
              .            (        . R    )    19 .
              .            (        .R    )    20 .
              .            (        .R    )    21 .
              .            (        .R    )    22 .
              .            (       . R    )    23 .
              .            (       . R    )    24 .
              .            (       . R    )    25 .
              .            (       . R    )    26 .
              .            (       . R    )    27 .
              .            (       . R    )    28 .
              .            (       . R    )    29 .
              .            (      . R     )    30 .
              .            (      . R     )    31 .
              .            (      . R     )    32 .
              .            (      . R     )    33 .
              .            (      . R     )    34 .
              .            (      . R     )    35 .
              .            (      . R     )    36 .
```

FIGURE 4.3(b) Autocorrelation Function of the First Differences of a Random Walk Model.

```
SERIES WITH D = 1 DS = 0 S = 0
MEAN =  -0.1318      SD =    2.070    (NOBS = 199)

AUTOCORRELATIONS
  LAGS ROW SE
   1-12 .07   -.07 -.09  .16  .11 -.05  .06  .00  .11 -.04  .09  .04  .02
  13-24 .08   -.10 -.04  .00 -.07 -.10  .05 -.04 -.08 -.01 -.03 -.05 -.08
  25-36 .08   -.09 -.02 -.06 -.11  .11  .06 -.09  .02  .02  .05  .00 -.03

CHI-SQUARE TEST           P-VALUE
Q(12)  =   16.2      12 D.F.   .181
Q(24)  =   26.4      12 D.F.   .331
Q(36)  =   38.8      36 D.F.   .343


            AUTOCORRELATION FUNCTION
            WITH   D = 1 DS = 0 S = 0


          -0.8  -0.4   0.0   0.4   0.8
        * . * . . . * . * . * . . * . *
          .              (R.  )           1  .
          .              (R.  )           2  .
          .              ( .  R           3  .
          .              ( .  R           4  .
          .              (R.  )           5  .
          .              ( .R)            6  .
          .              ( R  )           7  .
          .              ( .  R           8  .
          .              (R.  )           9  .
          .              ( .R)           10  .
          .              ( .R)           11  .
          .              ( R  )          12  .
          .              (R.  )          13  .
          .              (R.  )          14  .
          .              ( R  )          15  .
          .              (R.  )          16  .
          .              (R.  )          17  .
          .              ( .R)           18  .
          .              (R.  )          19  .
          .              (R.  )          20  .
          .              ( R  )          21  .
          .              ( R  )          22  .
          .              (R.  )          23  .
          .              (R.  )          24  .
          .              (R.  )          25  .
          .              ( R  )          26  .
          .              (R.  )          27  .
          .              R .  )          28  .
          .              ( .  R          29  .
          .              ( .R)           30  .
          .              (R.  )          31  .
          .              ( R  )          32  .
          .              ( R  )          33  .
          .              ( .R)           34  .
          .              ( R  )          35  .
          .              ( R  )          36  .
```

FIGURE 4.3(c)   Autocorrelation Function of the Second Differences of a Random Walk Model.

```
SERIES WITH D = 2 DS = 0 S = 0
MEAN  =  0.1117E-01  SD =    3.027    (NOBS = 198)

AUTOCORRELATIONS
  LAGS ROW SE
   1-12 .07  -.49 -.12   .13   .06 -.12   .08 -.08   .12 -.13   .09 -.02   .05
  13-24 .09  -.08  .01   .05 -.03 -.08   .12 -.02 -.06   .04   .00   .01 -.02
  25-36 .09  -.04  .05   .01 -.13   .13   .05 -.12   .05 -.01   .04 -.01 -.06

CHI-SQUARE TEST           P-VALUE
Q(12) =  69.9     12 D.F.  .000
Q(24) =  77.8     24 D.F.  .000
Q(36) =  92.8     36 D.F.  .000


               AUTOCORRELATION FUNCTION
               WITH  D = 2 DS = 0 S = 0


            -0.8  -0.4   0.0   0.4   0.8
          * . .* . .* . .* . .* . .* . .*
          .        R    ( . )              1 .
          .             (R .   )           2 .
          .             (  . R)            3 .
          .             (  .R )            4 .
          .             (R .   )           5 .
          .             (  .R )            6 .
          .             ( R.   )           7 .
          .             (  . R)            8 .
          .             (R .   )           9 .
          .             (  .R )           10 .
          .             (  R   )          11 .
          .             (  .R )           12 .
          .             ( R.   )          13 .
          .             (  R   )          14 .
          .             (  .R )           15 .
          .             (  R   )          16 .
          .             ( R.   )          17 .
          .             (  . R)           18 .
          .             (  R   )          19 .
          .             ( R.   )          20 .
          .             (  .R )           21 .
          .             (  R   )          22 .
          .             (  R   )          23 .
          .             (  R   )          24 .
          .             ( R.   )          25 .
          .             (  .R )           26 .
          .             (  R   )          27 .
          .             (R .   )          28 .
          .             (  . R)           29 .
          .             (  .R )           30 .
          .             (R .   )          31 .
          .             (  .R )           32 .
          .             (  R   )          33 .
          .             (  .R )           34 .
          .             (  R   )          35 .
          .             ( R.   )          36 .
```

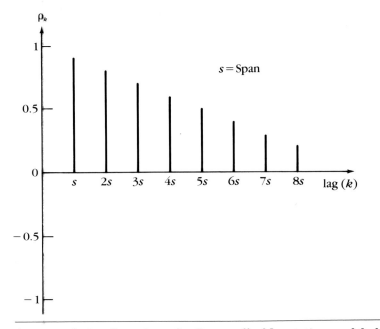

FIGURE 4.4   Autocorrelation Function of a Seasonally Nonstationary Model.

out quickly, then more than likely seasonal differencing with a span of 12 will be needed to induce stationarity in the series.

In Figure 4.5 we have shown the autocorrelations of a seasonally nonstationary series generated as

$$(1 - \phi B)(1 - B^6)z_t = a_t,$$

with $\phi = 0.7$. Although the autocorrelations initially die out quickly, the autocorrelation becomes large again at lag 6. Indeed we notice substantial swings in the autocorrelation function with autocorrelations at multiple of the seasonal span not decreasing rapidly. Again, this is strong evidence that the series is nonstationary and that seasonal differencing could induce stationarity.

For a series to be stationary, it is necessary for the autocorrelations to die out quickly as the number of lags increases. We do not claim that we should difference the series until there are no spikes in the *acf*. As soon as the spikes in the *acf* die out rapidly, there is no need for additional differencing. Specifically, we would want to carefully examine the stationarity of a series if the autocorrelations at lag k (or sk for a seasonal series with span s), for k greater than 5, are still large in absolute value, say larger than 0.7. This is, however, a rule of thumb that should not be taken at face value and assumed to hold for all situations.

FIGURE 4.5 **Autocorrelation Function of a Seasonally Nonstationary AR(1)×SAR(1) Model.**

```
SERIES WITH D = 0 DS = 0 S = 0
MEAN  =  -6.562      SD =    9.242    (NOBS = 200)

AUTOCORRELATIONS
  LAGS ROW SE
   1-12 .07    .41 -.05 -.34 -.07   .36   .93   .37 -.07 -.37 -.12   .31   .85
  13-24 .17    .33 -.10 -.39 -.16   .26   .78   .29 -.11 -.40 -.19   .22   .73
  25-36 .22    .27 -.11 -.39 -.20   .20   .69   .26 -.11 -.37 -.19   .19   .66

CHI-SQUARE TEST              P-VALUE
Q(12) =   504.     12 D.F.    .000
Q(24) =   917.     24 D.F.    .000
Q(36) = .128E+04   36 D.F.    .000
```

```
                AUTOCORRELATION FUNCTION
              WITH  D = 0 DS = 0 S = 0

              -0.8  -0.4   0.0   0.4   0.8
              * . . . . . * . . . * . . . * . . . . . *
              .                 ( . )    R             1 .
              .                 (R. )                  2 .
              .              R  ( . )                   3 .
              .                 ( R. )                  4 .
              .                 ( . ) R                 5 .
              .                 ( . )          R        6 .
              .                 ( . ) R                 7 .
              .                 ( R. )                  8 .
              .              R  ( . )                   9 .
              .                 ( R. )                 10 .
              .                 ( . )R                 11 .
              .                 ( . )          R       12 .
              .                 (  .   R               13 .
              .                 ( R. )                 14 .
              .               R(  .   )                15 .
              .                 ( R. )                 16 .
              .                 ( . R)                 17 .
              .                 ( . )      R           18 .
              .               ( . R )                  19 .
              .               ( R. )                   20 .
              .               R .   )                  21 .
              .               ( R ..  )                22 .
              .               ( . R )                  23 .
              .               ( . )    R               24 .
              .               ( . R )                  25 .
              .               ( R. )                   26 .
              .               R .   )                  27 .
              .             ( R .   )                  28 .
              .             ( . R )                    29 .
              .             ( . )  R                   30 .
              .             ( . R )                    31 .
              .             ( R. )                     32 .
              .             (R .   )                    33 .
              .             ( R. )                     34 .
              .             ( . R )                    35 .
              .             ( . )  R                   36 .
```

4.2.3 The Autocorrelation Function of Stationary Series

Once it has been reasonably ascertained that the series is stationary, the next step in identifying a model is to determine what type of autoregressive or moving average or mixed model adequately fits the data. The actual structure of the ARIMA$(p,d,q) \times (P,D,Q)_s$ model is obtained by comparing the sample *acf* of a stationary series with theoretical population *acf*'s.

In comparing the sample *acf* with population *acf*'s, we should keep in mind that there is no rule that stipulates that after differencing a series consecutively or seasonally to induce stationarity, we must now include nonseasonal or seasonal parameters. Indeed, it is quite possible that after appropriate stationarity transformations have been made, the new series is simply a white noise series and there is no need to include any parameter whatsoever. This is specifically the case for a random walk series. Only if autocorrelations are large at lags corresponding to the span and possibly multiplies thereof should we include seasonal parameters.

The balance of this section will be devoted to the derivation of the *acf* of some ARIMA models typically encountered in applied work.

AUTOREGRESSIVE MODELS

First-order autoregressive models   In Section 3.1 we indicated that a first-order autoregressive model, AR(1), could be represented as

$$z_t = \phi_1 z_{t-1} + a_t, \tag{4.5}$$

with a_t being white noise (i.e., independently distributed with mean 0, constant variance σ_a^2, and independent of z_{t-1}), and we showed that the stationarity condition requires that

$$|\phi_1| < 1. \tag{4.6}$$

Next we indicated that the autocovariances of an AR(1) also follow an AR(1) process; that is,

$$\lambda_k = \phi_1 \lambda_{k-1}, \tag{4.7}$$

and that as a result the autocorrelations of an AR(1) can be expressed as

$$\rho_k = \phi_1^k, \qquad k > 0. \tag{4.8}$$

A stationary series which has an *acf* that decays to 0 in a fashion similar to equation (4.8) [that is, exponentially $(0 < \phi_1 < 1)$ or oscillating in sign $(-1 < \phi_1 < 0)$], can very likely be construed as being governed by an AR(1) process.

In order to become familiar with the identification approach of comparing an observed *acf* with a theoretical *acf* in the process of selecting an appropriate ARIMA model, and at the same time to realize the implication of dealing with sample rather than theoretical *acf*'s, we have presented in Figure 4.6 the *acf* of a simulated AR(1) model with $\phi_1 = 0.8$. We therefore know the true ARIMA process which generated the data. The calculations of the *acf* in Figure 4.6 are based on 200 data points generated by an AR(1) process. According to (4.8) the first ten autocorrelations should have the following theoretical values:

$$0.80 \quad 0.64 \quad 0.51 \quad 0.41 \quad 0.33 \quad 0.26 \quad 0.21 \quad 0.17 \quad 0.13 \quad 0.11.$$

As can be seen from Figure 4.6, the calculated autocorrelations are not exactly equal, but are quite close to these figures.

Purely random models   A discrete time series, z_t, is called a *purely random model* if the data form a sequence of mutually independent, identically distributed random variables,

$$z_t = a_t. \tag{4.9}$$

Given that (4.9) is a special case of an AR(1) model with $\phi_1 = 0$, it follows immediately from (4.8) that

$$\rho_k = 0, \qquad k > 0. \tag{4.10}$$

Sample autocorrelations for such a model in practice will not be *exactly* equal to zero, but will be very small or statistically insignificant.[3] Figure 4.3(b) showed the *acf* of a purely random model based on simulated data. Again, most of the autocorrelations are quite small. In Section 4.5 we will discuss ways for evaluating the significance of autocorrelations; that is, for deciding when small sample autocorrelations can be accepted as evidence that the population autocorrelations are zero.

Second-order autoregressive models   A second-order autoregressive model, AR(2), can be represented as[4]

$$z_t = \phi_1 z_{t-1} + \phi_2 z_{t-2} + a_t. \tag{4.11}$$

---

[3] We will later show how exactly to apply statistical tests.

[4] Necessary conditions for (weak) stationarity of an AR(2) model are

$$\phi_2 + \phi_1 < 1$$
$$\phi_2 - \phi_1 < 1$$
$$|\phi_2| < 1. \tag{4.12}$$

Note that for suitable values of ϕ_2, ϕ_1 can take on values between -2 and $+2$, and the series will still be stationary.

The rule about the size of the roots of the polynomial still hold, i.e, the roots should lie outside the unit circle, or the size of the roots should be larger than one. See also Section 5.3.1. For a second- and higher-order polynomial the roots are however complicated functions of the underlying parameters.

FIGURE 4.6   Autocorrelation Function of an AR(1) Model.

```
SERIES WITH D = 0 DS = 0 S = 0
MEAN =  -0.4991      SD =    3.090    (NOBS = 200)

AUTOCORRELATIONS
  LAGS ROW SE
  1-12  .07   .76  .62  .47  .37  .27  .19  .13  .02 -.04 -.12 -.17 -.26
  13-24 .14  -.24 -.25 -.26 -.27 -.33 -.36 -.36 -.31 -.28 -.21 -.20 -.15
  25-36 .17  -.08 -.08 -.07 -.09 -.03  .00  .03  .04  .05  .07  .10  .08

CHI-SQUARE TEST           P-VALUE
Q(12) =  323.    12 D.F.   .000
Q(24) =  524.    24 D.F.   .000
Q(36) =  537.    36 D.F.   .000


             AUTOCORRELATION FUNCTION
             WITH  D = 0 DS = 0 S = 0

        -0.8  -0.4   0.0   0.4   0.8
        * . .* . .* . .* . .* . .* . .*
        .              ( . )        R      1 .
        .              ( .  )        R     2 .
        .              ( .   )  R           3 .
        .              ( .   )  R           4 .
        .              ( .   R               5 .
        .              ( .   R)              6 .
        .              ( . R )               7 .
        .              (  R  )               8 .
        .              ( R.  )               9 .
        .              ( R . )              10 .
        .              (R .  )              11 .
        .              R .  )               12 .
        .              R .  )               13 .
        .              R .  )               14 .
        .              R .  )               15 .
        .              R .  )               16 .
        .             R( .  )               17 .
        .              R .   )              18 .
        .              R .   )              19 .
        .              R .   )              20 .
        .             (R .   )              21 .
        .             ( R .   )             22 .
        .             ( R .   )             23 .
        .             ( R .   )             24 .
        .             ( R.   )              25 .
        .             ( R.   )              26 .
        .             ( R.   )              27 .
        .             ( R.   )              28 .
        .             ( R   )               29 .
        .             ( R   )               30 .
        .             ( R   )               31 .
        .             ( .R   )              32 .
        .             ( .R   )              33 .
        .             ( .R   )              34 .
        .             ( . R  )              35 .
        .             ( .R  )               36 .
```

The variance of such a process is

$$\lambda_0 = E(z_t z_t) = E[\phi_1 z_{t-1} + \phi_2 z_{t-2} + a_t) z_t]$$
$$= \phi_1 E(z_{t-1} z_t) + \phi_2 E(z_{t-2} z_t) + E(a_t z_t)$$
$$= \phi_1 \lambda_1 + \phi_2 \lambda_2 + \sigma_a^2. \tag{4.13}$$

In calculating $E(a_t z_t) = \sigma_a^2$, recognize in (4.11) that z_t is a function of z_{t-1}, z_{t-2}, and a_t, and that a_t is independent of past z_t's.

The autocovariances at lag k are defined as $E(z_t z_{t-k})$, and using (4.11) we immediately have

$$\lambda_k = \phi_1 \lambda_{k-1} + \phi_2 \lambda_{k-2}, \qquad k > 0. \tag{4.14}$$

In (4.14) there is no σ_a^2 term because all the terms in the expression for z_{t-k}, $k > 0$; that is, the terms z_{t-k-1}, z_{t-k-2}, and a_{t-k} are independent of a_t. The autocorrelations for an AR(2) model are calculated by scaling the autocovariances by λ_0:

$$\rho_k = \phi_1 \rho_{k-1} + \phi_2 \rho_{k-2}, \qquad k > 0, \tag{4.15}$$

with $\rho_k = \lambda_k / \lambda_0$.

Using (4.15) for $k = 1$ and $k = 2$,[5] that is,

$$\rho_1 = \phi_1 + \phi_2 \rho_1$$
$$\rho_2 = \phi_1 \rho_1 + \phi_2, \tag{4.16}$$

we can now solve explicitly for ρ_1 and ρ_2 in terms of ϕ_1 and ϕ_2, and obtain

$$\rho_1 = \frac{\phi_1}{1 - \phi_2} \tag{4.17}$$

$$\rho_2 = \frac{\phi_1^2}{1 - \phi_2} + \phi_2. \tag{4.18}$$

Note also that the relationship expressing the autocorrelations of an AR(1) and AR(2) process as a function of the AR parameters, equations (4.7) and (4.15) respectively, follows the same basic dynamic relationship as the processes themselves, except that the white noise term is missing. In fact it can be shown that for an AR(p) process, the autocorrelations at lag k will follow a pth-order autoregressive process[6]

[5] The equations (4.16) are the Yule–Walker equations of an AR(2) model, see Yule (1927) and Walker (1931).

[6] The Yule–Walker equations of an AR(p) process are defined as equation (4.19) for $k = 1, \ldots, p$:

$$\rho_1 = \phi_1 + \phi_2 \rho_1 + \cdots + \phi_p \rho_{p-1}$$
$$\rho_2 = \phi_1 \rho_1 + \phi_2 + \cdots + \phi_p \rho_{p-2}$$
$$\vdots$$
$$\rho_p = \phi_1 \rho_1 + \phi_2 \rho_2 + \cdots + \phi_p. \tag{4.20}$$

These p equations make it possible to solve for $\rho_1, \ldots, \rho_p$ in terms of the autoregressive parameters $\phi_1, \ldots, \phi_p$.

$$\rho_k = \phi_1 \rho_{k-1} + \phi_2 \rho_{k-2} + \cdots + \phi_p \rho_{k-p}, \qquad k > 0. \tag{4.19}$$

Figure 4.7 shows the *acf* calculated from data simulated according to the following AR(2) process

$$z_t = 0.75 z_{t-1} - 0.50 z_{t-2} + a_t. \tag{4.21}$$

Based on (4.17) and (4.18) the first two autocorrelations should have the theoretical values $\rho_1 = 0.50$, $\rho_2 = -0.33$, and based on (4.15), $\rho_3 = -0.50$, $\rho_4 = -0.21$, $\rho_5 = 0.09$, and $\rho_6 = 0.17$. Although the autocorrelations calculated from the simulated data are not exactly equal to the above values, the pattern of these first six sample autocorrelations is very similar.

Summary We evaluated the population autocorrelations for a purely random process, an AR(1), and an AR(2) process, and compared these with autocorrelations estimated from data simulated according to a known process. We found that the calculated *acf* resembles the population function quite well. We may therefore feel confident that, based on sample autocorrelations, we will be able to detect the pattern of the underlying theoretical autocorrelations.

We now present in Figure 4.8 some typical population *acf*'s for these models. Figure 4.8(a) shows the *acf* of a purely random process. From (4.10) we know that all the ρ_k's equal 0. Figures 4.8(b) and 4.8(c), respectively, contains the *acf* of an AR(1) process with positive and negative values for ϕ_1. The autocorrelations of an AR(1) process with positive ϕ_1 decay exponentially to 0, while for ϕ_1 negative their decay oscillates in sign.

Figures 4.8(d), (e), (f), and (g) show the *acf* of four typical AR(2) models. Note that for all practical purposes Figures 4.8(d) and 4.8(b), and similarly Figures 4.8(e) and 4.8(c), are the same. Differentiation between certain forms of AR(1) and AR(2) processes, therefore, cannot be made solely on the basis of autocorrelations. In Section 4.3 we will discuss another characteristic of ARIMA models, namely the partial autocorrelation which, when used in conjunction with the autocorrelations, will help in differentiating between the ARIMA models.

MOVING AVERAGE MODELS

First-order moving average models In Section 3.2 we indicated that an MA(1) process can be represented as

$$z_t = a_t - \theta_1 a_{t-1}, \tag{4.22}$$

with a_t white noise. We also showed in (3.42) that the autocorrelation can be derived as

$$\rho_1 = -\theta_1/(1 + \theta_1^2)$$
$$\rho_k = 0, \qquad k \geq 2. \tag{4.23}$$

Therefore, the first autocorrelation will be of opposite algebraic sign as the value of the moving average parameter θ_1.

FIGURE 4.7   Autocorrelation Function of an AR(2) Model.

```
SERIES WITH D = 0 DS = 0 S = 0
MEAN = -0.1156     SD =    2.507    (NOBS = 200)

AUTOCORRELATIONS
  LAGS ROW SE
   1-12  .07    .48 -.11 -.35 -.20  .02  .14  .13  .01 -.02 -.04 -.10 -.19
  13-24  .10   -.08  .05  .11  .08 -.09 -.20 -.19 -.06 -.01 -.01 -.06  .01
  25-36  .10    .13  .08 -.05 -.17 -.06  .04  .07  .00 -.03  .04  .15  .05


CHI-SQUARE TEST          P-VALUE
Q(12) =  99.9     12 D.F.   .000
Q(24) =  126.     24 D.F.   .000
Q(36) =  148.     36 D.F.   .000


            AUTOCORRELATION FUNCTION
            WITH   D = 0 DS = 0 S = 0


          -0.8  -0.4   0.0   0.4   0.8
          *  *  *  *  *  *  *  *  *  *  *  *  *
          .  .  .  .  .  .  .  .  .  .  .  .  .
          .                   ( . )       R          1  .
          .                   (R .   )               2  .
          .                R ( .   )                 3  .
          .                   R  .   )               4  .
          .                   (  R  )                5  .
          .                   (  . R)                6  .
          .                   (  . R)                7  .
          .                   (  R  )                8  .
          .                   (  R  )                9  .
          .                   ( R.  )               10  .
          .                   (R .  )               11  .
          .                   R  .  )               12  .
          .                   ( R.  )               13  .
          .                   (  .R )               14  .
          .                   (  . R)               15  .
          .                   (  .R )               16  .
          .                   ( R.  )               17  .
          .                   R  .  )               18  .
          .                   R  .  )               19  .
          .                   ( R.  )               20  .
          .                   (  R  )               21  .
          .                   (  R  )               22  .
          .                   ( R.  )               23  .
          .                   (  R  )               24  .
          .                   (  . R)               25  .
          .                   (  .R )               26  .
          .                   ( R.  )               27  .
          .                   (R .  )               28  .
          .                   ( R.  )               29  .
          .                   (  .R )               30  .
          .                   (  .R )               31  .
          .                   (  R  )               32  .
          .                   ( R.  )               33  .
          .                   (  .R )               34  .
          .                   (  . R)               35  .
          .                   (  .R )               36  .
```

FIGURE 4.8   Autocorrelation Functions of Autoregressive Models.

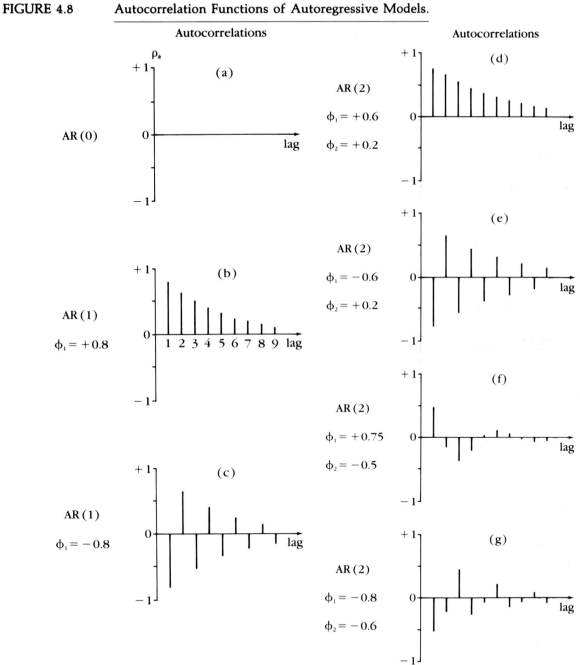

In Figure 4.9 we represent the autocorrelations calculated from data simulated according to the following MA(1) process with $\theta_1 = 0.7$:

$$z_t = a_t - 0.7a_{t-1}. \tag{4.24}$$

From (4.23) it follows that the population values of the autocorrelations should be $\rho_1 = -0.47$ and all other $\rho_k = 0$, $k > 1$. Again the results in Figure 4.9 show that the calculated autocorrelations are indeed close to the theoretical ones.

Second-order moving average models A second-order moving average process, MA(2), can be represented as

$$z_t = a_t - \theta_1 a_{t-1} - \theta_2 a_{t-2}. \tag{4.25}$$

First, express lagged values of z_t as a function of lagged error terms:

$$
\begin{aligned}
z_t &= a_t - \theta_1 a_{t-1} - \theta_2 a_{t-2} \\
z_{t-1} &= \quad\quad a_{t-1} - \theta_1 a_{t-2} - \theta_2 a_{t-3} \\
z_{t-2} &= \quad\quad\quad\quad a_{t-2} - \theta_1 a_{t-3} - \theta_2 a_{t-4} \\
z_{t-3} &= \quad\quad\quad\quad\quad\quad a_{t-3} - \theta_1 a_{t-4} - \theta_2 a_{t-5}.
\end{aligned} \tag{4.26}
$$

Now squaring the right-hand side of the first line of (4.26) we find that

$$\lambda_0 = E(z_t^2) = (1 + \theta_1^2 + \theta_2^2)\,\sigma_a^2. \tag{4.27a}$$

Similarly, $E(z_t z_{t-k})$, $k > 0$ can be evaluated by multiplying the right-hand side of the first line of (4.26) with the right-hand side of the line expressing z_{t-k}. This leads to

$$
\begin{aligned}
\lambda_1 &= E(z_t z_{t-1}) = -\theta_1(1 - \theta_2)\sigma_a^2 \\
\lambda_2 &= E(z_t z_{t-2}) = -\theta_2 \sigma_a^2 \\
\lambda_k &= E(z_t z_{t-k}) = 0, \quad\quad \text{for } k > 2.
\end{aligned} \tag{4.27b}
$$

The autocorrelations of an MA(2) model, therefore, are

$$
\begin{aligned}
\rho_1 &= -\theta_1(1 - \theta_2)/(1 + \theta_1^2 + \theta_2^2) \\
\rho_2 &= -\theta_2/(1 + \theta_1^2 + \theta_2^2) \\
\rho_k &= 0, \quad\quad \text{for } k > 2.
\end{aligned} \tag{4.28}
$$

Thus the *acf* of an MA(2) model abruptly *cuts off* after lag 2, and the sign of ρ_1 and ρ_2 is the opposite of θ_1 and θ_2, respectively. Note that the invertibility conditions for an MA(2) process require $|\theta_2| < 1$ [see (4.32), footnote 7]. In general the last nonzero autocorrelation has always the opposite sign of the highest order moving average parameter.

In Figure 4.10 we again present sample autocorrelations based data simulated according to an MA(2) process with the following parameter values: $\theta_1 = 1.5$ and $\theta_2 = -0.6$:

$$z_t = a_t - 1.5a_{t-1} + 0.6a_{t-2}. \tag{4.29}$$

FIGURE 4.9   Autocorrelation Function of an MA(1) Model.

```
SERIES WITH D = 0 DS = 0 S = 0
MEAN =  -0.3403E-01  SD =    2.567     (NOBS = 200)

AUTOCORRELATIONS
  LAGS ROW SE
  1-12   .07   -.56   .15 -.11   .07 -.02  -.05   .14 -.15   .10 -.08   .16 -.24
 13-24   .10    .14 -.02 -.04   .10 -.07   .03 -.12   .13 -.12   .14 -.09 -.06
 25-36   .11    .16 -.14   .17 -.22   .14 -.07   .05 -.01   .00 -.08   .12 -.02

CHI-SQUARE TEST          P-VALUE
Q(12) =  104.     12 D.F.   .000
Q(24) =  130.     24 D.F.   .000
Q(36) =  170.     36 D.F.   .000
```

```
               AUTOCORRELATION FUNCTION
               WITH  D = 0 DS = 0 S = 0


          -0.8  -0.4   0.0   0.4   0.8
          * . * . * . * . * . * . * . *
          .         R    (  .  )              1 .
          .              (   .  R )           2 .
          .              ( R  .    )          3 .
          .              (   . R  )           4 .
          .              (   R    )           5 .
          .              ( R .    )           6 .
          .              (   .  R )           7 .
          .              ( R .    )           8 .
          .              (   . R  )           9 .
          .              ( R .    )          10 .
          .              (   .  R )          11 .
          .            R (   .    )          12 .
          .              (   . R  )          13 .
          .              (   R    )          14 .
          .              ( R .    )          15 .
          .              (   .  R )          16 .
          .              ( R .    )          17 .
          .              (   R    )          18 .
          .              ( R .    )          19 .
          .              (   . R  )          20 .
          .              ( R .    )          21 .
          .              (   . R  )          22 .
          .              (  R .    )         23 .
          .              (  R .    )         24 .
          .              (   . R  )          25 .
          .              ( R .    )          26 .
          .              (   .    R           27 .
          .            R (   .    )          28 .
          .              (   . R  )          29 .
          .              ( R .    )          30 .
          .              (   . R  )          31 .
          .              (   R    )          32 .
          .              (   R    )          33 .
          .              ( R .    )          34 .
          .              (   . R  )          35 .
          .              (   R    )          36 .
```

FIGURE 4.10   Autocorrelation Function of an MA(2) Model.

```
SERIES WITH D = 0 DS = 0 S = 0
MEAN =  -0.3352E-01  SD =   4.156   (NOBS = 200)

AUTOCORRELATIONS
  LAGS ROW SE
  1-12  .07  -.73  .32 -.14  .08 -.02 -.07  .16 -.18  .15 -.14  .20 -.24
 13-24  .12   .17 -.04 -.05  .10 -.09  .08 -.13  .17 -.17  .16 -.08 -.06
 25-36  .12   .17 -.20  .23 -.25  .18 -.10  .06 -.02  .03 -.09  .11 -.04

CHI-SQUARE TEST          P-VALUE
Q(12) =  177.      12 D.F.   .000
Q(24) =  215.      24 D.F.   .000
Q(36) =  274.      36 D.F.   .000

              AUTOCORRELATION FUNCTION
              WITH   D = 0 DS = 0 S = 0

               -0.8  -0.4   0.0   0.4   0.8
               *  *  *  *  *  *  *  *  *  *  *
              . . . . . . . . . . . . . . . . .
               .     R          ( .  )            1 .
               .                ( .  ) R          2 .
               .                (R .  )           3 .
               .                (  .R )           4 .
               .                (  R  )           5 .
               .                ( R. )            6 .
               .                (  . R)           7 .
               .                R  .  )           8 .
               .                (  . R)           9 .
               .                (R .  )          10 .
               .                (  . R           11 .
               .                R( .  )          12 .
               .                (  . R           13 .
               .                ( R. )           14 .
               .                ( R. )           15 .
               .                (  . R)          16 .
               .                ( R. )           17 .
               .                (  .R )          18 .
               .                ( R .  )         19 .
               .                (  . R)          20 .
               .                (R .  )          21 .
               .                (  . R )         22 .
               .                ( R. )           23 .
               .                ( R. )           24 .
               .                (  . R )         25 .
               .                (R .  )          26 .
               .                (  . R           27 .
               .                R  .  )          28 .
               .                (  . R)          29 .
               .                ( R. )           30 .
               .                (  .R )          31 .
               .                ( R  )           32 .
               .                ( R  )           33 .
               .                ( R. )           34 .
               .                (  . R )         35 .
               .                ( R. )           36 .
```

From (4.28) it follows that the theoretical autocorrelations for an MA(2) model with these parameter values are $\rho_1 = -0.66$, $\rho_2 = 0.17$, and all other ρ_k's = 0. Again, the calculated autocorrelations are very close to these theoretical values.

Invertibility conditions In Chapter 3 we presented some heuristic arguments to show that the invertibility conditions imposed on the parameter of a moving average process correspond to the stationarity conditions imposed on the autoregressive parameters. We are now in a better position to show in a more concrete fashion why the invertibility conditions must be imposed. To do so, consider the following two MA(1) processes:

$$z_t = a_t + 0.5a_{t-1} \tag{4.30a}$$
$$z_t = a_t + 2a_{t-1}. \tag{4.30b}$$

It follows directly from (4.23) that both processes have *identical* autocorrelations:

$$\rho_1 = 0.4$$
$$\rho_k = 0, \qquad \text{for } k > 1. \tag{4.31}$$

That is, for two different processes, the *acf*'s are the same and we are unable to go back, to invert, uniquely from the *acf* (4.31) to the process (4.30a) or (4.30b). If we impose the invertibility condition[7] $|\theta_1| < 1$ we can then exclude model (4.30b). At the same time model (4.30a) allows us to give a sensible interpretation of the autoregressive rewriting of an MA process; see, for example, (3.35) for an MA(1) model.

Summary We have shown that for an MA(1) process, the *acf* has a cutoff after lag 1 and that for an MA(2) process, the *acf* has a cutoff after lag 2. This feature generalizes to higher-order moving average processes: an MA(q) process has a cutoff after lag q. This autocorrelation pattern is in sharp contrast with the patterns of the *acf*'s of autoregressive models which show a *gradual* decline in the values of the autocorrelations.

Figure 4.11 contains some typical *acf*'s of MA(1) and MA(2) processes. As can be seen, different values for θ_1 and θ_2 produce markedly different patterns in the *acf*'s, but the cutoff point is always equal to the order of the process. Notice also that the highest-order autocorrelation always has

[7] For an MA(2) process, it can be shown that the invertibility conditions require that
$$\theta_2 + \theta_1 < 1$$
$$\theta_2 - \theta_1 < 1$$
$$|\theta_2| < 1. \tag{4.32}$$
These conditions are similar to the stationarity conditions imposed on the parameters ϕ_1 and θ_2 of an AR(2) model. See also footnote 4.

FIGURE 4.11   Autocorrelation Functions of Moving Average Models.

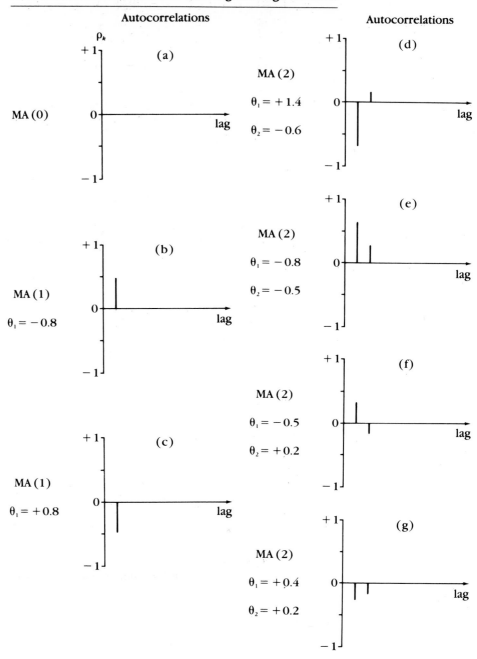

the opposite sign of the highest order MA parameter. In Figure 4.11(e), ρ_2 is positive, but θ_2 is negative.

MIXED MODELS: ARMA(1,1)

In Section 3.3 we specified the ARMA(1,1) model[8] as

$$z_t = \phi_1 z_{t-1} + a_t - \theta_1 a_{t-1} \qquad (4.33)$$

where ϕ_1 and θ_1 are respectively the autoregressive and moving average parameter and a_t is white noise. In that same section, we showed that the autocorrelations for such a process [see 3.56] equal

$$\rho_1 = \frac{(1 - \phi_1 \theta_1)(\phi_1 - \theta_1)}{1 + \theta_1^2 - 2\theta_1 \phi_1} \qquad (4.34a)$$

$$\rho_k = \phi_1 \rho_{k-1}, \qquad k > 1. \qquad (4.34b)$$

Thus, similar to the *acf* of an AR(1) process, the *acf* of the ARMA(1,1) process decays from the starting value ρ_1. However, the parameter ρ_1 now depends on both the autoregressive and moving average parameters of the model and not just the ϕ_1 parameter. Also, the sign of ρ_1 is determined by the sign of the difference $(\phi_1 - \theta_1)$.

In Figure 4.12 we graphed the *acf*'s of several typical ARMA(1,1) processes. A general conclusion from this figure is that it is very difficult to distinguish some of these *acf*'s from the *acf*'s of an AR(1) process as presented earlier in Figure 4.8. We are, however, now in a position to introduce another tool that should help us to narrow down the possible ARIMA models.

4.3 PARTIAL AUTOCORRELATION FUNCTIONS OF ARIMA MODELS

In Figure 4.8 we showed the *acf*'s of several AR(1) and AR(2) models, and pointed out that for all practical purposes there was hardly any difference between several of these. Furthermore, we stated that in order to distinguish certain AR(1) from certain AR(2) models we would have to use additional information. Similarly, the differences between Figures 4.12 and 4.8 are also not very obvious. In this section we will discuss one more characteristic of ARIMA models, namely the *partial autocorrelation function, (pacf)* which will help us to distinguish one model from another. We will begin by defining the term *partial autocorrelation*. Afterward we will discuss the partial autocorrelation function for some commonly encountered autoregressive, moving average, and mixed models.

---

[8] Stationarity and invertibility conditions require that $|\phi_1| < 1$ and $|\theta_1| < 1$, respectively.

4.3.1 Partial Autocorrelations

Any ARIMA(p,d,q) model, as we have seen, can always be expressed as a pure autoregressive model. These autoregressive models possess *acf*'s which, although they die out quickly, could stretch out to infinity. The partial autocorrelations constitute a device for summarizing all the information contained in the *acf* of an AR process in a small number of nonzero statistics. For an AR(p) process, only p such nonzero statistics are necessary, rather than an infinite number of nonzero autocorrelations. To see how this works, let us discuss this concept in the context of specific ARIMA models.

FIGURE 4.12   Autocorrelation Functions of ARMA(1,1) Models.

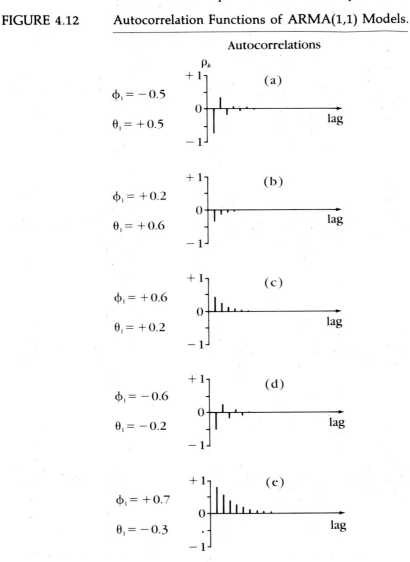

4.3.2 Autoregressive Models

In constructing autoregressive models we may want to check if we should include an additional lagged z_t in the model to represent the data more adequately. Suppose that after having fitted an $AR(k-1)$ model we want to see if the data should not be represented by an $AR(k)$ model. We therefore include an additional lagged variable in the model, namely z_{t-k}. If the value of the coefficient $|\phi_k|$ is 'large' we should include the z_{t-k}; otherwise we can omit the variable z_{t-k} and assume that the $AR(k-1)$ representation of the process is adequate. (The word 'large' will be qualified in Section 4.5.)

This coefficient ϕ_k measures the "excess" correlation not accounted for by the AR model of order $(k-1)$; that is, it measures the "partial" effect that z_{t-k} has in explaining the behavior of z_t in a model which already includes the terms $z_{t-1}, z_{t-2}, \ldots, z_{t-(k-1)}$. The highest-order autoregressive coefficient in the model, in this case ϕ_k, is defined as the partial autocorrelation at lag k,[9] and is denoted by ϕ_{kk}. The plot of ϕ_{kk} for different values of k, against k, is called the *partial autocorrelation function*, or *pacf*.

If the time series process is an $AR(k-1)$ process, then by successively calculating whether the parameters $\phi_1, \phi_2, \ldots, \phi_{k-1}, \phi_k$ should be included in the model we would find that $\phi_{11}, \phi_{22}, \ldots, \phi_{k-1,k-1}$ would have values which are different from 0, and that the value for ϕ_{kk} would be 0. One should keep in mind that the coefficient of any z_{t-j}, $j \leq k+1$, in an $AR(k+1)$ model will in general not be equal to ϕ_{jj}, the coefficient of $z_{t-j}, j \leq k$, in an $AR(k)$ model, because an $AR(k+1)$ model contains one additional variable, the lagged variable $z_{t-(k+1)}$.

We will now demonstrate how partial autocorrelations can be used to identify autoregressive models. Suppose that the true time series model is an $AR(1)$ model,

$$z_t = \phi_1 z_{t-1} + a_t. \qquad (4.35)$$

However, we also want to evaluate an $AR(2)$ model, namely

$$z_t = \phi_1 z_{t-1} + \phi_2 z_{t-2} + a_t. \qquad (4.36)$$

From (4.35) we obtain $\phi_{11} = \phi_1$, and from (4.36) $\phi_{22} = \phi_2$. If the true model is (4.35), we would expect that ϕ_{11} would be different from 0, and that ϕ_{22}, aside from sampling error,[10] would be equal to 0. Similarly, if we estimate

---

[9] If we represent the time series at different lags by the variables x_k, that is, $z_t = x_0$, $z_{t-1} = x_1, \ldots, z_{t-k} = x_k$, then the partial autocorrelations ϕ_{kk} of this time series are identical to the partial correlations between the variables x_0 and x_k, the influence of the variables $x_1, x_2, \ldots, x_{k-1}$ held constant.

[10] We will evaluate the uncertainty of the estimates of the partial autocorrelations in Section 4.5.

FIGURE 4.13 *Acf* and *Pacf* of Autoregressive Models.

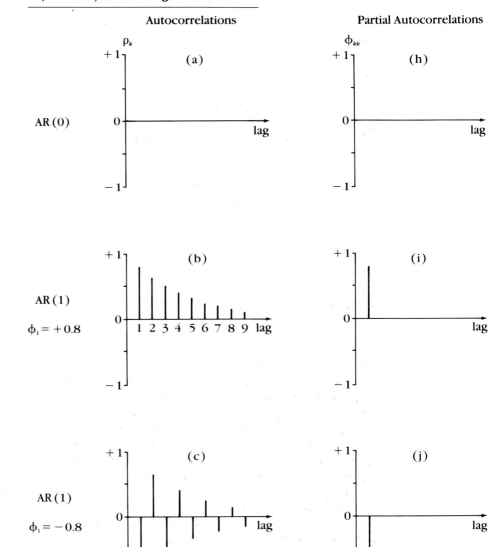

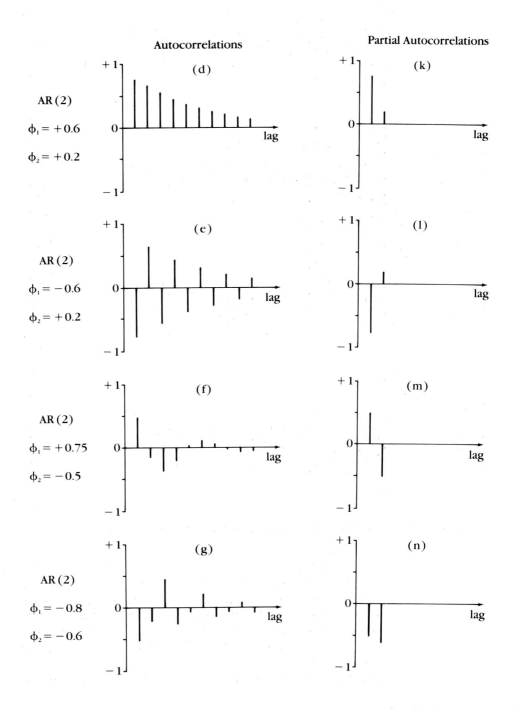

AR(3), AR(4), ... models, the third-, and fourth-, and higher-order partial autocorrelations should all be 0.

Thus, in constructing an AR(p) model we can use the partial autocorrelations to tell how many parameters to include in the model. We begin by entertaining the hypothesis that $p = 1$. If the partial autocorrelation ϕ_{11} is substantially different from 0, we conclude that we are dealing with an AR process *at least* of order 1. To see whether the process is of order 2 or greater, we evaluate ϕ_{22}. If ϕ_{22} differs from 0, we conclude that the process is *at least* of order 2. Notice that each time we are only looking at the highest order coefficients to obtain the partial autocorrelation. Similarly we could evaluate ϕ_{33}, ϕ_{44}, ...

In Figure 4.13 we have given the *pacf*'s of some autoregressive processes together with the *acf*'s. We now clearly see a difference between the *pacf* characteristics of an AR(1) process and an AR(2) process. The *pacf* of an AR(1) model has only one spike at lag 1, whereas the *pacf* of an AR(2) model contains two spikes, the first at lag 1 and the other at lag 2. Therefore, by combining the information contained in the *acf* and *pacf* it is possible to distinguish one AR process from another.

The partial autocorrelations described in this section are *population* partial autocorrelations. In Section 4.5 we will show how to use *sample* partial autocorrelations. These, of course, should be compared with the theoretical ones to identify tentative models.

4.3.3 Moving Average Models

To see how the partial autocorrelation function can be used in identifying a moving average process, we rewrite the MA model solely in terms of lagged z_t's, the inverted form, to obtain an infinite order autoregressive process. For an MA(1) process this inverted form takes the following form [see (3.35)],

$$z_t = -\theta_1 z_{t-1} - \theta_1^2 z_{t-2} - \theta_1^3 z_{t-3} - \cdots + a_t. \tag{4.37}$$

Since all the autoregressive parameters in (4.37) are necessary to represent the MA(1) process, the *pacf* of this process, contrary to that of an AR(p) process, will *not* possess a cutoff at lag p, but will be infinitely long. However, because of the invertibility condition, $|\theta_1| < 1$, the *pacf* will gradually die out.

It can easily be shown that any MA(q) process can be rewritten to form an AR process of infinite order. Consequently, the *pacf* of an MA process will be infinitely long. However, because of the invertibility conditions, the partial autocorrelations usually die out rather quickly. The graphs on the right side of Figure 4.14 contain the *pacf*'s of some typical MA(1) and MA(2) processes.

4.3.4 Mixed Models

Rewriting an ARMA(p,q) model in terms of autoregressive process, we will obtain a pure AR process of infinite order, if $q \neq 0$. For example, the autoregressive representation of an ARMA(1,1) model

$$z_t = \phi_1 z_{t-1} + a_t - \theta_1 a_{t-1} \tag{4.38}$$

is

$$z_t = a_t + (\phi_1 - \theta_1)z_{t-1} + \theta_1(\phi_1 - \theta_1)z_{t-2} + \theta_1^2(\phi_1 - \theta_1)z_{t-3} + \cdots . \tag{4.39}$$

If the ARMA(1,1) model is stationary and invertible ($|\phi_1| < 1$, $|\theta_1| < 1$), then the *pacf* will gradually die out. This general decaying pattern typifies the *pacf* of all mixed processes. Figure 4.15 contains the *pacf* of several ARMA(1,1) processes, together with the corresponding *acf* already given in Figure 4.12.

Table 4.1 on page 94 contains a summary of the patterns of the autocorrelation function and partial autocorrelation function of stationary and invertible autoregressive, moving average, and mixed processes.

Note, in particular, the duality between the AR and MA processes:

- the *acf* of an MA process behaves like the *pacf* of an AR process; and
- the *pacf* of an MA process behaves like the *acf* of an AR process.

4.4 AUTOCORRELATION FUNCTIONS AND PARTIAL AUTOCORRELATION FUNCTIONS OF SEASONAL MODELS

In Section 3.6 we introduced seasonal models, and indicated that the *acf* of pure seasonal autoregressive, moving average, and mixed models behaves in the same way as the *acf* of a nonseasonal model except that the values of the autocorrelations appear at multiples of the span. This generalization also extends to the *pacf* of seasonal ARIMA models.

Modeling seasonal time series often involves the use of nonseasonal as well as seasonal parameters resulting in multiplicative seasonal models as presented in equation (3.101). Unfortunately, the general characteristics of the *acf*'s and *pacf*'s of multiplicative seasonal models are somewhat harder to visualize. In this section we will discuss the *acf* and *pacf* of the two special models already discussed in Section 3.6.4 and provide a table describing the properties of some other multiplicative seasonal models.

4.4.1 Autoregressive Multiplicative Seasonal Models

In Section 3.6.4 we presented the following autoregressive multiplicative seasonal model [see equation (3.111)],

FIGURE 4.14 *Acf* and *Pacf* of Moving Average Models.

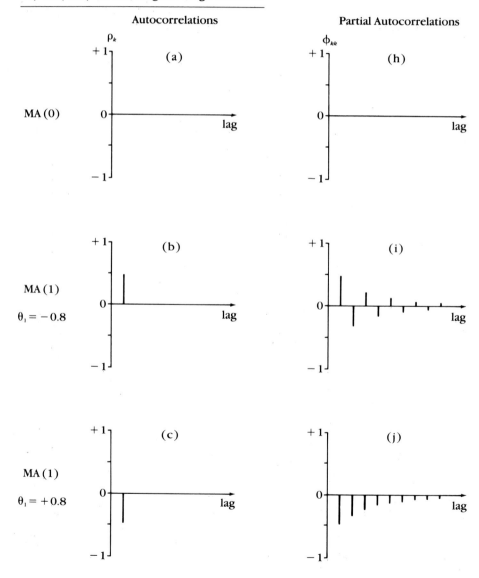

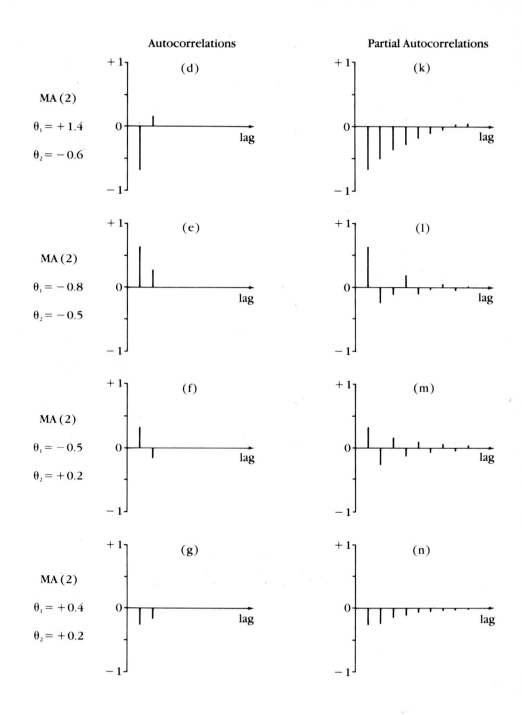

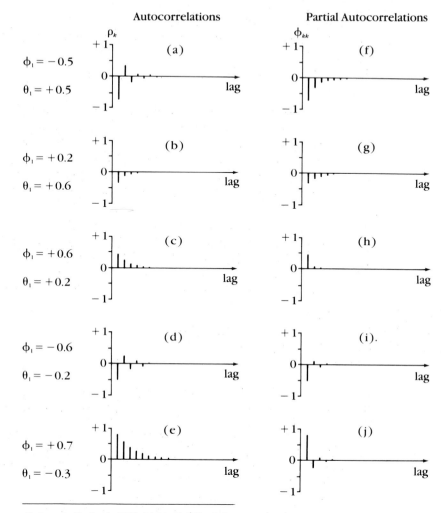

FIGURE 4.15   *Acf* and *Pacf* of ARMA(1,1) Models.

TABLE 4.1   Autocorrelation and Partial Autocorrelation Functions of Nonseasonal Models

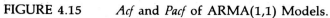

MODEL	AUTOCORRELATION FUNCTION	PARTIAL AUTOCORRELATION FUNCTION
AR(p)	Tails off	Cuts off after lag p
MA(q)	Cuts off after lag q	Tails off
ARMA(p,q)	Tails off	Tails off

$$(1 - \phi_1 B)(1 - \Phi_1 B^4)z_t = a_t. \tag{4.40}$$

We can also write (4.40) as

$$z_t = \phi_1 z_{t-1} + \Phi_1 z_{t-4} - \phi_1 \Phi_1 z_{t-5} + a_t. \tag{4.41}$$

From (4.41) it is clear that whatever the partial autocorrelations (ϕ_{kk}) are for $k \le 5$, the values of ϕ_{kk} for $k > 5$ will all be 0. That is, there will be a cutoff after lag 5.

Given that for an AR(2) model the autocorrelations are already complicated functions of the parameters in the model, it should not come as a surprise that for the model (4.41), a kind of fifth-order autoregressive process, this will certainly also be the case. However, given the stationarity conditions, $|\phi_1| < 1$ and $|\Phi_1| < 1$, the *acf* will still show a decaying behavior rather than a cutoff.

Without explicitly calculating the autocorrelations and partial autocorrelations, one cannot easily differentiate this multiplicative seasonal model from a regular autoregressive model of order 5. In Chapter 5 we will present some additional diagnostic checks that will help in differentiating between these two forms of models. Specifically we will recommend that if the *acf* dies out gradually and the *pacf* has a cutoff at lag 5, we should estimate an AR(5) and then evaluate the parameter estimates. From (4.41) it should be obvious that if we estimate an AR(5) model

$$z_t = \phi_1 z_{t-1} + \phi_2 z_{t-2} + \phi_3 z_{t-3} + \phi_4 z_{t-4} + \phi_5 z_{t-5} + a_t$$

and find that ϕ_2 and ϕ_3 are 0 and $\phi_5 = -\phi_1 \phi_4$, this should be evidence in favor of the autoregressive multiplicative seasonal model (4.41).

In Figures 4.16 and 4.17 we have presented some typical *acf*'s and *pacf*'s for multiplicative autoregressive and seasonal autoregressive model with span = 12. Specifically, we have reproduced the *acf*'s and *pacf*'s for the same autoregressive models used in Figure 4.13, but augmented with a seasonal AR(1) polynomial with span 12. Clearly all *pacf*'s in Figure 4.16 have a cutoff at lag 13. The beginning of each *acf* shows a similar pattern as a pure AR(1) process (see Figure 4.13). However, the seasonal AR model pulls the *acf* up, in absolute value, as the *acf* approaches the seasonal span lag. Then, starting from lag 13, the *acf* has a very similar pattern as the first 12 lags. Also, the *acf* at multiples of the span behaves very much as a pure AR(1) process. Let us concentrate on parts (b) and (f) of Figure 4.16. The AR(1) parameter has a value $\phi_1 = 0.8$ and the seasonal AR(1) has a value $\Phi_1 = -0.8$. We notice that the *acf* starts at 0.774 and then gradually declines. At lag 7 the effect of the seasonal parameter is felt which makes the *acf* increase in absolute value, until it reaches the value -0.774 at lag 12. Then, starting from lag 13, the *acf* decreases again in absolute value, then increases until it reaches the value 0.621 at lag 24. This value of the *acf* at lag 24 is positive, whereas at lags 12 and 36 the value is negative. Again this corresponds in general behavior with the *acf* of a pure AR(1) process with $\phi_1 = -0.8$ [see Figure 4.13(c)].

The corresponding *pacf* can also be decomposed. The *pacf* of a pure AR(1) model has a positive spike at lag 1, if the autoregressive parameter ϕ_1 is positive and has a negative spike at lag 1 if the parameter ϕ_1 is negative (see Figure 4.13). For the multiplicative model, we also find that the *pacf* contains a positive value at lag 1 corresponding to ϕ_1 positive and a negative value at lag 12 corresponding to Φ_1 negative. But we also observe a large

FIGURE 4.16   *Acf* and *Pacf* of AR(1)×SAR(1) Seasonal Models.

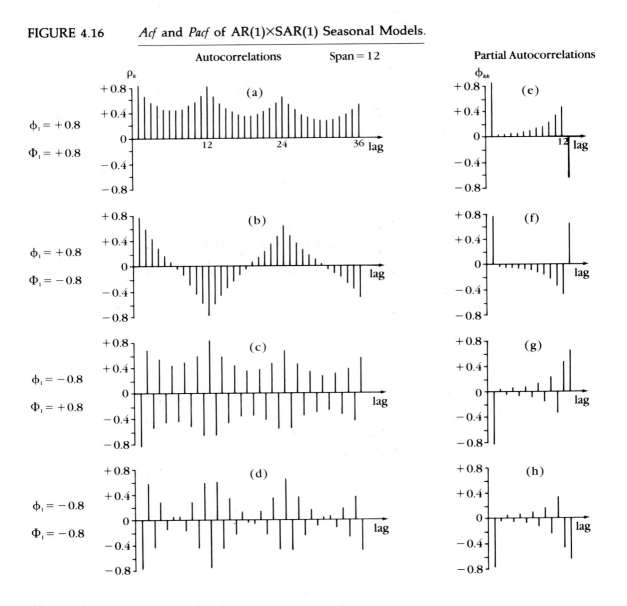

spike at lag 13, with sign the opposite of the sign of product of the lag 1 and lag 12 partial autocorrelations. Also, as we increase the lags of the *pacf* from 2 to 12, therefore include more and more term in the calculation of ϕ_{kk}, the terms gradually pick up the effect of the lag 12 parameter. Remember that the *pacf* represents partial correlations; that is, correlations between w_t and w_{t-k}, given that $w_{t-1}, \ldots, w_{t-k+1}$, are already included.

In Figure 4.17 we have represented the *acf*'s and *pacf*'s of multiplicative AR(2) and seasonal AR(1) model. Again it is instructive to compare the patterns with those of a pure AR(2) process represented in Figure 4.13. The basic characteristic that is easily recognized is that the *pacf* has a cutoff at lag 14. Also, all lag 12 partial autocorrelations in Figures 4.17(i)–(l) are positive, corresponding to $\Phi_1 = 0.8$, whereas all lag 12 partial autocorrelations in Figures 4.17(m)–(p) are negative corresponding to $\Phi_1 = -0.8$. Also, the sign of the lag 13 and lag 14 partial autocorrelation is the opposite of the sign of the product of the lag 1 and lag 12 and the lag 2 and lag 12 partial autocorrelation, respectively.

4.4.2 Moving Average Multiplicative Seasonal Models

The other frequently encountered seasonal model is the moving average multiplicative seasonal model discussed in Section 3.6, equation (3.112), namely

$$z_t = (1 - \theta_1 B)(1 - \Theta_1 B^4)a_t. \tag{4.42}$$

We assume that no differencing is needed to induce stationarity and that the model has a span of 4. Model (4.42) can also be expressed as

$$z_t = a_t - \theta_1 a_{t-1} - \Theta_1 a_{t-4} + \theta_1 \Theta_1 a_{t-5}, \tag{4.43}$$

which shows that the model can be interpreted as a moving average model of order $q + sQ = 1 + (4 \times 1) = 5$, with the parameter of a_{t-5} restricted to equal the negative of the product of the a_{t-1} and the a_{t-4} parameters.

Since the *acf* of moving average processes abruptly cuts off after the lag corresponding to the order of the process, the autocorrelations of this model will be 0 for lags greater than five. Indeed it can be shown that the autocovariances of (4.42) equal

$$\lambda_0 = (1 + \theta_1^2)(1 + \Theta_1^2)\sigma_a^2$$
$$\lambda_1 = -\theta_1 (1 + \Theta_1^2)\sigma_a^2$$
$$\lambda_2 = 0$$
$$\lambda_3 = \theta_1 \Theta_1 \sigma_a^2$$
$$\lambda_4 = -\Theta_1 (1 + \theta_1^2)\sigma_a^2$$
$$\lambda_5 = \theta_1 \Theta_1 \sigma_a^2$$
$$\lambda_k = 0, \quad k > 5, \tag{4.44}$$

FIGURE 4.17   *Acf* and *Pacf* of AR(2)×SAR(1) Seasonal Models.

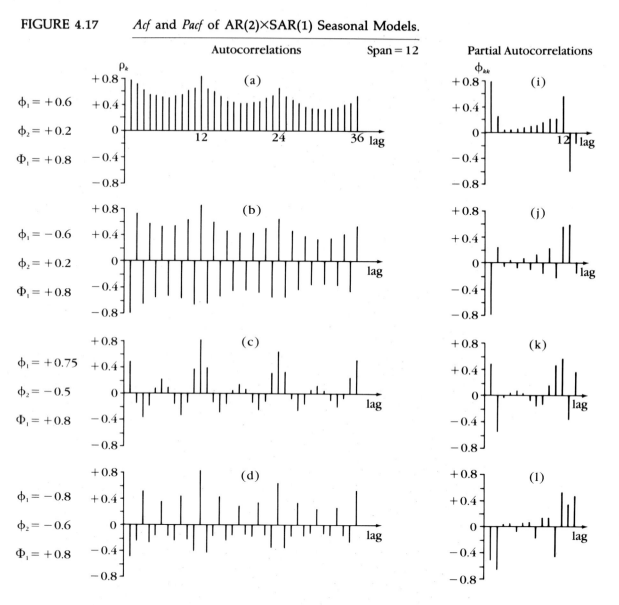

Autocorrelations

Partial Autocorrelations

$\phi_1 = +0.6$

$\phi_2 = +0.2$

$\Phi_1 = -0.8$

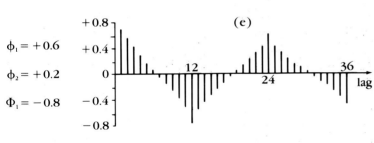

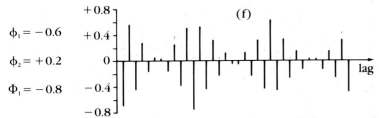

(e)

(m)

$\phi_1 = -0.6$

$\phi_2 = +0.2$

$\Phi_1 = -0.8$

(f)

(n)

$\phi_1 = +0.75$

$\phi_2 = -0.5$

$\Phi_1 = -0.8$

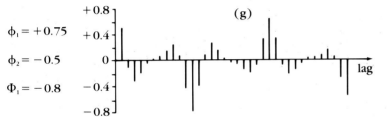

(g)

(o)

$\phi_1 = -0.8$

$\phi_2 = -0.6$

$\Phi_1 = -0.8$

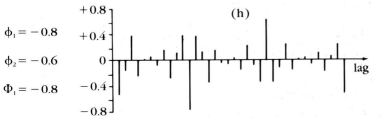

(h)

(p)

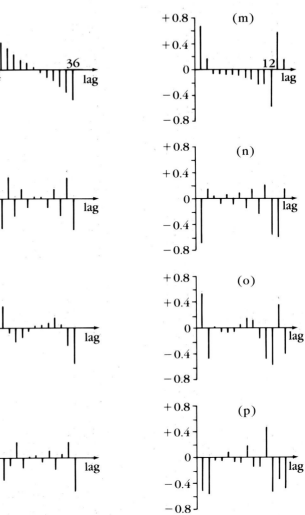

and the autocorrelations can be expressed as

$$\rho_1 = -\theta_1/(1 + \theta_1^2) \qquad (4.45a)$$
$$\rho_4 = -\Theta_1/(1 + \Theta_1^2) \qquad (4.45b)$$

and

$$\rho_2 = 0$$
$$\rho_3 = \rho_5 = \rho_1\rho_4$$
$$\rho_k = 0, \qquad k > 5. \qquad (4.45c)$$

The important features of the autocorrelations are: First, ρ_1 is only influenced by the regular MA parameter θ_1 and ρ_4 is only influenced by the seasonal MA parameter Θ_1; and second, the *acf* is symmetric in the neighborhood of the autocorrelation ρ_4 with ρ_3 equal to ρ_5. This symmetrical behavior is an important characteristic of the *acf* of moving average multiplicative seasonal models. The fact that ρ_3 is not 0 is contrary to what we might have expected since the error term a_{t-3} is not present in (4.43). This *acf* pattern (4.45) is typical of multiplicative seasonal models with both consecutive and seasonal moving average terms.

Model (4.42) can also be written as an autoregressive process with an infinite number of terms involving θ_1 and Θ_1. Since the invertibility conditions restrict the values of the parameters to be $|\theta_1| < 1$ and $|\Theta_1| < 1$, the values of the coefficients of the lagged z_t's will become smaller and smaller in absolute value as the order of the process increases. Therefore, the pattern of the *pacf* of this multiplicative seasonal model will be decaying. Apart from this feature, the exact behavior of the partial autocorrelation is difficult to work out in general.

In Figures 4.18 and 4.19 we have again represented the *acf* and *pacf* of some typical moving average multiplicative seasonal models with span = 12. The *pacf*'s certainly confirm that it will be very difficult to recognize the specific seasonal moving average models. The *acf*'s are, however, easier to recognize. In Figure 4.18, representing the multiplicative MA(1)×SMA(1) models, the *acf* has a cutoff at lag 13, with only ρ_1, ρ_{11}, ρ_{12}, and ρ_{13} nonzero. Furthermore, ρ_1 has the opposite sign of θ_1, ρ_{12} the opposite sign of Θ_1, and $\rho_{11} = \rho_{13} = \rho_1\rho_{12}$. These same general characteristics apply to Figure 4.19. In this figure we have the *acf*'s and *pacf*'s of some typical multiplicative MA(2)×SMA(1) model. Again there is now a cutoff at lag 14 in the *acf*, with only ρ_1, ρ_2, ρ_{10}, ρ_{11}, ρ_{12}, ρ_{13}, and ρ_{14} nonzero. In addition ρ_2 has the opposite sign of θ_2, ρ_{12} the opposite sign of Θ_1 and $\rho_{10} = \rho_{14} = \rho_2\rho_{12}$, $\rho_{11} = \rho_{13} = \rho_1\rho_{12}$. Formulas for the autocorrelations for these and other multiplicative seasonal models are given in Table 4.3, pages 104–105.

As mentioned, the *pacf*'s are difficult to interpret. However, we do notice that the shape of the first ten partial autocorrelations for the MA(1)×SMA(1) models and the first nine partial autocorrelations for the MA(2)×SMA(1) models are similar to the partial autocorrelations of the pure MA(1) and MA(2) model (see Figure 4.14).

4.4.3 General Characteristics of Multiplicative Models

There are very few generalizations about the *acf* and *pacf* that can be used to identify multiplicative seasonal models. Table 4.2 contains some of the basic characteristics of the *acf* and *pacf* of seasonal models. These characteristics can be used to see if perhaps a seasonal model may be worth considering to fit the data.

FIGURE 4.18   *Acf* and *Pacf* of MA(1)×SMA(1) Seasonal Models.

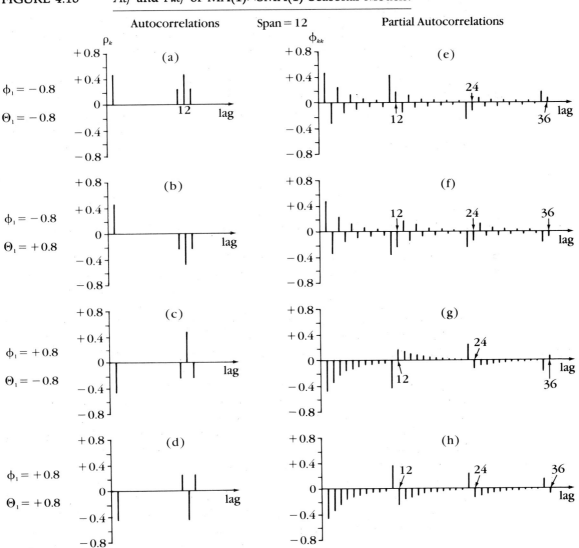

FIGURE 4.19   *Acf* and *Pacf* of MA(2)×SMA(1) Seasonal Models.

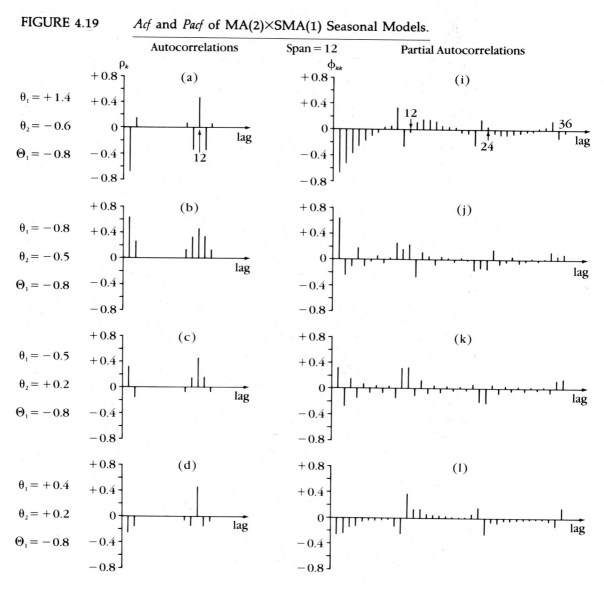

Autocorrelations   Partial Autocorrelations

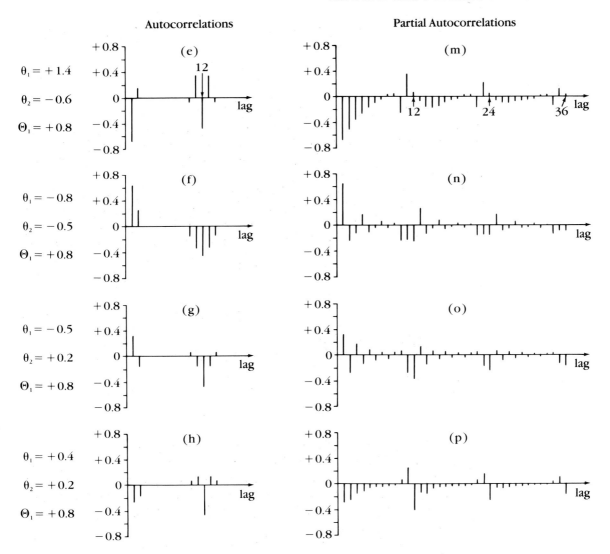

TABLE 4.2   Autocorrelation and Partial Autocorrelation Functions of Seasonal Models

MODEL SPAN = s	AUTOCORRELATION FUNCTION	PARTIAL AUTOCORRELATION FUNCTION
AR(p), Seas. AR(P)	Tails off	Cuts off after lag $p + sP$
MA(q), Seas. MA(Q)	Cuts off after lag $q + sQ$	Tails off
Mixed Models	Tails off	Tails off

TABLE 4.3 Autocovariances of Some Multiplicative Seasonal Models[a]

MODEL	(AUTOCOVARIANCES OF w_t)/σ_a^2	SPECIAL CHARACTERISTICS
(1) $w_t = (1 - \theta B)(1 - \Theta B^s)a_t$ $w_t = a_t - \theta a_{t-1} - \Theta a_{t-s} + \theta\Theta a_{t-s-1}$ $s \geq 3$	$\gamma_0 = (1 + \theta^2)(1 + \Theta^2)$ $\gamma_1 = -\theta(1 + \Theta^2)$ $\gamma_{s-1} = \theta\Theta$ $\gamma_s = -\Theta(1 + \theta^2)$ $\gamma_{s+1} = \gamma_{s-1}$ All other autocovariances are 0.	(a) $\gamma_{s-1} = \gamma_{s+1}$ (b) $\rho_{s-1} = \rho_{s+1} = \rho_1\rho_s$
(2) $w_t = (1 - \theta_1 B - \theta_s B^s - \theta_{s+1}B^{s+1})a_t$ $w_t = a_t - \theta_1 a_{t-1} - \theta_s a_{t-s}$ $\quad -\theta_{s+1}a_{t-s-1}$ $s \geq 3$	$\gamma_0 = 1 + \theta_1^2 + \theta_s^2 + \theta_{s+1}^2$ $\gamma_1 = -\theta_1 + \theta_s\theta_{s+1}$ $\gamma_{s-1} = \theta_1\theta_s$ $\gamma_s = \theta_1\theta_{s+1} - \theta_s$ $\gamma_{s+1} = -\theta_{s+1}$ All other autocovariances are 0.	(a) In general, $\gamma_{s-1} \neq \gamma_{s+1}$ $\gamma_1\gamma_s \neq \gamma_{s+1}$
(3) $w_t = (1 - \theta_1 B - \theta_2 B^2)(1 - \Theta B^s)a_t$ $w_t = a_t - \theta_1 a_{t-1} - \theta_2 a_{t-2} - \Theta a_{t-s}$ $\quad + \theta_1\Theta a_{t-s-1} + \theta_2\Theta a_{t-s-2}$ $s \geq 5$	$\gamma_0 = (1 + \theta_1^2 + \theta_2^2)(1 + \Theta^2)$ $\gamma_1 = -\theta_1(1 - \theta_2)(1 + \Theta^2)$ $\gamma_2 = -\theta_2(1 + \Theta^2)$ $\gamma_{s-2} = \theta_2\Theta$ $\gamma_{s-1} = \theta_1\Theta(1 - \theta_2)$ $\gamma_s = -\Theta(1 + \theta_1^2 + \theta_2^2)$ $\gamma_{s+1} = \gamma_{s-1}$ $\gamma_{s+2} = \gamma_{s-2}$ All other autocovariances are 0.	(a) $\gamma_{s-2} = \gamma_{s+2}$ (b) $\gamma_{s-1} = \gamma_{s+1}$
(4) $w_t = (1 - \theta B)(1 - \Theta_1 B^s - \Theta_2 B^{2s})a_t$ $w_t = a_t - \theta a_{t-1} - \Theta_1 a_{t-s} + \theta\Theta_1 a_{t-s-1}$ $\quad - \Theta_2 a_{t-2s} + \theta\Theta_2 a_{t-2s-1}$ $s \geq 3$	$\gamma_0 = (1 + \theta^2)(1 + \Theta_1^2 + \Theta_2^2)$ $\gamma_1 = -\theta(1 + \Theta_1^2 + \Theta_2^2)$ $\gamma_{s-1} = \theta\Theta_1(1 - \Theta_2)$ $\gamma_s = -\Theta_1(1 + \theta^2)(1 - \Theta_2)$ $\gamma_{s+1} = \gamma_{s-1}$ $\gamma_{2s-1} = \theta\Theta_2$ $\gamma_{2s} = -\Theta_2(1 + \theta^2)$ $\gamma_{2s+1} = \gamma_{2s-1}$ All other autocovariances are 0.	(a) $\gamma_{s-1} = \gamma_{s+1}$ (b) $\gamma_{2s-1} = \gamma_{2s+1}$

MODEL	(AUTOCOVARIANCES OF w_t)/σ_a^2	SPECIAL CHARACTERISTICS
(5) $\quad (1 - \Phi B^s)w_t = (1 - \theta B)(1 - \Theta B^s)a_t$ $w_t - \Phi w_{t-s} = a_t - \theta a_{t-1} - \Theta a_{t-s}$ $\qquad\qquad + \theta\Theta a_{t-s-1}$ $\qquad\qquad s \geq 3$	$\gamma_0 = (1+\theta^2)\left[1 + \dfrac{(\Theta-\Phi)^2}{1-\Phi^2}\right]$ $\gamma_1 = -\theta\left[1 + \dfrac{(\Theta-\Phi)^2}{1-\Phi^2}\right]$ $\gamma_{s-1} = \theta\left[\Theta - \Phi - \dfrac{\Phi(\Theta-\Phi)^2}{1-\Phi^2}\right]$ $\gamma_s = -(1+\theta^2)\left[\Theta - \Phi - \dfrac{\Phi(\Theta-\Phi)^2}{1-\Phi^2}\right]$ $\gamma_{s+1} = \gamma_{s-1}$ $\gamma_j = \Phi\gamma_{j-s} \qquad j \geq s+2$ For $s \geq 4$, $\gamma_2, \gamma_3, \ldots, \gamma_{s-2}$ are all 0.	(a) $\gamma_{s-1} = \gamma_{s+1}$ (b) $\gamma_j = \Phi\gamma_{j-s} \qquad j \geq s+2$
(6) $\quad (1 - \Phi B^s)w_t = (1 - \theta_1 B - \theta_s B^s$ $\qquad\qquad - \theta_{s+1}B^{s+1})a_t$ $w_t - \Phi w_{t-s} = a_t - \theta_1 a_{t-1} - \theta_s a_{t-s}$ $\qquad\qquad -\theta_{s+1}a_{t-s-1}$ $\qquad\qquad s \geq 3$	$\gamma_0 = 1 + \theta_1^2 + \dfrac{(\theta_s-\Phi)^2}{1-\Phi^2} + \dfrac{(\theta_{s+1}+\theta_1\Phi)^2}{1-\Phi^2}$ $\gamma_1 = -\theta_1 + \dfrac{(\theta_s - \Phi)(\theta_{s+1}+\theta_1\Phi)}{1-\Phi^2}$ $\gamma_{s-1} = (\theta_s - \Phi)\left[\theta_1 + \Phi\dfrac{(\theta_{s+1}+\Phi\theta_1)}{1-\Phi^2}\right]$ $\gamma_s = -(\theta_s - \Phi)\left[1 - \Phi\dfrac{(\theta_s - \Phi)}{1-\Phi^2}\right]$ $\qquad + (\theta_{s+1} + \theta_1\Phi)\left[\theta_1 + \Phi\dfrac{(\theta_{s+1} + \theta_1\Phi)}{1-\Phi^2}\right]$ $\gamma_{s+1} = -(\theta_{s+1} + \theta_1\Phi)\left[1 - \Phi\dfrac{(\theta_s - \Phi)}{1-\Phi^2}\right]$ $\gamma_j = \Phi\gamma_{j-s} \qquad j \geq s+2$ For $s \geq 4$, $\gamma_2, \gamma_3, \ldots, \gamma_{s-2}$ are all 0.	(a) $\gamma_{s-1} \neq \gamma_{s+1}$ (b) $\gamma_j = \Phi\gamma_{j-s} \qquad j \geq s+2$

* *Source:* Box and Jenkins (1976), pp. 329–333.

In many cases tentative identification of multiplicative models can be made by comparing the sample autocorrelations with the population autocorrelations. Box and Jenkins have summarized the major *acf* characteristics of some frequently encountered multiplicative seasonal models. Table 4.3 is based on their results.

4.5 SAMPLE AUTOCORRELATIONS AND SAMPLE PARTIAL AUTOCORRELATIONS[11]

In practice, one never knows the population values of autocorrelations and partial autocorrelations of the underlying stochastic process. Consequently, in identifying a tentative model, one must use the sample autocorrelation and partial autocorrelation functions to see if they are similar to those of typical models for which the parameters are known.

Since sample autocorrelations and partial autocorrelations are only estimates, they are subject to sampling errors, and as such will never correspond exactly to the underlying true autocorrelations and partial autocorrelations. Sample autocorrelations will be denoted by r_k, and sample partial autocorrelations by $\hat{\phi}_{kk}$.

4.5.1 Sample Autocorrelations

An estimate of ρ_k can be calculated using the formula

$$r_k = \frac{c_k}{c_0} \tag{4.46}$$

where c_k, defined as

$$c_k = \frac{1}{n} \sum_{t=1}^{n-k} z_t z_{t+k}, \qquad k \geq 0, \tag{4.47}$$

is the estimate of the autocovariance λ_k. Notice that for each increase in k one observation is lost. The statistic c_1 is estimated from $n-1$ pairs of observations; c_2 is estimated from $n-2$ pairs of observations, and so forth. Again, z_t represents deviations from the mean of the stationary data and n is the number of observations available after suitable differencing has been made.

To use the identification methods described in the earlier sections of this chapter, it is necessary to know when ρ_k is effectively 0. For this purpose we must use the standard error of the sample autocorrelations. For lags k greater than some value q beyond which the theoretical autocorrelation function may be deemed to have died out, Bartlett (1946, 1966) has shown that an approximate estimate of the autocorrelation variance is given by:

[11] In this section we assume that the reader is familiar with some basic principles of statistical inference in general and hypothesis testing in particular. For background reading we can refer to DeGroot (1975) and Snedecor and Cochran (1980).

$$\mathrm{Var}(r_k) \simeq \frac{1}{n}\,[1 + 2(\rho_1^2 + \rho_2^2 + \cdots + \rho_q^2)], \qquad k > q, \qquad (4.48)$$

where n is defined as in (4.47). In practice, however, one replaces the population autocorrelations ρ_k of a particular ARIMA model with the estimated autocorrelations r_k to obtain an estimate of the approximate variance of r_k as

$$\widehat{\mathrm{Var}}(r_k) = \frac{1}{n}\,(1 + 2\sum_{i=1}^{q} r_i^2), \qquad k > q. \qquad (4.49)$$

The square root of (4.49) is called the *large sample standard error*,[12] $\mathrm{SE}(r_k)$, of r_k.

EXAMPLE

The following autocorrelations were estimated based on a time series of length $n = 100$ observations, and generated by an MA(1) process for which it was *known* that $\rho_1 = -0.4$ and $\rho_k = 0$ for $k > 1$:

k	1	2	3	4	5	6	7	8	9	10
r_k	−0.30	−0.13	0.07	−0.10	0.06	0.04	0.04	−0.12	−0.02	0.12

Given these estimated autocorrelations, r_k, is it possible for all population autocorrelations, ρ_k, to be 0? To answer this question, let us calculate the estimate of the variance of the autocorrelations assuming that all the ρ_k's are 0. Setting all r_k's in (4.49) to 0 yields the following estimate of the variance:

$$\widehat{\mathrm{Var}}(r_k) = \frac{1}{n} = 0.01, \qquad k > 0. \qquad (4.50)$$

The standard error of r_k is thus

$$\mathrm{SE}(r_k) = \sqrt{\widehat{\mathrm{Var}}(r_k)} = 0.10, \qquad k > 0. \qquad (4.51)$$

Since the absolute value of r_1 is three times the standard error, we can conclude that ρ_1 is nonzero. However, in order to confirm that this series is really governed by an MA(1) process and not a higher-order process, we have to check if this series is compatible with ρ_1 being nonzero, and $\rho_k = 0$ for $k > 1$. Setting $r_1 = -0.30$ and $r_k = 0$ for $k > 1$ in (4.49), we obtain the following estimate of the variance for r_k, $k > 1$.

[12] Program IDENT which is part of the Time Series (TS) collection (see Appendix C), prints out the so-called "ROW SE," that is, the *standard errors* for r_k under the assumption that q takes on the values 0, 12, 24, . . . (the program prints twelve autocorrelations per row). In the plot of the autocorrelations, 95% large-sample confidence limits are plotted for $q = 0, 1, 2, \ldots$. These large-sample confidence limits are constructed using results of Anderson (1942) which show that for moderately large samples, and if the theoretical autocorrelation is 0, the distribution of the estimated autocorrelations is approximately normal. A similar result holds for the estimated partial autocorrelations.

$$\widehat{\mathrm{Var}}(r_k) = \frac{1}{100}[1 + 2(-0.30)^2] = 0.012, \qquad k > 1. \qquad (4.52)$$

The $\mathrm{SE}(r_k)$ is 0.109. Since the estimate autocorrelations for all lags greater than 1 are small compared with this SE, we can conclude that the series is compatible with an *acf* in which $\rho_1 \neq 0$ and $\rho_k = 0$ for $k > 1$.

Q STATISTIC

Rather than considering each autocorrelation individually, quite often we might want to see if, as a group, say the first 24 autocorrelations, show evidence of model inadequacies. Box and Pierce (1970) showed that for a purely random process, that is, a model with *all* $\rho_k = 0$, the statistic·

$$Q(K) = n(n+2) \sum_{k=1}^{K} \frac{1}{n-k} r_k^2 \qquad (4.53)$$

is distributed approximately as a χ^2(chi-square) distribution with K degrees of freedom, $\chi^2(K)$, where n is the number of data points available after differencing the series, and K is the number of autocorrelations used in the summation. This test using the Q statistic is sometimes called the *Portmanteau test.* If the computed value of Q is less than the table value of the χ^2 statistic with K degrees of freedom, given a prespecified significance level, the group of autocorrelations used to calculate the test can be assumed to be not different from 0. This indicates that the data generating the autocorrelations are random. If the computed Q statistic is larger than the χ^2 value from a χ^2 table, the autocorrelations are significantly different from 0, indicating the existence of some pattern. Fortunately, many computer programs will not only calculate the value of Q statistic but the probability of the null hypothesis as well. Therefore, there is less and less need to have access to these distribution tables.

The Q statistic in our example takes on the value $Q = 98.33$, and based on a χ^2 distribution with 10 degrees of freedom, we find a χ^2 statistic value of 25.188, given a significance level of 0.005. Therefore there is less than half a percent chance that the model is a purely random process.

Box and Pierce initially suggested the use of a different Q statistic, but later work by Ljung and Box (1978) showed that improved small sample performance was obtained with the form of the Q-test specified in (4.53). This form of the Q-test is sometimes referred to as the Ljung–Box Q-test.[13] It should be noted that the Q-test is not a very powerful tool for detecting specific departures from white noise.[14] Furthermore, the Q statistic is also

---

[13] Program IDENT calls this the CHI-SQUARE TEST and prints out the Q statistic value as well as the *P*-value defined as the probability that the correct model is a purely random process (i.e., the difference between 1 and the tail area probability). For an interpretation of the *P*-value, see DeGroot (1973), Gibbons and Pratt (1975), and Dickey (1977).

[14] Much research has been done on the actual performance of this Q statistic; see Davies et al. (1977) and Davies and Newbold (1979).

sensitive to the value of K, the number of autocorrelations used to calculate Q. For economic data, $K = 12$ and $K = 24$ have proven to be useful.

4.5.2 Sample Partial Autocorrelations

Unlike the autocorrelations, the partial autocorrelations cannot be estimated using a simple, straightforward formula. Indeed the partial autocorrelations $\hat{\phi}_{kk}$ are calculated from a solution of the Yule–Walker equation system, expressing the partial autocorrelations as a function of the autocorrelations.[15] We refer the reader to Box and Jenkins (1976, p. 64). The important question we have to address here is how to use the $\hat{\phi}_{kk}$ to evaluate the unknown population partial autocorrelations, and in particular to evaluate when population partial autocorrelations can be considered to be 0. Again, we therefore need to evaluate the SE of $\hat{\phi}_{kk}$.

Quenouille (1949) showed that the variance of the estimate of the partial autocorrelations, $\hat{\phi}_{kk}$, is approximately equal to

$$\mathrm{Var}(\hat{\phi}_{kk}) \simeq \frac{1}{n}, \qquad k > p, \tag{4.54}$$

where n equals the number of observations after suitable differencing, and p represents the first p partial autocorrelations that are assumed to be nonzero. Notice that, contrary to the variance of the autocorrelations, the variance of the partial autocorrelations only depends upon the sample size. Equation (4.54) provides a way, after having observed p nonzero partial autocorrelations, to evaluate if all other $\hat{\phi}_{kk}$'s are different from 0.

EXAMPLE

The following partial autocorrelations were estimated based on data generated by the AR(1) process $z_t = 0.5 z_{t-1} + a_t$, with $n = 100$:

k	1	2	3	4	5	6	7	8	9	10
$\hat{\phi}_{kk}$	0.40	−0.12	−0.14	0.01	0.15	−0.07	−0.05	0.10	0.06	0.02

For this process the theoretical value for ϕ_{11} is 0.5 and for $k > 1$, $\phi_{kk} = 0$. The approximate standard errors, based on (4.54), are equal to

$$\mathrm{SE}(\hat{\phi}_{kk}) = \frac{1}{\sqrt{n}} = 0.10 \qquad k > p.$$

Since the value of 0.40 is four times the value of the standard error, we can safely infer that ϕ_{11} is nonzero. Furthermore, since all other sample partial autocorrelations are in absolute value less than two SE and most of them even less than one SE, we can safely conclude that all other ϕ_{kk} are 0. This,

---

[15] Box and Jenkins (1976, p. 82 ff.) present a recursive method for calculating the *pacf* which is attributed to Durbin (1960). The TS discussed in Appendix C programs make use of this recursive method.

of course, is what we expected, since the data have been generated by an AR(1) process.

In evaluating sample autocorrelations and partial autocorrelations, one should avoid attaching too much significance to every detail which might appear in the estimates of these statistics. Keep in mind that the probability of having, say, one autocorrelation lie outside the 95% large-sample confidence intervals, given that the data are really random, increases with the number of autocorrelations which are evaluated. For example, if the first 20 values of r_k are plotted against k, then one should expect to find at least one significant value outside the 95% confidence limits, even if the data were truly random.

Another point to consider when deciding whether or not lags are significantly different from 0 is the actual interpretation that can or cannot be given to these spikes. If spikes appear at lags to which some physical interpretation can be attached, such as lag 1 or 2 or a lag corresponding to a seasonal variation, then these spikes may be considered worthwhile to be included

FIGURE 4.20   Sample *Acf*'s of an AR(1) Model.[a]

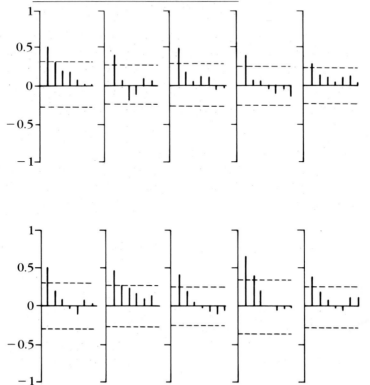

[a] Ten realizations of length $n = 100$ for the process $z_t = 0.5z_{t-1} + a_t$. The dotted lines are drawn at ±2 standard errors. Source: Nelson (1973), p. 73.

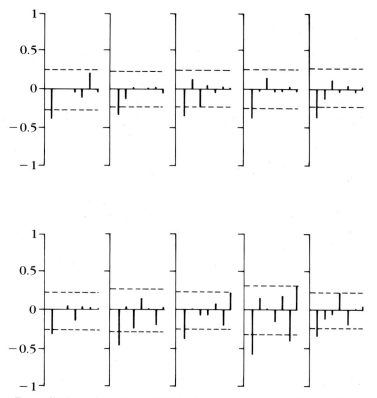

^a Ten realizations of length $n = 100$ for the process $z_t = a_t - 0.5\, a_{t-1}$. The
dotted lines are drawn at ± 2 standard errors. Source: Nelson (1973), p. 75.

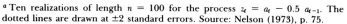

FIGURE 4.21   Sample *Acf*'s of an MA(1) Model.^a

in the model. Otherwise, some more careful analysis should be performed
before deciding that spikes really represent a higher-order model. Estimating
an AR(37) because the partial autocorrelation at lag 37 is large, whereas all
others are insignificant, except say the partial autocorrelation at lag 1, looks
to be quite odd. An AR(1) model would certainly be a very good alternative.

Furthermore, sample autocorrelations and partial autocorrelations will
not always exactly match population autocorrelations and partial autocorrelations. Judgment has to be exercised when deciding if the autocorrelations
are declining exponentially or if a spike is really significant. To illustrate
the need for exercising judgment and restraint when analyzing autocorrelations and partial autocorrelations, Nelson (1973) performed a series of interesting experiments. Using an AR(1) process with $\phi_1 = 0.5$ and an MA(1)
with $\theta_1 = 0.5$, he calculated sample autocorrelations for ten independent
artificial realizations, each containing 100 observations. These results are reported in Figures 4.20 and 4.21.

Note that although the sample autocorrelation functions for the AR(1) process resemble the general appearance of the theoretical autocorrelation function, $r_k \simeq \rho_k = (0.5)^k$, there are also substantial departures between specific sample *acf*'s and the theoretical *acf*. In theory, the *acf* of the MA(1) process should have only one spike at lag 1 with a value of -0.40. An examination of Figure 4.21 shows some differences in the values of the first autocorrelation. In one of the cases we even observed a large spike at lag 6.

4.6
SUMMARY

In this chapter, we have discussed that by comparing sample autocorrelation functions and partial autocorrelation functions with population autocorrelation and partial autocorrelation functions, one can identify tentative models to fit the data. Although this process does require a great deal of judgment which can be acquired with experience, these tools have proven to be quite valuable in narrowing down the class of ARIMA models that should be considered to represent a particular time series. In Chapters 7 and 8 we will actively make use of these tools to analyze some real data. First, however, we must estimate the parameters of the model and then show how forecasts can be generated with these ARIMA models.

In the next chapter we shall see how the parameters of the tentatively identified models are estimated and how diagnostic checks can be made on the models to evaluate their adequacy. Indeed, we will show how the residuals of misspecified models can be used to make logical alterations on the models in order to more adequately depict the process governing the series. Remember that the aim of the identification phase is not to really specify the correct model, but just to narrow down the number of models that should be seriously evaluated.

CHAPTER 5

ESTIMATION
AND
DIAGNOSTIC
CHECKING

In previous chapters we presented the general characteristics of the ARIMA models. In particular we calculated, evaluated, and generalized the behavior of the autocorrelation, the partial autocorrelation, and memory functions of several seasonal and nonseasonal ARIMA models. In Chapter 4 we discussed the identification of the ARIMA models and showed how the *acf* and *pacf* can be used to select appropriate models to tentatively represent the data.

With the set of tentative models narrowed down to just a few possibilities we can now estimate the parameters of these models and then evaluate which one truly represents the process governing the series.

In this chapter we will first briefly discuss the estimation process, give an example, and then discuss in Section 5.3 the very important diagnostic checks that should be performed in order to evaluate the reliability of any particular ARIMA model.

All the examples used in this and subsequent chapters will be analyzed using a collection of highly interactive computer time series programs[1] under the name TS. As mentioned above, in no way need the user have access to this particular collection of programs, although access to any of the many commercially available Box–Jenkins programs is highly recommended.

5.1 MODEL ESTIMATION

The next step after identifying a particular ARIMA model from the general class of multiplicative models,

$$\phi(B)\Phi(B^s)w_t = \theta(B)\Theta(B^s)a_t, \tag{5.1}$$

where

$$w_t = \nabla_s^D \nabla^d z_t \tag{5.2}$$

is to estimate the vectors of parameters $\underline{\phi} = (\phi_1, \phi_2, \dots, \phi_p)'$, $\underline{\Phi} = (\Phi_1, \Phi_2, \dots, \Phi_P)'$, $\underline{\theta} = (\theta_1, \theta_2, \dots, \theta_q)'$, and $\underline{\Theta} = (\Theta_1, \Theta_2, \dots, \Theta_Q)'$.

There are basically two methods available for estimating these parameters. One such method is the least squares method; the other is the maximum likelihood method.

Under the least squares method one chooses those values of the parameters which will make the sum of the squared residual as small as possible. That is, one chooses $\hat{\underline{\phi}}$, $\hat{\underline{\Phi}}$, $\hat{\underline{\theta}}$, and $\hat{\underline{\Theta}}$ as estimators of $\underline{\phi}$, $\underline{\Phi}$, $\underline{\theta}$, and $\underline{\Theta}$, respectively, so that the sum of squared residuals, SSR,

$$S(\hat{\underline{\phi}}, \hat{\underline{\Phi}}, \hat{\underline{\theta}}, \hat{\underline{\Theta}}) = \sum_{t=1}^{n} \hat{a}_t^2 \tag{5.3}$$

[1] Appendix C contains a primer that gives more detailed information about these programs.

is a minimum. As usual, n is the number of observations, possibly adjusted for the differencing required to make the series stationary.

Two difficulties arise in minimizing (5.3):

1. The model requires starting values for the data w_0, w_{-1}, w_{-2}, . . . , w_{1-p}, and for the errors a_0, a_{-1}, . . . , a_{1-q}, and possibly also for seasonal terms, if seasonal parameters are part of the model in which case, the subscript $1 - p$ would be $1 - (p + sP)$ and the subscript $1 - q$ would be $1 - (q + sQ)$.
2. In general, the model is nonlinear in the coefficients.

To understand the difficulties involved with the starting values consider an ARMA(1,1) model. This model written out for period $t = 1$ takes the form

$$w_1 = \phi_1 w_0 + a_1 - \theta_1 a_0. \tag{5.4}$$

From this equation it is clear that in order to use this model for all time periods $t = 1, 2, . . . , n$ we would need to know the values of w_0 and a_0.

There are two ways in which we can deal with this starting value problem. One way is to circumvent this problem and to calculate for this ARMA(1,1) model the sum of squares (5.3) only for the periods $t = 2, 3, . . . , n$ (i.e., drop the first observation).

Another approach is to recognize that we can forecast not only for periods in the future (i.e., beyond the last observation), but also for periods prior to the start of the sample period. For obvious reasons this latter type of forecasting is called 'back forecasting' or 'backcasting.'[2] Under the least squares options, the program that is part of the TS collection, program ESTIMA, uses the back forecasting approach to generate starting values.[3] One issue that arises with back forecasting is deciding how far back we want to go in order to make the influence of the starting values negligible. Unfortunately the answer depends on the particular process and the values of the coefficients. In particular, if the process is close to being nonstationary or noninvertible (i.e., the roots of the autoregressive or moving average polynomial lie close to the unit circle,[4] it may be necessary to backcast the data quite far.

Finally, an alternative approach is to calculate exact maximum likelihood estimates. From a number of algorithms proposed in the literature, the highly efficient computational approach discussed in Ansley (1979) has been imple-

---

[2] Minimizing the sum of squares (5.3) with the use of backcasting is sometimes called the unconditional approach. The conditional approach would not estimate the unknown starting values, but would replace these with some appropriate assumed values and the estimation would then be performed conditionally on these values.

[3] Many other commercially available computer packages allow for back forecasting.

[4] For a more detailed discussion of the use of the roots in analyzing stationarity, see Section 5.3.1. See also section 3.1.4, footnote 5 for the definition of a root.

mented in the TS program ESTIMA. Ansley has shown that a convenient computational form for maximum likelihood estimation (MLE) can be obtained by maximizing a quadratic form, similar to (5.3), but involving n transformed residuals. Therefore, after calculating these transformed residuals, we can again apply a standard nonlinear least squares optimizer.

Much more could be written about the conditional least squares and the maximum likelihood methods as well as the nonlinear estimation procedure. However, we would exceed very quickly the technical level at which we have aimed this book. For the reader who would want to study the nonlinear estimation method as it applies to the ARIMA model, we refer to the book of Box and Jenkins (1970, Chapter 7) and to the article of Ansley (1979), as well as to the references cited by these authors.

From a practical point of view we recommend that the user always estimate the process using the least squares (LS) method first and reverse to the maximum likelihood for the final model. This recommendation takes into account the fact that the LS method is certainly computationally more efficient and, second, that the user can then use the LS results as guess values to start the MLE option. In many cases the number of iterations of the MLE procedure is considerably reduced when initiating the algorithm with starting values close to the optimum.

Why an ARIMA Model is in general nonlinear in the coefficients can be seen as follows. Consider an ARMA(1,1) model and solve for the current error term to obtain

$$a_t = w_t - \phi_1 w_{t-1} + \theta_1 a_{t-1}.$$

Notice that in order to evaluate a_t we must not only know ϕ_1 and θ_1, and the data, w_t, w_{t-1}, but we must also have access to the value of a_{t-1}. But this value a_{t-1}, itself a function of the coefficients ϕ_1 and θ_1, is only known after the optimization. Therefore we are not in a position to express the residuals as a linear function of the coefficients, given known values for w_t, w_{t-1} and a_{t-1}. It should also be clear that the nonlinearity disappears as soon as the ARIMA model contains no moving average terms.

The computer program ESTIMA relies on the Marquardt nonlinear optimizer (see Marquardt, 1963)[5] to minimize the sum of squared residuals SSR (5.3), or the maximum likelihood modified term of squared residuals. The implementation of this nonlinear algorithm requires that the user specify starting values for the parameters to initiate the nonlinear estimation procedure. These preliminary values, or *guess values,* of the parameters can be obtained by comparing the values of the sample autocorrelations with the parameter representation of the population autocorrelations and then solving

---

[5] Other computer packages may make use of other nonlinear least squares optimizers or algorithms. In the ARIMA models I have encountered, I have found that the selection of a particular optimizer does not significantly affect the estimates.

for each individual parameter. For example, if we are estimating an AR(1) process, and $r_1 = 0.7$, then a good guess value for ϕ_1 is also 0.7 because we know that $\rho_1 = \phi_1$. Experience, however, has shown that the number of iterations necessary to minimize (5.3) and so to estimate the parameters of the ARIMA models is not overly sensitive to the initial guess values.[6]

Similarly, experience indicates that if the iterations fail to converge to the correct answer, this can depend both on the guess values used to start the nonlinear optimization and on the extent of overspecification in the model. Overspecification will be discussed in Section 5.3.2. When convergence fails, one should try to find better guess values or use a model representation with fewer parameters. A check on the accuracy of the numerical solution obtained is to try several reasonable guess values and see if the iterations converge to the approximate same answer.

5.2 AN EXAMPLE

Table 5.1 and Figure 5.1 contain a typical output of the estimation of an ARIMA Model. The series analyzed is the chemical batch yield data presented earlier in Figure 1.2. (The data are also given in Appendix A.) To illustrate the estimation we selected an AR(2) process to represent the data. The estimation, using the first 65 data points, yields the following results:

$$z_t = 56.962 \ -0.324z_{t-1} + 0.219z_{t-2}, \hat{\sigma}_a = 10.598. \qquad (5.5)$$
$$(10.849) \quad (0.125) \quad (0.124)$$

The numbers within the parentheses are estimates of the standard error of the estimates, which from here on will be denoted as SE. Also, $\hat{\sigma}_a$ is an estimate of the process residual standard deviation (EST. RES. SD) and is defined as the square root of an unbiased estimate of the variance. When we use the term unbiased we imply that we have used a degrees of freedom correction. Specifically, this estimate is defined as

$$\hat{\sigma}_a^2 = \frac{1}{\nu} \sum_{t=1}^{n} \hat{a}_t^2, \qquad (5.6)$$

where ν = degrees of freedom [i.e., the number of differenced observations (n) minus the number of coefficients estimates (3)]. The information contained in Table 5.1 and Figure 5.1 will be discussed below. However Appendix C Section C.4.2 has a detailed explanation of each quantity.

---

[6] There are other computer packages that can calculate guess values based on some characteristics of the data at hand. Nevertheless, the program should always display these starting values and the user should routinely inspect these values and possibly recalculate certain results using alternative guess values to evaluate the convergences.

TABLE 5.1   Chemical Batch Yield Data—AR(2) Model

```
NOBS = 65

INITIAL VALUES
 AR( 1)  -0.4000E+00
 AR( 2)   0.2100E+00
 CONST    0.5100E+02

MODEL WITH D = 0 DS = 0 S = 0
MEAN =   51.48     SD =   11.61    (NOBS = 65)

INIT SSR =   0.1402E+05
REL. CHANGE IN SSR <=   0.1000E-05
FINAL SSR =  0.6978E+04
5 ITERATIONS

LEAST SQUARES METHOD
PARAMETER ESTIMATES
====================
          EST       SE      EST/SE      95% CONF LIMITS
 AR( 1)  -0.324    0.125    -2.601    -0.568    -0.080
 AR( 2)   0.219    0.124     1.770    -0.024     0.463
 CONST   56.962   10.849     5.250    35.698    78.226

EST.RES.SD =   1.0598E+01
EST.RES.SD(WITH BACK FORECAST) =     1.0609E+01

R SQR =  .205
DF = 62
F =    7.989    (2,62 DF)  P-VALUE =  .001

CORRELATION MATRIX
        AR( 1)  AR( 2)
 AR( 2)   .406
 CON( 3) -.834   -.831

ROOTS   REAL   IMAGINARY  SIZE
 AR    -1.520    0.000    1.520
 AR     2.997    0.000    2.997
```

```
CHEMICAL BATCH YIELD

AUTOCORRELATIONS OF RESIDUALS
  LAGS ROW SE
  1-12  .12   .01 -.01  .00 -.06 -.16 -.05  .04 -.11 -.02  .10  .12 -.10
 13-24  .13   .07  .08  .07  .16 -.07 -.05 -.07 -.01  .02 -.07 -.08 -.10

CHI-SQUARE TEST          P-VALUE
Q(12) =  6.33     10 D.F.  .787
Q(24) = 13.3      22 D.F.  .925

AUTOCORRELATIONS OF FIRST DIFFERENCED RESIDUALS
  LAGS ROW SE
  1-12  .13  -.48 -.03  .04  .04 -.12  .01  .10 -.11 -.01  .05  .12 -.19
 13-24  .16   .08  .00 -.05  .17 -.12  .01 -.03  .01  .06 -.05  .02 -.06

CHI-SQUARE TEST          P-VALUE
Q(12) = 22.5      10 D.F.  .013
Q(24) = 28.3      22 D.F.  .167

AUTOCORRELATION FUNCTION
            RESIDUALS                      FIRST DIFFERENCED RESIDUALS

   -0.8  -0.4   0.0   0.4   0.8        -0.8  -0.4   0.0   0.4   0.8
 * . . * . . * . * . * . * . .  * . * . * .   * . * . * . * . . * . * . * . * . * . . * .
  .            (   R    )      1  .        R (     .     )
  .            (   R    )      2  .         (    R.     )
  .            (   R    )      3  .         (     .R    )
  .            (  R.    )      4  .         (     .R    )
  .            ( R .    )      5  .         (R    .     )
  .            (  R.    )      6  .         (    R      )
  .            (   .R   )      7  .         (          R)
  .            ( R .    )      8  .        (  R    .     )
  .            (   R    )      9  .         (    R      )
  .            (    .  R )    10  .         (     .R    )
  .            (    .  R )    11  .         (     .   R )
  .            ( R .    )     12  .        (R     .     )
  .            (   .R   )     13  .         (     .R    )
  .            (   .R   )     14  .         (    R      )
  .            (   .R   )     15  .         (    R.     )
  .            (   .  R  )    16  .         (     .    R)
  .            (  R.    )     17  .        ( R     .     )
  .            (  R.    )     18  .         (    R      )
  .            (  R.    )     19  .         (    R      )
  .            (   R    )     20  .         (    R      )
  .            (   R    )     21  .         (     .R    )
  .            (  R.    )     22  .         (    R.     )
  .            (  R.    )     23  .         (    R      )
  .            ( R .    )     24  .         (    R.     )
```

FIGURE 5.1 Chemical Batch Yield Data—Residuals from AR(2) Model.

```
NOBS = 65      MIN =   -28.35    MAX =    24.57
MEAN = -3.4581E-02 SD =   10.35      STUDENTIZED RANGE =   5.113
SKEWNESS = -0.2699    KURTOSIS =    3.337

                    INCREMENT =    1.0
                  ..........................................0.........................
   1  .  -2.68    .                                        *  .
   2  .  10.0     .            *                              .              *
   3  . -23.5     .    *                                      .        *
   4  .  7.45     .                                           .      *
   5  . -0.996    .                                        *. .
   6  .  3.77     .                                           . *
   7  .  10.4     .                                           .             *
   8  . -12.2     .                      *                    .
   9  .  3.26     .                                           . *
  10  .  1.16     .                                           .*
  11  .  16.6     .                                           .                   *
  12  . -9.48     .                          *                .
  13  . -4.20     .                               *           .
  14  . -6.17     .                             *             .
  15  .  1.49     .                                           .*
  16  . -2.94     .                                     *     .
  17  .  24.6     .                                           .                        *
  18  .  14.3     .                                           .            *
  19  . -19.7     .            *                              .
  20  .  17.0     .                                           .              *
  21  .  9.90     .                                           .          *
  22  .  0.327    .                                           .*
  23  .  0.320    .                                           .*
  24  .  6.73     .                                           .       *
  25  . -3.49     .                                    *      .
  26  .  1.45     .                                           .*
  27  .  1.64     .                                           .*
  28  . -8.27     .                        *                  .
  29  . -28.4     .*                                          .
  30  .  0.264    .                                           .*
  31  .  6.68     .                                           .      *
  32  .  17.3     .                                           .                *
  33  .  11.1     .                                           .           *
  34  .  19.6     .                                           .                    *
  35  .  4.73     .                                           .   *
  36  .  1.00     .                                           .*
  37  . -4.14     .                                 *         .
  38  . -1.11     .                                         *.
  39  . -13.3     .                     *                     .
  40  . -3.15     .                                   *       .
  41  .  0.640    .                                           .*
  42  .  1.74     .                                           .*
  43  . -4.67     .                                 *         .
  44  .  2.55     .                                           . *
  45  .  1.64     .                                           .*
  46  .  8.73     .                                           .        *
  47  . -3.84     .                                 *         .
  48  .  7.69     .                                           .       *
  49  . -2.87     .                                      *    .
  50  . -5.07     .                              *            .
  51  . -11.5     .                       *                   .
  52  .  2.94     .                                           . *
  53  .  8.82     .                                           .        *
  54  . -11.1     .                      *                    .
  55  . -2.74     .                                     *     .
  56  .  0.218    .                                           .*
  57  . -18.7     .                  *                        .
  58  . -21.7     .        *                                  .
  59  .  0.920    .                                           .*
  60  . -2.14     .                                     *     .
  61  .  13.8     .                                           .            *
  62  . -6.80     .                            *              .
  63  . -9.57     .                         *                 .
  64  .  10.9     .                                           .           *
  65  . -9.49     .                          *                .
                  ..........................................  .........................
```

5.3 DIAGNOSTIC CHECKS

Once a model has been identified and its parameters estimated, it is necessary to verify whether the model can be improved upon or not. First, however, we should discuss some of the results obtained from estimating an AR(2) model.

For those familiar with a standard regression output, most of the quantities contained on Table 5.1 will be self-explanatory. However, it still will be useful to define all the quantities. In the output it is clearly indicated what the initial guess values are for the nonlinear estimation procedure and that no consecutive differencing ($D = 0$) or seasonal differencing ($DS = 0$) has been specified. Furthermore, it is indicated that we estimate the AR(2) model by least squares (LS). We also applied MLE to this data set and found no real differences in the estimation results. We will for now only use the LS method.

For any coefficient estimate, we find the following statistics:

1. EST: The point ESTimate of the true coefficients of the ARIMA model;
2. SE: The estimate of the Standard Error of the estimates (the square root of an unbiased estimate of the sampling variance);
3. EST/SE: The ratio of the point ESTimate to its Standard Error; and
4. 95% CONF LIMITS: The limits of an appropriate 95% large-sample confidence interval.

In addition, several measures of goodness are also given:

1. EST.RES.SD: An ESTimate of the Standard Deviation of the RESiduals in the ARIMA model. This estimate is the square root of an unbiased estimate of the corresponding variance as defined in (5.6).
2. EST.RES.SD(WITH BACK FORECAST): An ESTimate of the true or process RESidual Standard Deviation. This estimate also uses the back forecasted residuals and is the square root of an unbiased estimate of the process residual variance.
3. R SQR: An estimate of the fraction of the variance of the data z_t or, if differencing has been applied, the variance of w_t accounted for by the fitted ARIMA model. The R^2(R SQR) is also called the coefficient of determination.

For a model with an intercept the R^2 statistic is calculated as

$$R^2 = 1 - SSR / \sum_{t=1}^{n} (w_t - \bar{w})^2,$$

where SSR is the sum of the squared residuals excluding the back forecasted residuals. If the model does not contain an intercept, R^2 is calculated as specified above, except that in the program ESTIMA we now use the sum of the transformed data squared, Σw_t^2, with no subtraction of the mean.

Although the R^2 in the ARIMA model has the same basic interpretation as the R^2 in the linear regression model, the special character of the time series models offers the opportunity of additional insight which can be of considerable value in actual data analysis. It can be shown that the population R^2 is related to the coefficients ϕ_i and θ_i and does not depend on σ_a^2. This is one point where intuition based on the multiple regression is misleading. Nelson (1976) analyzed some special ARMA models and found that for an AR(1) model $R^2 = \phi_1^2$ so that if $\phi_1 = 0.5$, we can anticipate an R^2 of at most 0.25. This preliminary assessment of R^2 can save the model builder a good deal of disappointment. For an MA(1) model it can be shown that $R^2 = \theta_1^2/(1 + \theta_1^2)$, and therefore approaches a maximum of 0.50. Unfortunately, the case of mixed ARMA models does no longer produce as clear results as the pure MA models. For more detailed information we refer to Nelson (1976).

We also warn the user against the use of the R^2 measure for selection among alternative models. As in the case for linear regression, using R^2 to select a model is erroneous if the dependent variable, here the variable w_t, is not the same. Specifically, we cannot use the R^2 to select among models with different stationarity transformations. As we will show in the next chapter, the variance of the forecast errors is determined by the variance of the error term. Therefore, for forecasting purposes the variance of the error term is the criterion one is most interested in.

The program ESTIMA prints out the number of degrees of freedom (DF), (i.e., the difference between the number of observations after differencing and the number of parameters estimated). As an additional overall goodness-of-fit measure of the complete model, the value of the F statistic[7] is given as well as its "P-value,"[8] which is equal to the 'tail area' probability of the hypothesis that all coefficients are zero, except the intercept.

Finally, ESTIMA provides an estimate of the sampling correlation matrix of the coefficient estimates. This correlation matrix expresses the degree of correlation that exists between the different coefficient estimates. In Table 5.1 we have estimated the correlation between the ϕ_1 and ϕ_2 coefficients to be 0.406. High correlation between estimates is pointing in the direction of model simplification. We will return to this in Sections 5.3.3 and 5.3.4.

We will now describe four groups of tests or diagnostic checks which can next be used to evaluate the model adequacy. If inadequacies are detected,

---

[7] The derivation of the F statistic can be found in any statistic inference book, e.g. DeGroot (1975), Snedecor and Cochran (1980). In testing the overall goodness-of-fit measure, large values of the F statistic and accompanying small P-value indicate that the parameters in the model significantly contribute to the overall explanation of the variance in the dependent variable.

[8] For an interpretation of the P-value and its relationship to the probability that the null hypothesis is true, see DeGroot (1973), Gibbons and Pratt (1975), and Dickey (1977).

then more than likely a different specification of the model will be required to fit the data. The four groups of diagnostic checks are:

1. Stationarity analysis;
2. residual analyses;
3. fitting extra parameters—the underspecified model; and
4. omitting parameters—the overspecified model.

Although we present the diagnostic checks in the above order we do not imply by this that the diagnostic checks should necessarily be applied sequentially. Indeed, in applying the diagnostic checks one must compare the results of various checks to see if as a whole there is sufficient evidence of model inadequacy. Making a decision based on just one test may lead to erroneous conclusions. If the model is inadequate, then some of these diagnostic checks will provide clues for the modifications necessary to improve the model specifications.

5.3.1 Stationarity Analyses

The program ESTIMA prints out the roots of the estimated polynomials as well as the size of these roots. The size is defined as the square root of the sum of the squares of the real and the imaginary part of each root. For stationarity and invertibility to hold, the size of each root should be larger than one (should be outside the unit circle).

For the chemical batch yield data the sizes of both roots of the AR(2) polynomial are larger than one in absolute value and therefore there is no evidence from these numbers that the model is nonstationary. If, for example, one of the roots were close to one, this would be an indication that additional differencing would be required to induce stationarity. Roots close to one for the MA polynomial indicates that the model is possibly overdifferenced.

This can be made clear with a simple example. Suppose that we estimate the following AR(2) model:

$$(1 - \phi_1 B - \phi_2 B^2)z_t = a_t$$

and find that one root is close to one. For expositional purposes we will assume that the root exactly equals one. We can therefore rewrite the model as

$$(1 - B)(1 - \phi B)z_t = a_t$$

or as

$$(1 - \phi B)w_t = a_t$$

with

$$w_t = (1 - B)z_t.$$

Alternatively, suppose we estimate the following MA(2) model:

$$w_t = (1 - \theta_1 B - \theta_2 B^2) a_t$$

and again find that one of the roots equals one. We can then again rewrite the model as

$$w_t = (1 - B)(1 - \theta B) a_t.$$

If $w_t = (1 - B) z_t$, this analysis shows that the model can be reduced to

$$z_t = (1 - \theta B) a_t.$$

5.3.2 Residual Analyses

If a model adequately depicts the ARIMA process governing the series, then the errors of the model should be white noise as defined in Chapter 3; that is, they should have mean 0, constant variance, and be uncorrelated over time. Consequently, an examination of the properties of the errors should allow us to evaluate the model adequacy. However, one should always keep in mind that the analyses of the errors are based on estimates; that is, on residuals of the model, defined as the differences between the observed and the fitted values, and not on population residuals. For example, for an AR(1) process the errors are defined as

$$a_t = z_t - \phi_1 z_{t-1} \tag{5.7}$$

and the residuals as

$$\hat{a}_t = z_t - \hat{\phi}_1 z_{t-1}, \tag{5.8}$$

where $\hat{\phi}_1$ is an estimate of ϕ_1 and $\hat{\phi}_1 z_{t-1}$ is the fitted value of z_t.

As mentioned in Chapter 3, the analyst should always start the time series model building with a time series plot. Such a plot would very quickly reveal a number of data and model formulation problems that other analyses might not be able to detect. A careful inspection of the time series plot could reveal sources of nonstationarity, such as variance nonstationarity, which are difficult to detect by inspection of the *acf* or *pacf*. Such a plot would also help in evaluating whether there are outliers in the series which could really be part of the data or could possibly be due to gross errors such as incorrect key punching. Of course, the latter kind are easily fixed. Real outliers, however, are more difficult to handle, although solutions have been proposed in the literature. [References include Fox (1972), Kleiner et al. (1979), Martin (1980), and Martin et al. (1983).] Again, the *acf* and *pacf* of time series containing outliers can be quite misleading and would therefore not reveal the underlying ARIMA process.

Figure 5.1 contains the residual plot[9] of the AR(2) model applied to the chemical batch data. A visual inspection of this plot does not immediately reveal any unusual problems. The large residuals at observations 17 and 29 could not, off hand, be explained. We therefore proceeded with the analysis.

It seems reasonable that the autocorrelations of the residuals would also yield valuable information about possible model inadequacies. The analysis of the residuals is, however, further hindered by the fact that residuals, though estimates of the errors, are necessarily correlated with each other even when the true errors are independent.

If the residuals are truly white noise, then their *acf* should have no spikes, and sample autocorrelations should all be small. Based on Bartlett's formula, equation (4.48), the estimate of the approximate large-sample standard error for individual autocorrelations under the assumption that the errors follow a white noise process is $1/\sqrt{n}$. Thus, residual autocorrelations, r_k, which lie, say, outside the range $\pm 1.96/\sqrt{n}$ (that is, outside the approximate 95% large-sample confidence limits), are significantly different from 0. Box and Pierce (1970), however, have shown that this standard error estimate is only a good approximation for high lagged residual autocorrelations. The standard error of the first few residual autocorrelations can be much less than $1/\sqrt{n}$. Consequently, values of r_k at low lags which are inside the approximate confidence limits may still be significantly different from 0, and may need further analysis.

The residual autocorrelations of the chemical batch yield data calculated in Table 5.1 are all within the approximate 95% large-sample confidence limit. (The standard error under the hypothesis that the residuals are white noise is listed under the heading ROW SE. For the first row that estimate is 0.12.) Furthermore, the first few autocorrelations are substantially smaller than the SE. Consequently, even allowing for the fact that for small lags the SE can overestimate the true standard error, one cannot reject the null hypothesis that the errors follow a white noise process. Based on this diagnostic check there seems to be no evidence for doubting the model adequacy.

A second approach for analyzing the residual autocorrelations is to rely on the Ljung–Box Q statistic, defined in Section 4.5, equation (4.53) as

$$Q(K) = n(n + 2) \sum_{k=1}^{K} \frac{1}{n - k} r_k^2(\hat{a}). \tag{5.9}$$

[9] In the heading of this figure, the TS program has printed out some summary information about the series plotted. Most of these statistics have been discussed above. The studentized range is defined as the range of the data divided by the standard deviation (SD). The coefficient of skewness is the scaled third moment about the mean, calculated as $m_3/(var \times SD)$, with $m_3 = \Sigma (x_i - \bar{x})^3/n$; and $\bar{x}$ is the mean of the data and *var* is the variance of the data. If the data comes from a normal population, the coefficient of skewness is 0. The kurtosis is the scaled fourth moment about the mean calculated as m_4/var^2, with $m_4 = \Sigma (x_i - \bar{x})^4/n$. For normally distributed data the kurtosis has the value 3. For more information about these statistics, see Snedecor and Cochran (1980).

In the literature this is also known as applying the *Portmanteau* test on the residual. If the fitted model is appropriate, (i.e., if the errors are white noise), Q is approximately distributed as a χ^2 distributed variable with $K-p-q-P-Q$ degrees of freedom. Notice that the symbol K in $Q(K)$ refers to the number of terms used in the summation. The symbols p, q, P, and Q are the numbers of parameters in the ARIMA model, respectively the autoregressive, the moving average, the seasonal autoregressive and the seasonal moving average parameters. Note that the degrees of freedom associated with the Q statistic, as defined in Section 4.5, and used for the model identification do not involve a correction factor but is just K, the number of terms used in the summation. This is so because at the identification stage no parameters are estimated. To make this dependence on the number of parameters used to estimate $\hat{a}$ explicit, we have included $\hat{a}$ in the autocorrelation notation in equation (5.9). The hypothesis that the errors are white noise is rejected when values of the Q statistic are large relative to the value from a χ^2 distribution table. As already mentioned in Section 4.5, the Q statistic is sensitive to the value of K, the number of residual autocorrelations used to calculate Q. Indeed it is quite possible that a Q statistic based on $K = 12$, that is, based on the first twelve residual autocorrelations, could lead toward rejection of the hypothesis that all these autocorrelations, as a whole, are 0, whereas for the same data a Q statistic based on $K = 24$ could lead towards not rejecting the hypothesis that all these 24 autocorrelations, as a whole, are 0. Again this should not upset the user. But once more we stress that no single test statistic should be the sole basis for accepting or rejecting a model.

The estimation results in Table 5.1 for the chemical batch yield data show that the Ljung–Box Q statistic based on the first 24 autocorrelations has a value of 13.3. With 22 degrees of freedom the value corresponds to a *P*-value, a 'tail area' probability or significance level, of 0.925. Therefore the evidence is overwhelmingly in favor of the null hypothesis that the errors are uncorrelated.

A third approach for determining if the errors are white noise is to evaluate the autocorrelations of the first differenced residuals. If the errors, a_t, form a white noise process, say,

$$a_t = e_t, \tag{5.10}$$

then the first differences of (5.10) should follow an MA(1) process, with the moving average parameter θ_1 equal to 1; that is,

$$a_t - a_{t-1} = e_t - \theta_1 e_{t-1}, \ \theta_1 = 1. \tag{5.11}$$

The first autocorrelation of an MA(1) process with $\theta_1 = 1$ equals -0.5, $\rho_1 = -\theta_1/(1 + \theta_1^2)$. The estimate of the first autocorrelation of the first differenced residuals for the chemical batch yield data has a value $r_1 = -0.48$ (see Table 5.1), strongly suggesting that the errors may indeed follow a white noise process.

It is important to recognize that the residuals of misspecified models can be used to make logical alterations on the models in order to more adequately depict the process governing the series.

Suppose that we estimated the following MA(1) process:

$$w_t = e_t - \theta_1 e_{t-1}$$
$$= (1 - \theta_1 B) e_t, \tag{5.12}$$

and found evidence that the residuals rather than being white noise were also governed by an MA(1) process,

$$e_t = a_t - \lambda a_{t-1}$$
$$= (1 - \lambda B) a_t. \tag{5.13}$$

The evidence for such a conclusion might be a large value for the first residual autocorrelation. By combining (5.12) with (5.13) we obtain

$$w_t = (1 - \theta_1 B)(1 - \lambda B) a_t$$
$$= (1 - \theta_1 B - \lambda B + \theta_1 \lambda B^2) a_t$$
$$= a_t - (\theta_1 + \lambda) a_{t-1} + \lambda \theta_1 a_{t-2}, \tag{5.14}$$

which indicates that an MA(2) process might be more appropriate than an MA(1) process. Residual tests and some of the other tests to be described in the next section should be made before concluding that the MA(2) representation is really better than the MA(1) representation.

Finally, in some situations, particularly those involving seasonal models, it might be possible that not all periodic characteristics of the series have been filtered out. A way to detect if periodicities are present in the residuals is to evaluate the *normalized cumulative periodogram*.[10]

The normalized cumulative periodogram is defined as[11]

$$C(f_j) = \frac{\sum\limits_{i=1}^{j} R^2(f_i)}{ns^2}, \tag{5.15}$$

where

$$R^2(f_i) = \frac{2}{n} \left[\left(\sum_{t=1}^{n} a_t \cos 2\pi f_i t \right)^2 + \left(\sum_{t=1}^{n} a_t \sin 2\pi f_i t \right)^2 \right],$$

$f_i = i/n$ (the frequency) and $1/f_i =$ the period, $a_t =$ the errors, and $s^2 =$ a large-sample estimate of σ_a^2, the error variance.

---

[10] The balance of this section could be skipped without affecting the understanding of the remaining material covered in this book.

[11] See Box and Jenkins (1976, Section 8.2.4).

The quantity $R^2(f_i)$ is a measure of how closely the trigonometric functions (cos, sin) with frequency f_i fit the residuals.[12] It is a device for correlating the a_t's with sine and cosine waves of different frequencies. For example, a seasonal residual pattern with period 12 (= span 12) or frequency $1/12 = 0.08$ would register large values for $R^2(1/12)$, $R^2(1/24)$, $R^2(1/36)$, etc. As a result the cumulative periodogram $C(f_j)$ would show jumps at these frequencies. The large-sample estimate s^2 of σ_a^2 does not involve any degrees of freedom adjustment.

A useful visual device is a plot of the cumulative periodogram against the frequency f_j or the period $1/f_j$. For a white noise series, the plot of $C(f_j)$ against f_j or $1/f_j$ should be scattered about a straight line joining the points (0, 0) and (0.5, 1). Model inadequacies would produce nonrandom errors whose cumulative periodogram would show systematic deviations from this line. Pronounced departures at low frequencies or high periodicity may be evidence of a cycle with a long period which in itself could be interpreted as evidence of trend, and so of insufficient differencing. Figure 5.2 shows an example of such a cumulative periodogram. If the bumps occur near the periods s, $2s$, $3s$, etc., then it is quite probable that seasonal effects have not been adequately modeled.

The deviations from the straight line can be evaluated using the large-sample Kolmogorov–Smirnov confidence limits. In Figure 5.2 these limits are depicted by two straight lines drawn parallel above and below the white noise straight line. The distance between the lines is determined by the confidence coefficient. The dark lines in Figure 5.2 denote the 95% large-sample confidence limits and the dashed lines denote the 75% large-sample confidence limits. For a 95% confidence interval, the confidence lines are drawn at distances $\pm 1.36/\sqrt{m}$ above and below the theoretical 45° line, where $m = (n - 2)/2$ for n even and $m = (n - 1)/2$ for n odd, and n is the number of observations possible after suitable differencing.[13]

In calculating the cumulative periodogram we have to rely on residuals as estimates of the true underlying errors; that is, a_t in (5.15) is replaced by $\hat{a}_t$. Even if the model were adequate, the residuals might still show some deviations from the white noise line. We should therefore be on the look out only for gross deviations from the 45° line, and should not worry about strict adherence to the Kolmogorov–Smirnov limits.

Figure 5.3 contains the cumulative periodogram of the residuals of the AR(2) model fitted for the chemical batch yield series. All the points are within the approximate 90% large-sample confidence limits,[14] a sign that

---

[12] The normalized cumulative periodogram is calculated using the fast Fourier transform, which results in a great reduction in computation time. For references see Singleton (1967) and Brigham (1974).

[13] For more information about the Kolmogorov–Smirnov test see Box and Jenkins (1976, p. 297).

[14] The TS program only calculated the 90% large-sample confidence limits.

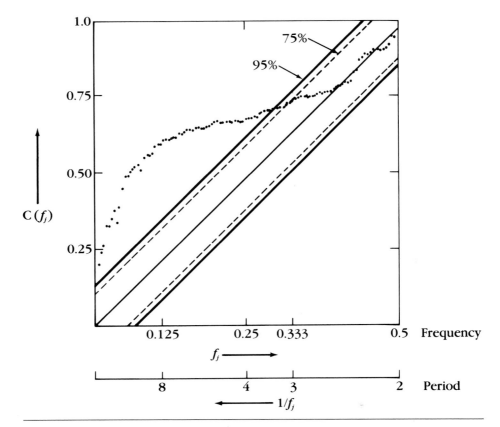

C (f_j)

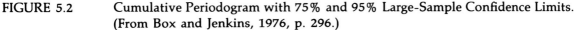

FIGURE 5.2   Cumulative Periodogram with 75% and 95% Large-Sample Confidence Limits. (From Box and Jenkins, 1976, p. 296.)

the model is more than likely adequate. Notice that in the computer output we do not directly indicate the confidence limit lines, but use a + sign each time the cumulative periodogram exceeds the 90% large-sample confidence limits.

A very important point to keep in mind when analyzing residual autocorrelations is that even if an occasional autocorrelation at some obscure lag appears to be significant, it is highly likely that the residuals are still white noise, and that further elaboration of the model is unnecessary.

5.3.3 Overspecified Model: Omitting Parameters

Another very useful check on the model adequacy is to evaluate whether the current model does not contain redundant parameters. Redundant parameters can be spotted by a careful use of the estimate of the large-sample standard error of the coefficient estimates (SE) and the estimate of the large-sample correlations between these coefficient estimates.

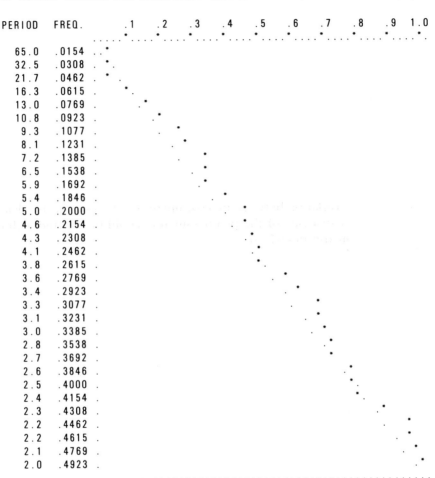

```
NORMALIZED CUMULATIVE PERIODOGRAM (CP) OF RESIDUALS
EXPECTED CP PLOTTED WITH (.); ACTUAL CP PLOTTED WITH (*)
IF ACTUAL EXCEEDS 90% KOLMOGOROV-SMIRNOV LIMITS, PLOTTED WITH (+)

PERIOD  FREQ.       .1   .2   .3   .4   .5   .6   .7   .8   .9  1.0
                    . . . . . . . . . . . . . . . . . . . . . . .
                    *       *       *       *       *       *    *
 65.0  .0154  . . *
 32.5  .0308  .  * .
 21.7  .0462  .  *  .
 16.3  .0615  .    * .
 13.0  .0769  .     . *
 10.8  .0923  .        . *
  9.3  .1077  .        .  *
  8.1  .1231  .         . *
  7.2  .1385  .            *
  6.5  .1538  .            *
  5.9  .1692  .            *
  5.4  .1846  .             *
  5.0  .2000  .              *
  4.6  .2154  .               *
  4.3  .2308  .               *
  4.1  .2462  .               *
  3.8  .2615  .                *
  3.6  .2769  .                 *
  3.4  .2923  .                  *
  3.3  .3077  .                   *
  3.1  .3231  .                    *
  3.0  .3385  .                    *
  2.8  .3538  .                     *
  2.7  .3692  .                      *
  2.6  .3846  .                       *
  2.5  .4000  .                       *
  2.4  .4154  .                        *
  2.3  .4308  .                         *
  2.2  .4462  .                          *
  2.2  .4615  .                          *
  2.1  .4769  .                           *
  2.0  .4923  .                            *
              . . . . . . . . . . . . . . . . . . . . . . . . . .
```

FIGURE 5.3   Chemical Batch Yield Data—Cumulative Periodogram.

We can use the SE to evaluate the statistical significance of a single coefficient. As a general rule of thumb one may claim that a coefficient is significant, more specifically, significantly different from zero, if the absolute value of the point estimate is at least twice the value of the standard error. If the point estimate of a parameter satisfies this rule, then the probability of obtaining a coefficient with a sign opposite from the point estimate is small and negligible.

An insignificant parameter is an indication that the model may be overspecified and a simplification of the model may be possible. If this insignificant parameter happens to be the one of the highest order, then, in general, the ARIMA model could be simplified by removing that parameter from the model.

Suppose the estimates we obtain of an AR(2) model are as follows:

$$z_t = 0.76z_{t-1} + 0.08z_{t-2} + a_t,$$
$$(0.03) \qquad (0.10)$$

(5.16)

with standard errors within parentheses. We notice immediately that the estimate of the second-order autoregressive parameter ϕ_2 is very small relative to its standard error, the ratio equals 0.8, and hence we would advocate reducing the model to an AR(1) process.

If the insignificant parameter is not the highest-order parameter, then examining the large-sample correlations between the parameter estimates is required to determine which parameter to delete from the model. A high correlation between two parameter estimates indicates that there is a possibility that one of the two parameters could be omitted without changing much on the model adequacy. The included parameter could be able to pick up the contribution of the omitted parameter because of the strong correlation between the two.

If the insignificant parameter is not the highest-order parameter but is strongly correlated with the highest-order parameter, we would be inclined to evaluate the model without the highest-order parameter. If no such correlation exists between the parameters, then we could reestimate the ARIMA model with the insignificant parameter suppressed. If ϕ_1 in (5.16) was insignificant and the correlation between ϕ_1 and ϕ_2 was small, then we should evaluate the estimation results of the following restricted model:

$$z_t = \phi_2 z_{t-2} + a_t.$$

(5.17)

For the chemical batch yield data the AR(2) model estimated in Table 5.1 reveals that the highest-order autoregressive coefficient ϕ_2 is relatively small in comparison with its standard error EST/SE = 1.551. Therefore we should certainly evaluate if the yield series could not be as adequately modeled by an AR(1) model. Table 5.2 contains the estimates of this AR(1) model. Comparing the results in Tables 5.1 and 5.2, we must conclude that the AR(1) model fits the data just as well as the more elaborate AR(2) model. The estimated residual standard error is only increased from 10.60 to 10.77; the Q statistics on the residuals have also increased, but the P-value for $Q(24)$ is still 0.802.

If the large-sample correlations between the parameter estimates indicate a high correlation between coefficients of different polynomials the situation requires even more careful analysis. For example, the correlation matrix might indicate a high correlation between ϕ_2 and θ_3 with or without any of these coefficients being insignificant. In such a situation we can only resort to an

TABLE 5.2 **Chemical Batch Yield Data—AR(1) MODEL**

```
MODEL WITH D = 0 DS = 0 S = 0
MEAN =   51.48     SD =    11.61     (NOBS = 65)

LEAST SQUARES METHOD
PARAMETER ESTIMATES
====================
              EST       SE      EST/SE      95% CONF LIMITS
   AR( 1)   -0.410    0.116    -3.533     -0.637    -0.182
   CONST    72.685    6.146    11.826     60.639    84.731

EST.RES.SD =   1.0774E+01
EST.RES.SD(WITH BACK FORECAST) =      1.0777E+01

R SQR =   .165
DF = 63
F =     12.44     (1,63 DF) P-VALUE =   .001

CORRELATION MATRIX
         AR( 1)
  CON( 2)  -.976

ROOTS    REAL   IMAGINARY   SIZE
  AR    -2.440   0.000     2.440

AUTOCORRELATIONS OF RESIDUALS
   LAGS ROW SE
    1-12  .12   .09  .17 -.07 -.04 -.19 -.07 -.01 -.09  .00  .07  .13 -.05
   13-24  .14   .09  .10  .10  .15 -.06 -.04 -.09 -.03 -.01 -.09 -.10 -.11

CHI-SQUARE TEST         P-VALUE
Q(12) =  8.58     11 D.F.  .661
Q(24) = 17.1      23 D.F.  .802

AUTOCORRELATIONS OF FIRST DIFFERENCED RESIDUALS
   LAGS ROW SE
    1-12  .13  -.53  .16 -.14  .11 -.15  .03  .06 -.09  .02  .01  .13 -.17
   13-24  .17   .07  .00 -.03  .15 -.12  .02 -.04  .02  .05 -.04  .01 -.06

CHI-SQUARE TEST         P-VALUE
Q(12) = 29.4      11 D.F.  .002
Q(24) = 34.1      23 D.F.  .064
```

evaluation of the model specifications where one or the other parameter has been omitted.

Finally, we must warn the user that the rules of thumb presented in this and the next section are just that, and should therefore never be followed blindly.

5.3.4 Underspecified Model: Fitting Extra Parameters

In order to verify that the tentatively identified model contains the appropriate number of parameters to represent the data, one can include an additional parameter in the ARIMA model to see if the addition results in an improvement over the original model.

In Section 5.3.3 we discovered that an AR(1) also adequately depicted the behavior of the chemical batch yield series. We could now evaluate a more elaborate model and, for example, refit this model after having included a moving average term; that is, we could now estimate an ARMA(1,1) model.

Table 5.3 contains the estimation results for this ARMA(1,1) model. Observe that the value of the MA(1) parameter is rather insignificant and that the correlation between the AR(1) and the MA(1) parameters is very large. From these results we can conclude that the MA(1) parameter should be dropped. The AR(1) adequately depicts the behavior of the data.

In fitting extra parameters to a time series model there is one serious pitfall, called the *parameter redundancy*. Parameter redundancy occurs when we add at the same time an autoregressive parameter and a moving average parameter to the model. Suppose that the correct model for the data is an AR(1) model written as

$$z_t = \phi z_{t-1} + a_t. \tag{5.18}$$

Next, write (5.18) for period $t - 1$:

$$z_{t-1} = \phi z_{t-2} + a_{t-1}. \tag{5.19}$$

Finally, subtract (5.19) from (5.18) and solve for z_t to obtain

$$z_t = (1 + \phi)z_{t-1} - \phi z_{t-2} + a_t - a_{t-1} \tag{5.20}$$

or

$$z_t = \phi_1 z_{t-1} + \phi_2 z_{t-2} + a_t - \theta_1 a_{t-1}. \tag{5.21}$$

The values for parameters ϕ_1, ϕ_2, θ_1 are appropriately defined in (5.20). The ARMA(2,1) model (5.21) can however also be derived from (5.18) by simply adding an additional autoregressive term and an additional moving average term. Since (5.21) was derived directly from (5.18), the difference between these two models is therefore illusory. As a matter of fact, (5.18) is to be preferred over (5.21) because it uses the data more efficiently. First,

TABLE 5.3   **Chemical Batch Yield Data—ARMA(1,1) Model**

```
MODEL WITH D = 0 DS = 0 S = 0
MEAN =    51.48     SD =    11.61     (NOBS = 65)

LEAST SQUARES METHOD
PARAMETER ESTIMATES
====================
            EST       SE      EST/SE      95% CONF LIMITS
AR( 1)    -0.760    0.156    -4.856     -1.066     -0.453
MA( 1)    -0.422    0.223    -1.892     -0.859      0.015
CONST     90.750    8.266    10.979     74.549    106.951

EST.RES.SD =    1.0607E+01
EST.RES.SD(WITH BACK FORECAST) =     1.0630E+01

R SQR =  .203
DF = 62
F =     7.919     (2,62 DF) P-VALUE =  .001

CORRELATION MATRIX
          AR( 1)   MA( 1)
  MA( 1)   .851
  CON( 3) -.974    -.827

ROOTS    REAL   IMAGINARY   SIZE
  AR    -1.316    0.000     1.316
  MA    -2.371    0.000     2.371

AUTOCORRELATIONS OF RESIDUALS
  LAGS ROW SE
  1-12   .12   .03   .05   .02  -.08  -.15  -.07   .03  -.11  -.01   .08   .13  -.09
  13-24  .13   .09   .09   .07   .16  -.07  -.04  -.07  -.02   .01  -.09  -.07  -.11

CHI-SQUARE TEST              P-VALUE
Q(12) =  6.26      10 D.F.   .793
Q(24) = 14.1       22 D.F.   .899

AUTOCORRELATIONS OF FIRST DIFFERENCED RESIDUALS
  LAGS ROW SE
  1-12   .13  -.50   .01   .04  -.01  -.08  -.01   .11  -.12   .02   .02   .13  -.19
  13-24  .16   .09   .00  -.05   .18  -.13   .02  -.03   .00   .07  -.06   .03  -.07

CHI-SQUARE TEST              P-VALUE
Q(12) = 23.7       10 D.F.   .008
Q(24) = 30.7       22 D.F.   .103
```

the observations on z_t for the period $t = 2, \ldots, n$ can be used, whereas (5.21) can only use z_t for $t = 3, \ldots, n$; second, (5.18) has less parameters than (5.21), although it could be observed that the parameters ϕ_1 and ϕ_2 are functionally related.

In more general terms the parameter redundancy problem can also be explained as follows. Rewrite (5.18) as

$$(1 - \phi B)z_t = a_t \tag{5.22}$$

and multiply (5.22) by $(1 - B)$ to obtain an equivalent form of (5.20) as

$$(1 - B)(1 - \phi B)z_t = (1 - B)a_t. \tag{5.23}$$

This equation makes it clear that it is not difficult to generate complicated or elaborate models as we can always multiply the left and right sides of an equation by the same factor. In (5.23) we use the common factor $(1\eta - B)$, but a more general form $(1 - \eta B)$ for arbitrary η or any polynomial could as easily be used.

5.4 SUMMARY

In this chapter we have presented the second link in the Box–Jenkins iterative identification–estimation/diagnostic checking–forecasting strategy. As you will recall after examining a time series plot of the data, patterns of the *acf* and *pacf* are evaluated in the *identification* phase to suggest some possible ARIMA models that could represent the data. Next, the parameters of these models are *estimated.* The parameters should be evaluated for over- or under-differencing using the size of the roots as a tool. Next, the significance of the estimates should be examined to see if model simplification is possible. At the same time the residuals must be analyzed to see if they can be accepted as conforming with white noise error. Otherwise a new model may have to be identified and/or estimated, restarting the iterative cycle, leading to a model that is statistically adequate and yet parsimonious for a given time series.

The final diagnostic check depends on the model's ability to forecast. In the next chapter we will discuss how forecasts are calculated and what kind of checks can be made to ascertain which model generates reliable forecasts.

In Chapters 7 and 8 the complete model-building strategy will be applied to the analysis of two time series typical of those encountered in economic and business practice.

CHAPTER 6

━━━━━

FORECASTING

6.1 OVERVIEW AND NOTATION

Once a fitted model has been judged as adequately representing the process governing the series, it can be used to generate forecasts for future periods. In this chapter we will show how ARIMA models can be used to generate forecasts for a time series. In order to make this discussion as succinct as possible, it is, however, necessary first to introduce some new notation.

Let the current period, the *origin* date, be period n, and suppose we want to forecast h time periods ahead to period $n + h$; that is, we want to know the value of the yet unrealized observation z_{n+h}. The time interval h is called the forecast *horizon*. The forecast for z_{n+h}, made at time period n for h periods ahead, is denoted by $z_n(h)$. That is, for $h = 1$, $z_n(1)$ is the one step ahead forecast of z_{n+1}; for $h = 2$, $z_n(2)$ is the forecast made at time n for period $n + 2$, using only observations z_1 through z_n.

Since the variable to be forecasted, z_{n+h}, is a random variable, it can only be fully described in terms of its forecast distribution, a probability distribution which is conditional on past and present data as well as on the specification of the ARIMA model. We will denote the forecast distribution of z_{n+h} by $f_{n,h}(z)$. In the next sections we will show how to obtain a point forecast which in some way is optimal, and how to construct forecast confidence intervals around this point forecast. Figure 6.1 illustrates a forecast distribution with point and interval forecasts.

FIGURE 6.1   The Forecast Distribution $f_{n,h}(z)$.

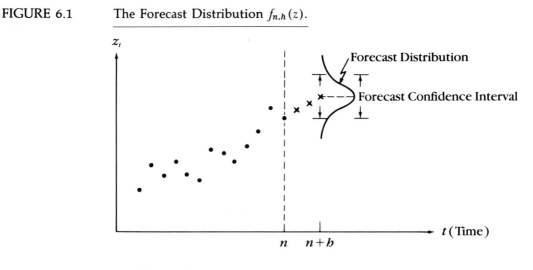

• = Observations

× = Point Forecasts

6.2. OPTIMAL FORECASTS AND COST FUNCTIONS

Optimality is a criterion which requires a yardstick to make comparisons with. We need a specific criterion by which forecasting methods can be ordered. A forecast can only be optimal with respect to something. The ideal situation is the one for which the yardstick is a *cost function.*

Consider the following example: Suppose that the owner of a grocery store orders some perishable commodity, such as fruit, based on an estimate of the demand in his store. Suppose also that he can evaluate the costs associated with being under- or overstocked; that is, he can calculate the costs involved in making a forecast error. If he is overstocked, the actual demand is therefore less than forecasted and he will have to throw away the leftover perishable fruit, so he will lose whatever he paid for it (the cost of overage); if he is understocked, the actual demand is above that forecasted and there will not be enough fruit, so he will lose goodwill and possibly the opportunity of selling other items in his store (the cost of underage).

A cost function, therefore, could be a function of both the actual demand, d^a, and the forecasted demand, d^f; that is

$$\text{Cost} = C(d^a, d^f). \tag{6.1}$$

Alternatively, this cost function could be expressed as a function of only the forecast error, $a = d^a - d^f$, or

$$\text{Cost} = C(a). \tag{6.2}$$

Let us continue our simplified example. Suppose that the fruit costs exactly \$1.00 per case and that the grocery owner is selling this fruit at \$1.25. If actual demand is above that forecasted, then the owner will lose at least 25¢, since 25¢ profit is made per case. However, for every case left over, the loss is \$1.00, as the fruit cannot be sold the next day. In this example the cost function of making an error, a, is given by

$$C(a) = \begin{cases} \$\ 0.25a & \text{if } a > 0 \\ \$\ 0 & \text{if } a = 0 \\ \$-1.00a & \text{if } a < 0, \end{cases} \tag{6.3}$$

where C is expressed in dollars. Figure 6.2 contains a plot of this cost function. Notice that this cost function is not symmetric, takes the value 0 when no error is made, is never negative, and increases in size as the errors become larger. In many applied situations, the cost functions will have similar properties, although they could be symmetric about $a = 0$ and need not consist of straight lines. Note that since the forecast error is random, so is the cost function $C(a)$.

If we define the error in making a forecast at time period n for future time period $n + h$, $z_n(h)$, as

$$a_n(h) = z_{n+h} - z_n(h), \tag{6.4}$$

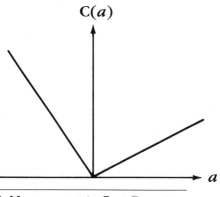

$$C(a)$$

$$a$$

FIGURE 6.2   A Nonsymmetric Cost Function.

then the *optimal* forecast for period $n + h$ is that value of $z_n(h)$ which minimizes the expected cost of $C[a_n(h)]$. When it is clear from the context, we will use the symbol a to denote the forecast error. In practice it is unlikely that the decision maker will know precisely the cost function. For this reason often a particular cost function is chosen for convenience and it is hoped that this cost function is a sufficiently good approximation to the true cost function. Quite often the following quadratic cost function is selected:

$$C(a) = \alpha a^2, \tag{6.5}$$

where α is some positive constant. It will be shown that the particular value of α is irrelevant, and so can be taken to be unity. This cost function defines the forecast error costs proportional to the forecast error squared. This cost function has the nice property that the solution for the optimal forecast is the mean of the forecast distribution. In the next section we will prove this statement and show how to calculate the mean of the forecast distribution.

The quadratic cost function implies that the costs are symmetric. Therefore, if the costs of being over- or understocked are generally not identical, relying on a quadratic cost function can be suboptimal. For example, suppose that a company has, on the basis of its prediction of future sales volume, decided to purchase additional machinery. If the forecast of future sales is higher than actual demand, the machines will be idle. Conversely, if the forecast is lower than actual demand, then the company will not be able to accept all orders. Regardless of what actually happens it is highly unlikely that the cost of overage and underage will be the same. Thus, the choice of a quadratic cost function is justifiable only on pragmatic grounds as this cost function leads to solutions that can easily be calculated.

In a few special cases, an optimal predictor can still be found for specific nonsymmetric cost functions. An example is the following cost function:

$$C(a) = \begin{cases} \alpha a & a > 0 \\ 0 & a = 0 \\ -\beta a & a < 0 \end{cases} \tag{6.6}$$

with α and $\beta > 0$, α being the unit cost of underage, and β being the unit cost of overage. In the above grocery store example $\alpha = \$0.25$ and $\beta = \$1.00$. To solve for the point forecast which minimizes the expected value of this cost function we must set the value of $z_n(h)$ such that the cumulative forecasting distribution[1] at $z_n(h)$ takes on the value[2]

$$F_{n,h}[z_n(h)] = \alpha/(\alpha + \beta). \tag{6.7}$$

In doing so we equalize the expected cost of overage and the expected cost of underage.

6.3 MINIMUM MEAN SQUARED ERROR FORECAST

We now will show that the mean of the forecast distribution minimizes the expected value of a quadratic cost function. That is, there is no other forecast which will produce errors whose squares have smaller expected values than the mean of the forecast distribution.

Let m_h be the expected value of z_{n+h} forecasted at time period n, $m_h = Ez_{n+h}$. Also, let m be any other forecast of z_{n+h} defined as

$$m = m_h + d, \tag{6.8}$$

where d is the difference between m and m_h. Using the point forecast m, the expected value of the forecast error squared is then

$$E[(z_{n+h} - m)^2] = E\{[z_{n+h} - (m_h + d)]^2\}. \tag{6.9}$$

Rearranging the right-hand side term of (6.9) we obtain

$$E[(z_{n+h} - m)^2] = E[(z_{n+h} - m_h)^2] - 2dE(z_{n+h} - m_h) + d^2. \tag{6.10}$$

Since $m_h = Ez_{n+h}$, the second term on the right-hand side of (6.10) is equal to 0 and since d^2 is nonnegative, in order to minimize (6.10) we set $d = 0$. The term $E[(z_{n+h} - m_h)^2]$ is the mean squared error of the forecast m_h. Therefore the optimal mean squared error forecast of z_{n+h} is obtained for $m = m_h = Ez_{n+h}$.

---

[1] The cumulative distribution $F(x')$ is defined as

$$F(x') = \int_{-\infty}^{x'} f(x)\,dx.$$

[2] We observe that this is the solution to the "critical fractile" problem. For a reference, see Vatter et al. (1978), p. 184.

The mean of the forecast distribution, $E(z_{n+h})$, can be calculated as follows. Let z_t be a stationary and invertible ARMA(p,q) process. For the time period $t = n + h$, this process can be expressed as

$$z_{n+h} = \phi_1 z_{n+h-1} + \cdots + \phi_p z_{n+h-p} + a_{n+h} - \theta_1 a_{n+h-1} - \cdots - \theta_q a_{n+h-q}. \quad (6.11)$$

The expected value of z_{n+h} in (6.11), calculated using information up to period n, is obtained as follows:

1. Replace the current and past errors a_{n+j}, $j \leq 0$, with actual residuals;
2. replace each future error a_{n+j}, $0 < j \leq h$, with its expectation, which, since a_{n+j} is white noise, is just 0;
3. replace current and past observations z_{n+j}, $j \leq 0$, with the actual observed values;
4. replace each future value of z_{n+j}, $0 < j < h$, with their appropriate forecast $z_n(j)$; therefore we should first forecast $z_{n+1}, z_{n+2}, \ldots, z_{n+h-1}$ in order to forecast z_{n+h}.

In addition, all the parameters in the model must be replaced with estimates. Methods for calculating these estimates were the subject of Chapter 5. In this chapter we will derive the forecasting properties assuming that these parameters are known.

The approach discussed above is not restricted to just ARMA models. This procedure can easily be extended to generate minimum mean squared error forecasts for any nonseasonal as well as multiplicative seasonal model, stationary as well as nonstationary.

In the next sections we shall use this approach to generate minimum mean squared error forecasts for some specific stationary and nonstationary models. We also will discuss the general properties of these forecasts in the context of nonseasonal models and present methods for calculating forecasts confidence intervals. The results explained there carry over to seasonal models in a straightforward fashion.

6.4 STATIONARY MODELS

6.4.1 First-Order Autoregressive Model

Consider the following AR(1) model with mean[3] $Ez_t = \mu$,

$$z_t - \mu = \phi_1(z_{t-1} - \mu) + a_t$$

or

$$z_t = (1 - \phi_1)\mu + \phi_1 z_{t-1} + a_t. \quad (6.12)$$

---

[3] In this chapter we will explicitly reintroduce the mean of the series and so analyze the raw data directly rather than the deviations from the mean.

TABLE 6.1　　　Forecast Profiles of an AR(1) Model ($\mu = 50$, $z_n = 30$).

h	1	2	3	4	5	6	7	8
$\phi_1 = 0.5$ $\quad z_n(h)$	40	45	47.5	48.8	49.4	49.7	49.8	49.9
$\phi_1 = -0.5$ $\quad z_n(h)$	60	45	52.5	48.8	50.6	49.7	50.2	49.9

Using the approach discussed in the previous section, the mean of the one period ahead forecast distribution is given by

$$Ez_{n+1} = z_n(1) = (1 - \phi_1)\mu + \phi_1 z_n. \qquad (6.13)$$

For h periods ahead the mean of the forecast distribution is

$$z_n(h) = (1 - \phi_1)\mu + \phi_1 z_n(h - 1), \qquad h > 1. \qquad (6.14)$$

In deriving these means we made use of the fact that all future error terms have mean 0; that is, $E(a_{n+h}) = 0$, $h > 0$.

EXAMPLE

Suppose that in (6.12) $\mu = 50$, $\phi_1 = 0.5$, and $z_n = 30$; then, an optimal point forecast[4] can be calculated for z_{n+1} by substituting these values in (6.13), that is,

$$z_n(1) = (1 - 0.5)(50) + (0.5)(30) = 40. \qquad (6.15)$$

Similarly, the forecast for z_{n+2} is

$$\begin{aligned} z_n(2) &= (1 - \phi_1)\mu + \phi_1 z_n(1) \\ &= (1 - 0.5)(50) + (0.5)(40) \\ &= 45. \end{aligned} \qquad (6.16)$$

The first eight forecasts of the AR(1) model with above parameters are given in the first row of Table 6.1.

　　As can be seen from the first line of Table 6.1, the forecast profile of the stationary AR(1) process used in the example is geometrically increasing from the last observed value of $z_n = 30$ to the mean of the series $\mu = 50$. If $z_n > \mu$, the forecast profile will show a geometric decrease towards the mean. Therefore, with $\phi_1 > 0$ there is no over- or undershooting of the mean value. As can be seen from the second line of Table 6.1, however, when $\phi_1 < 0$, the forecast will again approach the mean, but in an oscillating fashion.

---

[4] The term *optimal point forecast* always refers to a minimum mean squared error forecast.

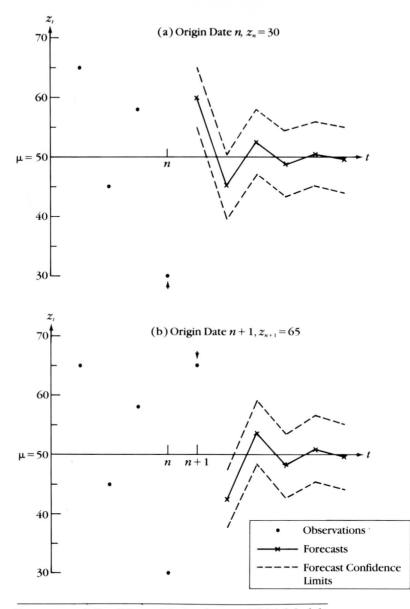

FIGURE 6.3   Forecast and Confidence Limits for an AR(1) Model.

The fact that the forecasts have to approach the mean of the series can be shown for an AR(1) model by solving for $z_n(h)$ in terms of z_n. Substituting (6.13) in (6.14) for $h = 2$, we obtain

$$z_n(2) = (1 - \phi_1)(1 + \phi_1)\mu + \phi_1^2 z_n. \qquad (6.17)$$

Continuing the substitution we can write for $z_n(h)$

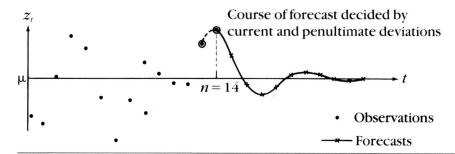

Course of forecast decided by
current and penultimate deviations

• Observations

—*— Forecasts

FIGURE 6.4   Forecast Profile of an AR(2) Model $[(1 - 0.75B + 0.52B^2)z_t = a_t$; Origin Date $n = 14]$. (From Box and Jenkins, 1976.)

$$z_n(h) = (1 - \phi_1)(1 + \phi_1 + \cdots + \phi_1^{h-1})\,\mu + \phi_1^h z_n. \qquad (6.18)$$

If we now forecast farther and farther out in the future, that is, if we let $h \to \infty$, and given the stationarity condition imposed on the model, $|\phi_1| < 1$, the best forecast will be the mean[5] of the series μ.

In Figure 6.3(a), we have drawn the forecast profile of the AR(1) process whose forecasts are given in the second row of Table 6.1. Figure 6.3(b) shows the forecast profile of this same AR(1) process starting at period $n + 1$, that is after the value $z_{n+1} = 65$ has been recorded. A discussion of the confidence limits which appear in these figures will take place in Section 6.6.

For an AR(1) process the forecast trajectory is determined only by the current deviation from the mean. For example, note that the forecast of the AR(1) process of the first row in Table 6.1 for the first period is 10 units away from the mean, $|40 - 50|$, that the next forecast is just five units from the mean, the third forecast 2.5 units, and so forth. For an AR(p) process the shape of the forecast profile is determined by the current and last ($p - 1$) deviations. Figure 6.4 shows the forecast profile of an AR(2) model.

In Section 3.1.3 we derived the memory function of an AR(1) process and showed that although the influence of a shock declined gradually it nevertheless persisted for an infinite number of periods. In analyzing the forecast profile of an AR(1) we observe a similar pattern of behavior. If the last observed value z_n is different from the mean of the series, then it will take an infinite number of periods before the mean will become the optimal predictor, although the optimal forecast can approach the mean quite rapidly.

[5] In (6.18) the terms in equation $1 + \phi_1 + \cdots + \phi_1^{h-1} + \cdots$ form a geometric series, and with $|\phi_1| < 1$ the terms sum to $1/(1 - \phi_1)$.

So

$$z_n(h) = \frac{1 - \phi_1}{1 - \phi_1}\,\mu = \mu.$$
$$(h \to \infty)$$

6.4.2 First-Order Moving Average Model

The forecasts for an MA(1) model can be generated in a similar fashion. For the period $n + 1$ an MA(1) process can be expressed as

$$z_{n+1} = \mu + a_{n+1} - \theta_1 a_n. \tag{6.19}$$

As was the case with the AR(1) process discussed above, z_t represents the actual data and not the deviations from the mean of the series.

At time n, the residual a_{n+1} has not been observed and is therefore replaced by its expected value 0. The value for the residual a_n can be calculated from[6]

$$a_n = z_n - \mu + \theta_1 a_{n-1}. \tag{6.20}$$

Therefore, the one period ahead forecast for an MA(1) model is just

$$z_n(1) = \mu - \theta_1 a_n. \tag{6.21}$$

For two periods ahead the forecast is simply equal to

$$z_n(2) = \mu \tag{6.22}$$

since both a_{n+2} and a_{n+1} are unknown and are therefore replaced by their mean, which is 0. Equations (6.21) and (6.22) clearly resemble the findings of Section 3.2.4, in which we saw that the effect of a shock in an MA(1) model only lasts for one period in the future.

The memory function of an MA(q) model [see equation (3.43)] shows that the effect of a shock in the model only lasts for q future periods. Similarly, only the first q values of the forecast profile of an MA(q) process will be determined by past disturbances. All other future values will be equal to the mean of the process. Consequently, for an MA(1) process, all forecasts except the current one period ahead forecast will be equal to the mean of the series; that is,

$$z_n(h) = \mu, \qquad h > 1. \tag{6.23}$$

EXAMPLE

Suppose that the parameter of the MA(1) specified in (6.19) have the following values: $\mu = 50$, $\theta_1 = 0.5$; suppose, furthermore, that from (6.20) we found that $a_n = 6$. The optimal one step ahead forecast can be derived by replacing those values in (6.21); that is,

$$z_n(1) = 50 - (0.5)(6)$$
$$= 47.$$

---

[6] It should be apparent that to solve for a_n we ultimately have to know a_0. To calculate this initial value we could pursue the following lines of attack: Use the conditional approach, in which we replace this value with its mean 0; or, use the unconditional approach, which is more precise and which employs the method of back forecasting discussed in Chapter 5. Program FRCAST in the TS collection uses the back forecasting technique.

For two or more periods ahead, the forecasts equal the mean of the series

$$z_n(h) = 50, \qquad h \geq 2.$$

6.4.3 An ARMA(1,1) Model

Finally, consider the following ARMA(1,1) process with mean μ

$$z_t - \mu = \phi_1(z_{t-1} - \mu) + a_t - \theta_1 a_{t-1} \qquad (6.24)$$

which, when written out for period $n + 1$, takes on the form

$$z_{n+1} = (1 - \phi_1)\mu + \phi_1 z_n + a_{n+1} - \theta_1 a_n. \qquad (6.25)$$

The forecasts for this model can be generated sequentially using the equations

$$z_n(1) = (1 - \phi_1)\mu + \phi_1 z_n - \theta_1 a_n$$
$$z_n(h) = (1 - \phi_1)\mu + \phi_1 z_n(h-1), \qquad h > 1. \qquad (6.26)$$

Observe that only the one period ahead forecast is directly influenced by past residuals. As in an AR(1) process, forecasts with longer horizons will again approach μ, the mean of z_n. Assuming that $\phi_1 > 0$ and that $z_n(1)$ is above the mean, the forecasts will decrease geometrically to that mean; similarly, if $\phi_1 > 0$ and if $z_n(1)$ is below the mean, then the forecast profile will approach the mean from below. If $\phi_1 < 0$, then the forecast will oscillate above and below the mean and gradually approach the mean. The difference with an AR(1) model is that for an ARMA(1,1) model the current error a_n is used to determine the one step ahead forecast $z_n(1)$.

6.5 NONSTATIONARY MODELS

The principles outlined in Section 6.4 for forecasting stationary time series models can easily be extended to include nonstationary models. The only change that needs to be made for these models is first to replace z_{n+j}, $j \leq 0$, and $j > 0$, with the appropriate differenced values w_{n+j}, and then solve explicitly for z_{n+h} in terms of other z_{n+j}'s and a_{n+j}'s, as our ultimate goal is to forecast z_{n+h} and not w_{n+h}. However, although the method of obtaining forecasts is similar, the forecast profiles of nonstationary time series models are quite different from the ones discussed in Section 6.4.

6.5.1 An IMA(1,1) Model

Consider the following nonstationary IMA(1,1) model written as

$$z_t - z_{t-1} = \mu + a_t - \theta_1 a_{t-1}, \qquad (6.27)$$

with μ the mean of the first differences $w_t = z_t - z_{t-1}$. Note that the model (6.27) is stationary in the first differences w_t but not in z_t.

Writing the model for the period $n+1$, and solving for z_{n+1}, we obtain

$$z_{n+1} = \mu + z_n + a_{n+1} - \theta_1 a_n. \qquad (6.28)$$

The optimal mean squared error forecast for z_{n+1} made at period n is obtained by replacing a_{n+1} by its mean value 0, using the actual value of z_n, and calculating a_n from

$$a_n = z_n - z_{n-1} - \mu + \theta_1 a_{n-1}. \qquad (6.29)$$

Note that in evaluating (6.29) the start-up problem is again present, as mentioned in Section 6.4.2, footnote 6.

The optimal forecasts for this model can be generated from the following equations:

$$z_n(1) = \mu + z_n - \theta_1 a_n$$
$$z_n(2) = \mu + z_n(1)$$

and

$$z_n(h) = \mu + z_n(h-1), \qquad h > 1. \qquad (6.30)$$

Substituting for $z_n(h-1)$, the last formula ultimately reduced to

$$z_n(h) = z_n(1) + \mu(h-1), \qquad h > 1. \qquad (6.31)$$

Equation (6.31) indicates that the forecast profile is a straight line with slope μ and with an intercept value equal to the one period ahead forecast $z_n(1)$. Table 6.2 contains the forecast profiles of an IMA(1,1) model made at time period n and at period $n+1$.

The availability of a new observation may produce changes in the forecasts. Note how the forecasts in Table 6.2 vary when the starting period is changed. The forecasts on the second line of the table were calculated after the actual value for z_{n+1} was recorded. Using the observed value of 215 for z_{n+1} together with (6.29), we found that $a_{n+1} = -50$. This information along with (6.30) yields $z_{n+1}(1) = 300$. As this new forecast for z_{n+2} is 25 units

TABLE 6.2   Forecast Profiles of an IMA(1,1) Model ($\mu = 60$, $\theta_1 = 0.5$, and $a_n = 8$).

	n	$n+1$	$n+2$	$n+3$	$n+4$	$n+5$	$n+6$	$n+7$
z_n	209	265	325	385	445	505	565	625
z_{n+1}		215	300	360	420	480	540	600

The actual value of z_n is 209. The second row shows forecasts made after having observed an actual value of z_{n+1} of 215.

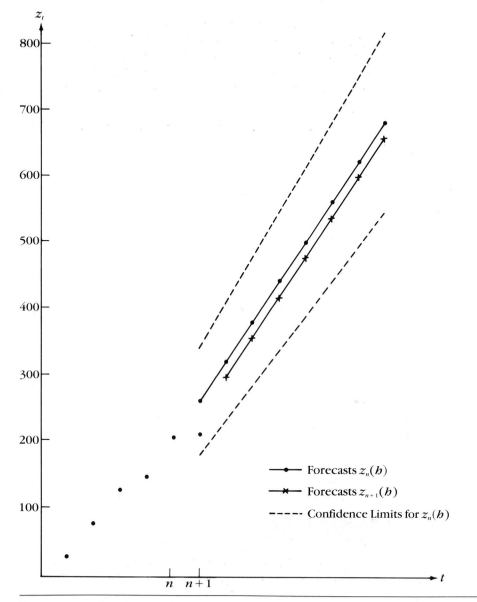

FIGURE 6.5   Forecast and Confidence Limits for an IMA(1,1) Model ($z_t - z_{t-1} = 50 + a_t - 0.5a_{t-1}$).

lower than the earlier forecast, it follows directly from (6.31) that all other forecasts will also be lowered by 25 units. Figure 6.5 shows the forecasts for the two origin dates. In Section 6.7 we will present an efficient way for updating existing forecasts.

6.5.2 An ARIMA(1,1,0) Model

Let us evaluate the forecast profile of the following ARIMA(1,1,0) model

$$(1 - \phi_1 B)(1 - B)z_t = \delta + a_t \tag{6.32}$$

with $\delta = (1 - \phi_1)\mu$, and μ is the mean of the first differences $w_t = (1 - B)z_t$. Solving for current z_t we obtain

$$z_t = \delta + (1 + \phi_1)z_{t-1} - \phi_1 z_{t-2} + a_t. \tag{6.33}$$

The minimum mean squared error forecasts for different horizons are given by

$$z_n(1) = \delta + (1 + \phi_1)z_n - \phi_1 z_{n-1}$$
$$z_n(2) = \delta + (1 + \phi_1)z_n(1) - \phi_1 z_n$$
$$z_n(h) = \delta + (1 + \phi_1)z_n(h-1) - \phi_1 z_n(h-2), \qquad h > 2. \tag{6.34}$$

Inferences about the forecast profile of z_{n+h} can be made by examining the forecast profile of the differences first. Rewriting (6.32) in terms of first differences we obtain

$$(1 - \phi_1 B)w_t = \delta + a_t$$

or

$$w_t = \phi_1 w_{t-1} + \delta + a_t. \tag{6.35}$$

This is the equation of an AR(1) model using first differences. In Section 6.4.1 we derived the forecast profile of these forecasted first differences as

$$w_n(1) = \phi_1 w_n + \delta$$
$$w_n(h) = \phi_1 w_n(h-1) + \delta, \qquad h > 1. \tag{6.36}$$

Rewriting (6.36), by substitution for $w_n(h-1)$, we have

$$w_n(h) = (1 + \phi_1 + \phi_2 + \cdots + \phi_1^{h-1})\delta + \phi_1^h w_n.$$

With $\delta = (1 - \phi_1)\mu$, we immediately find that for large h, the forecast of the first differences will approach its mean, μ (see footnote 5). In most economic applications μ is positive or zero. If $\phi_1 > 0$ and if w_n is below the mean μ, the forecast profile of the differences will approach this mean from below. Similarly, for $\phi_1 > 0$ and w_n above the mean μ, the forecasted changes will decrease to that mean.

To facilitate our discussion on what can be inferred about the forecast profile of the original series z_t using the w_t series, let us assume that $\phi_1 > 0$. If the forecasted differences decrease to the mean μ of the w_t series, and if we assume $\mu > 0$, then the forecasts of the z_t's will increase at *a decreasing rate* and eventually the *increase* will approach the mean μ of the differenced series. The forecast profile of z_t therefore approaches a straight line, with

TABLE 6.3   Forecast Profiles of an ARIMA(1,1,0) Model ($\phi_1 = 0.5$, $\delta = 5$).

	$n-1$	0	$n+1$	$n+2$	$n+3$	$n+4$	$n+5$	$n+6$	$n+7$	$n+8$
w_n		-12	-1	4.5	7.3	8.6	9.3	9.7	9.8	9.9
z_n	42	30	29	33.5	40.8	49.4	58.7	68.3	78.2	88.1
w_n			5	7.5	8.8	9.4	9.7	9.8	9.9	10.0
z_n			35	42.5	51.3	60.6	70.3	80.2	90.1	100.0

The last two rows show the forecasts of w_n and z_n starting at period $n+1$, given $z_{n+1} = 35$.

positive slope μ. If the forecast profile of w_t approaches a positive μ from below, then, for sufficiently negative w_n, the forecast profile of z_t will first show a decrease and then will increase at a rate gradually approaching μ. This is an example of a *turning point forecast*.

In Table 6.3 we have presented forecasts starting at time period n for w_{n+h} and z_{n+h}, based on model (6.32) with $\phi_1 = 0.5$, $\delta = 5$, $z_n = 30$, and $z_{n-1} = 42$. Based on these numbers we can calculate that $\mu = 10$. Figure 6.6 contains the forecast profiles for this model. Also included in the table and the figure are the forecasts starting at period $n+1$, assuming that we observed $z_{n+1} = 35$.

6.6 INTERVAL FORECASTS

In many situations, in addition an optimal point forecast, we may want to measure the uncertainty around this point forecast. In this section we will evaluate the standard error of the forecast error and then construct forecast confidence limits.

In order to calculate the standard errors of the forecast errors we first express the ARIMA process in the *error-shock* form; that is, by successive substitution for z_{t-1}, z_{t-2}, . . . , we write the model in terms of current and past errors only as

$$z_t = a_t + \psi_1 a_{t-1} + \psi_2 a_{t-2} + \cdots . \qquad (6.37)$$

The values of the parameters (ψ_1, ψ_2, . . .) depend upon the particular ARIMA model and are called the *error learning coefficients*.[7] In Chapter 3 we already used this error-shock form representation to calculate the memory function of several stationary processes. For an AR(1) process, for example, we found that $\psi_1 = \phi_1$, $\psi_2 = \phi_1^2$ and, in general [see equation (3.25)], that $\psi_k = \phi_1^k$.

The optimal forecast $z_n(h)$ can also be expressed using (6.37) in terms of current and past errors

---

[7] In the TS collection, the program FRCAST calls an error learning coefficient a PSI WEIGHT.

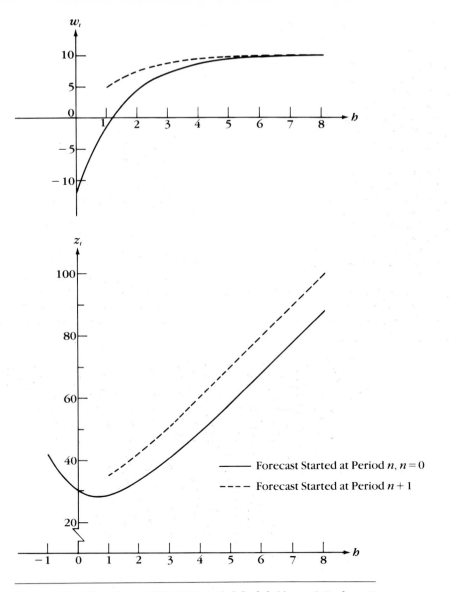

FIGURE 6.6   Forecast Profile of an ARIMA(1,1,0) Model ($\phi_1 = 0.5$, $\delta = 5$).

$$z_n(h) = \psi_h a_n + \psi_{h+1} a_{n-1} + \cdots .\tag{6.38}$$

As a result, the h step ahead forecast error

$$a_n(h) = z_{n+h} - z_n(h)\tag{6.39}$$

can be written as

$$a_n(h) = a_{n+h} + \psi_1 a_{n+h-1} + \cdots + \psi_{h-1} a_{n+1}. \qquad (6.40)$$

Because the errors a_t are independent, it follows from (6.40) that $a_n(h)$ is an MA($h-1$) process, *regardless* of the form of the ARIMA process being analyzed. In particular, the one step ahead errors

$$a_n(1) = a_{n+1} \qquad (6.41)$$

form a white noise series.

From (6.40) it follows directly that the forecast errors $a_n(h)$ have mean 0 and variance equal to

$$\text{Var}[a_n(h)] = E[a_n^2(h)] = \sigma_a^2 \sum_{j=0}^{h-1} \psi_j^2, \qquad \text{with } \psi_0 = 1. \qquad (6.42)$$

Observe that as the forecast horizon is lengthened, the error variances are monotonically nondecreasing:

$$\text{Var}[a_n(h)] - \text{Var}[a_n(h-1)] = \sigma_a^2 \, \psi_{h-1}^2 \geq 0. \qquad (6.43)$$

Equation (6.43) proves the intuitive notion that we can never know more as we look further in the future.

If we assume that the error terms a_t are normally distributed, then we can characterize the whole forecast distribution $f_{n,h}(z)$. The forecast distribution of z_{n+h}, $f_{n,h}(z)$, will be distributed as a normal random variable with mean $z_n(h)$ and variance $\text{Var}[a_n(h)]$. Under this error distribution it is straightforward to make probability statements about future observations. Indeed, the 95% large-sample[8] confidence interval for z_{n+h} is

$$z_n(h) \pm 1.96 \, \text{SE}[a_n(h)]. \qquad (6.44)$$

The *SE* stands for the standard error of the forecast error, defined as the square root of the variance (6.42). In the actual calculation of the confidence limits, we replace the error learning coefficients with their estimates and σ_a^2 with its estimate (5.6).

The above development has concentrated upon obtaining a confidence interval for the individual z_{n+1} value based on current and past values of z_t. Sometimes we may be interested in obtaining a confidence interval for the *mean* of z_{n+1}, based on current and past values of z_t. Intuitively, it should be clear that the point forecast will remain unchanged, but that we will be able to forecast the mean with more certainty than an individual observation. Indeed it can be shown that the variance of the *mean* of $z_n(h)$ will be

$$\text{Var}[Ez_n(h)] = E\{[z_n(h) - E(z_{n+h})]^2\}$$
$$= \sigma_a^2 \sum_{j=1}^{h-1} \psi_j^2. \qquad (6.45)$$

---

[8] The qualifier *large-sample* is added because, even if the errors are normally distributed, we still have to use estimates of the parameters of the model rather than their true values.

The difference with the variance (6.42) is that the variance of the mean of $z_n(h)$ does not include ψ_0 in the summation.

6.6.1 First-Order Autoregressive Model

The error-shock form for an AR(1) process with mean μ can be obtained directly from (3.25) as

$$z_t = \mu + a_t + \phi_1 a_{t-1} + \phi_1^2 a_{t-2} + \phi_1^3 a_{t-3} + \cdots.$$

The error learning coefficients are $\psi_j = \phi_1^j$, and as a result the variance of the forecast errors is given by

$$\text{Var}[a_n(h)] = \sigma_a^2 (1 + \phi_1^2 + \phi_1^4 + \cdots + \phi_1^{2(h-1)}). \tag{6.46}$$

As h becomes large the forecasting variance approaches $\sigma_a^2/(1 - \phi_1^2)$, the variance of the AR(1) data [see equation (3.10)]. In Figure 6.3 presented on page 144 we have indicated the confidence intervals of the AR(1) model.

6.6.2 First-Order Moving Average Model

It is apparent that the ARIMA representation of an MA(1) model,

$$z_t = \mu + a_t - \theta_1 a_{t-1}, \tag{6.47}$$

is also already in the error-shock form. Consequently, we have $\psi_1 = -\theta_1$ and $\psi_j = 0$, $j > 1$. The variances of the h step ahead forecast errors are therefore given by

$$\text{Var}[a_n(1)] = \sigma_a^2$$
$$\text{Var}[a_n(h)] = \sigma_a^2(1 + \theta_1^2), \qquad h > 1. \tag{6.48}$$

Notice that the variance of forecast errors for two or more steps ahead equals the variance of the MA(1) data [see equation (3.37)] and that this variance remains constant beyond the two step ahead forecasts.

The fact that the forecast error variance approaches a constant and is equal to the variance of the model as h becomes large is a result that extends to *all* seasonal and nonseasonal *stationary* models. As we will demonstrate with a specific model in the next section, for a *nonstationary* model this variance will increase without limit, an indication that far enough in the future we know very little about the future values of nonstationary series.

6.6.3 An IMA(1,1) Model

Let us now calculate the error variance of the following IMA(1,1) model:

$$z_t - z_{t-1} = \mu + a_t - \theta_1 a_{t-1}. \tag{6.49}$$

The error-shock representation of this process can be obtained as

$$z_t = \mu + a_t + (1 - \theta_1)a_{t-1} + (1 - \theta_1)a_{t-2} + \cdots. \qquad (6.50)$$

Therefore, the error learning coefficients are all constant and are equal to $\psi_i = (1 - \theta_1)$, $i = 1, 2, \ldots$. The forecast error variance can be calculated by

$$\text{Var}[a_n(h)] = \sigma_a^2[1 + (h-1)(1-\theta_1)^2]. \qquad (6.51)$$

Clearly, as h becomes large, this variance increases without limit. The confidence limits of an IMA process will become wider and wider the more steps ahead we forecast. Figure 6.5 on page 149 shows the confidence intervals of an IMA(1,1) model.

6.7 AN EXAMPLE

As an example we will forecast the chemical batch yield data analyzed in the previous chapter. In Section 5.2 we used the first 65 data points to estimate the model by least squares and obtained the following results:

$$z_t = 56.962 - 0.324z_{t-1} + 0.220z_{t-2}, \quad \hat{\sigma}_a = 10.598. \qquad (5.5)$$
$$\quad (10.849) \ (0.125) \qquad (0.124)$$

Notice in particular that the intercept term 56.962 is an estimate of $(1 - \phi_1 - \phi_2)\mu$. From this we obtain an estimate of μ as $\hat{\mu} = 51.60$, a quantity very close to the data sample mean of 51.48.

Using these results, we now propose to estimate the next 15 data points. These results are presented in Table 6.4, and a plot of the forecast together

TABLE 6.4   Chemical Batch Yield Data—Forecasts

OBS	ψ_i	95% LOWER CONF. LIMIT	FORECAST	95% UPPER CONF. LIMIT	ACTUAL
66	1.0000	36.72	57.49	78.26	59.00
67	−0.3241	25.05	46.89	68.72	40.00
68	0.3246	31.53	54.38	77.24	57.00
69	−0.1764	26.48	49.63	72.77	54.00
70	0.1284	29.51	52.81	76.11	23.00
71	−0.0803	27.38	50.74	74.09	
72	0.0542	28.72	52.11	75.49	
73	−0.0352	27.81	51.21	74.60	
74	0.0233	28.40	51.80	75.20	
75	−0.0153	28.01	51.41	74.81	
76	0.0101	28.26	51.67	75.07	
77	−0.0066	28.09	51.50	74.90	
78	0.0044	28.20	51.61	75.02	
79	−0.0029	28.13	51.54	74.94	
80	0.0019	28.18	51.58	74.99	

with the last 20 actual data points is contained in Figure 6.7. Notice the very large forecast error for period 70. However, under a more careful examination of the original data (see Figure 1.3), this last actual point represents quite a large drop in the data. From Figure 6.7 we notice that this data point is even outside the 95% large-sample confidence interval.

FIGURE 6.7   Chemical Batch Yield Forecasts.

```
FORECAST PLOTTED WITH (*), ACTUAL PLOTTED WITH (+)
95% CONF. LIMITS WITH ( AND )
ORIGIN = 65

                          INCREMENT =    1.1
                     . . . . . . . . . . . . . . . . . . . . . . . . . . . . . . . . . . . . . .
        45   .    50.0    .                                        +
        46   .    62.0    .                                                   +
        47   .    44.0    .                             +
        48   .    64.0    .                                                    +
        49   .    43.0    .                        +
        50   .    52.0    .                                   +
        51   .    38.0    .                   +
        52   .    59.0    .                                              +
        53   .    55.0    .                                         +
        54   .    41.0    .                      +
        55   .    53.0    .                                        +
        56   .    49.0    .                                     +
        57   .    34.0    .               +
        58   .    35.0    .                 +
        59   .    54.0    .                                       +
        60   .    45.0    .                          +
        61   .    68.0    .                                                       +
        62   .    38.0    .                  +
        63   .    50.0    .                                    +
        64   .    60.0    .                                            +
        65   .    39.0    .                   +
        66   .    57.5    .            (                             * +                    )
        67   .    46.9    . (                  +           *                    )
        68   .    54.4    .        (                          * +                    )
        69   .    49.6    .  (                             *    +               )
        70   .    52.8    .+       (                        *                      )
        71   .    50.7    .    (                      *                      )
        72   .    52.1    .     (                     *                      )
        73   .    51.2    .     (                     *                      )
        74   .    51.8    .      (                    *                      )
        75   .    51.4    .     (                     *                      )
        76   .    51.7    .     (                     *                      )
        77   .    51.5    .     (                     *                      )
        78   .    51.6    .     (                     *                      )
        79   .    51.5    .     (                     *                      )
        80   .    51.6    .     (                     *                      )
                     . . . . . . . . . . . . . . . . . . . . . . . . . . . . . . . . . . . . . .
```

From this figure we also clearly see that after ten forecasts the remaining forecasts are virtually constant, with value equal to 51.6, the estimate of the mean of data. Also, the confidence limits no longer increase, a reflection of the stationarity of the raw data. In Table 6.4 we also presented the ψ-values, the error learning coefficients. These coefficients are used to calculate the forecast error variances [see equation (6.42)]. Again, after ten periods these values have become quite small, reflecting the fact that no uncertainty is added as we increase the forecast horizon.

6.8 UPDATING THE FORECASTS

As soon as the value of the next data point, z_{n+1}, becomes available, we may want to adjust all forecasts for future periods. The new set of forecasts can be generated in two different ways:

1. *Sequential updated forecasting*   Using the additional data point, we reestimate the parameters of the current model and then proceed, making forecasts as usual.
2. *Adaptive forecasting*   We leave the parameter values of the ARIMA model unchanged but change the origin date to incorporate the new data point.

The sequential updating method only differs from what we have explained thus far by the fact that each time a new observation becomes available, the model is reestimated, reevaluated to see if it is still adequate, and then used to generate forecasts. In Chapters 7 and 8 we present examples of sequential updating. We would expect that the parameter values do not change very much when one new data point is added. If the parameter values were to change very much we would be rather skeptical about the stability of the ARIMA model. We do expect that as more and more data become available from the same process, the point estimates will be more precise.

Adaptive forecasting allows for the updating of h periods ahead forecasts of any ARIMA model without explicit need to access all past data. In fact, all that is needed to update forecasts made in a previous time period is the current one step ahead forecast error $a_n(1)$ and the old forecasts $z_n(l)$, $l = 2, 3, \ldots, h$. To show how adaptive forecasting works, we express the ARIMA model in the error-shock form [see equation (6.37)] and express the forecasts for period $n + h$ made at time n and $n + 1$, respectively, as

$$z_n(h) = \psi_h a_n + \psi_{h+1} a_{n-1} + \psi_{h+2} a_{n-2} + \cdots; \qquad (6.52)$$

$$z_{n+1}(h-1) = \psi_{h-1} a_{n+1} + \psi_h a_n + \psi_{h+1} a_{n-1} + \psi_{h+2} a_{n-2} + \cdots. \qquad (6.53)$$

Note that the future errors $a_{n+h}, a_{n+h-1}, \ldots, a_{n+2}$ are replaced by their mean value, 0. Then, subtract (6.52) from (6.53) to obtain an expression for the revision in the forecast of z_{n+h} as

$$z_{n+1}(h-1) - z_n(h) = \psi_{h-1} \, a_{n+1}$$
$$= \psi_{h-1}[z_{n+1} - z_n(1)] \qquad (6.54)$$

or

$$z_{n+1}(h-1) = z_n(h) + \psi_{h-1}[z_{n+1} - z_n(1)]. \qquad (6.55)$$

Equation (6.55) shows a remarkable result. Suppose that we currently have h forecasts made at origin n for the periods $n + 1, \ldots, n + h$, and after we observe the actual value for z_{n+1} we calculate the forecast error[9] $a_{n+1} = z_{n+1} - z_n(1)$. We can then update all other forecasts by adding to the forecasts made at origin n a fixed quantity proportional to the current one period ahead forecast error a_{n+1}, with the proportionality constant equal to the error learning coefficient ψ_{h-1}. The forecast mechanism can be so viewed as an *error learning process;* forecasts of future values are modified in the light of past forecasting errors. The ψ-weights are therefore appropriately called *error learning coefficients.*

Also, it is not necessary to recompute estimates of the widths of appropriate confidence intervals. As shown in equation (6.42), the variance for an h step ahead forecast is only determined by the error variance σ_a^2 and the error learning coefficients, which, since we do not reestimate the model, remain constant.

Finally, the forecast h periods ahead made at period $n + 1$, $z_{n+1}(h)$, cannot be calculated using equation (6.55), but is easily obtained from the forecasts at shorter lead times using the procedure explained in equation (6.11).

EXAMPLE
In Section 6.5 (see Table 6.2) we used an IMA(1,1) model to generate forecasts for $z_n(h)$, $h = 1, 2, \ldots, 7$. We also calculated new forecasts as soon as the actual value of z_{n+1} became available and found that all forecasts were decreased by 25 units. This corresponds to the fact that all error learning coefficients are constant and equal $(1 - \theta_1)$ and therefore the adaptive forecasting formula becomes

$$z_{n+1}(h-1) - z_n(h) = (1 - \theta_1)\,[z_{n+1} - z_n(1)]. \qquad (6.56)$$

The original forecast for z_{n+1} was 265, and so we made a forecast error, $a_{n+1} = -50$. With a parameter value $\theta_1 = 0.5$, equation (6.56) shows that all forecasts have to be equally revised downward by 25 units.

---

[9] Note that once we have observed the data at period $n + 1$, the forecast error $a_n(1)$ is written as a_{n+1}.

6.9. EXPONENTIAL SMOOTHING

In Section 6.5 we derived the forecast profile for an IMA(1,1) model. Let us now derive the forecast profile for the same model in a different way. The IMA(1,1) with mean $\mu = 0$ can be represented as

$$z_t - z_{t-1} = a_t - \theta_1 a_{t-1}. \tag{6.57}$$

Similarly, this model can be rewritten for time periods $n + 1$ and earlier as

$$
\begin{aligned}
z_{n+1} - z_n &= a_{n+1} - \theta_1 a_n \\
z_n - z_{n-1} &= a_n - \theta_1 a_{n-1} \\
z_{n-1} - z_{n-2} &= a_{n-1} - \theta_1 a_{n-2} \\
&\ \ \vdots
\end{aligned}
\tag{6.58}
$$

On multiplying both sides of the second equation in (6.58) by θ_1, both sides of the third equation by θ_1^2, and so on, and then adding all these transformed equations together, we can solve for z_{n+1} to obtain

$$
\begin{aligned}
z_{n+1} = a_{n+1} &+ (1 - \theta_1)z_n + \theta_1(1 - \theta_1)z_{n-1} + \theta_1^2(1 - \theta_1)z_{n-2} \\
&+ \theta_1^3(1 - \theta_1)z_{n-3} + \cdots .
\end{aligned}
\tag{6.59}
$$

The minimum mean squared error forecast for z_{n+1} is obtained from equation (6.59) after replacing the future error a_{n+1} by its mean value 0, that is

$$z_n(1) = (1 - \theta_1)z_n + \theta_1(1 - \theta_1)z_{n-1} + \theta_1^2(1 - \theta_1)z_{n-2} + \cdots . \tag{6.60}$$

The coefficients in (6.60) can be graphically represented as in Figure 6.8. Equation (6.60) represents the *exponentially weighted moving average process.* This equation shows that for an IMA(1,1) process, recent observations receive larger weights than observations more distant in the past. The observations

FIGURE 6.8   Exponential Smoothing Forecast Profile.

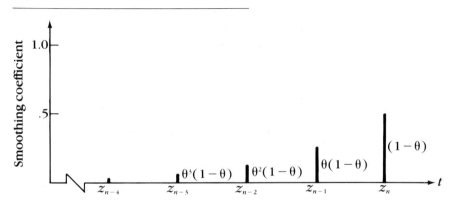

z_n, z_{n-1}, . . . are weighted in an exponentially decreasing fashion, hence the name *exponential smoothing*.

The weighting pattern may be adjusted by selecting different values of θ_1. If θ_1 is small, then the weight given to the current observation will be large and the successive weights will decline rapidly. However, if θ_1 is large, then the weight given to the current observation will be small and subsequent weights will decay slowly.

Exponential smoothing has been used for routine forecasting. It should, however, be recognized that exponential smoothing is only correct for an IMA(1,1) model, a model which is only one of the many ARIMA processes that could have generated a time series. In the light of this finding we have to label the method of always using exponential smoothing to forecast as an *ad hoc* method.

6.10 SUMMARY

In this chapter we have shown how to choose an optimal forecast based on specific cost functions. Then we showed that the mean of the forecast distribution minimizes the expected value of any quadratic cost function and is therefore a minimum mean squared error forecast. For an ARIMA model such a forecast can be expressed as a linear function of past data and past errors and as such can easily be calculated.

In Section 6.4 and 6.5 we derived the forecast and its properties for some commonly encountered stationary and nonstationary ARIMA models. In the next section we complemented the point forecast with their forecast confidence limits. Then in Section 6.7 we presented a simple example using the chemical batch yield data analyzed in earlier chapters.

In the next two sections we concluded this chapter with some information about updating of forecasts and with an explanation of exponential smoothing.

CHAPTER 7

UNITED STATES RESIDENTIAL CONSTRUCTION

Note: This chapter was written jointly with Sergio Koreisha.

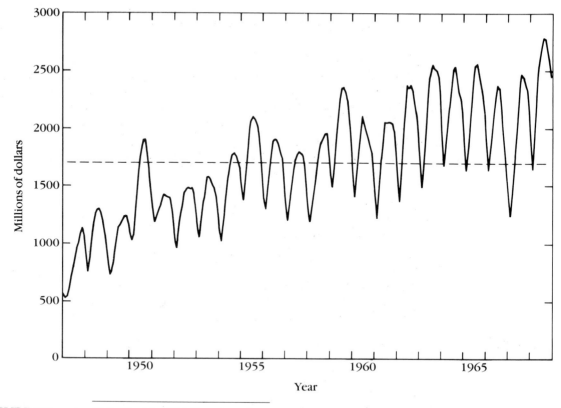

FIGURE 7.1   U.S. Residential Construction.

The objective of this chapter is to acquaint you with the type of judgment that is required to formulate time series models using Box–Jenkins methodology. Throughout this chapter we shall ask for your opinions and suggestions for model improvements. Space is provided for you to jot down some notes. After you write down your ideas, check them with ours. Please keep in mind, however, that the evaluations which follow were made using judgment, and should not be construed as being "the answer." You should view our comments as educated opinions. We hope that by checking your opinions back and forth with ours, you will become more conversant with this forecasting methodology.

The time series we shall analyze is new United States residential construction (non-farm). The data consist of 264 monthly observations covering the period from January 1947 to December 1968. The data are in current dollars (millions), and are unadjusted for seasonal variations.[1] This example is reason-

[1] *Source:* United States Department of Commerce, Office of Business Economics, *1971 Business Statistics,* The Biennial Supplement to the *Survey of Current Business,* pp. 49–50, 226. In Appendix A, Table A–6, we have listed the data.

ably advanced but at the same time confronts the reader with the typical problems encountered in using the Box–Jenkins methodology to forecast a time series. We feel that the interaction between questions and answers should allow the reader to tackle such an example without the further need to introduce additional made-up situations.

7.1 PRELIMINARY ANALYSIS

7.1.1 Data Examination

As a first step in analyzing the residential construction data, we recommend that you study the data to see what insights they can provide. Figure 7.1 contains the plot of the monthly construction data for the period January 1947 to December 1968. What conclusions can you draw from examining this time series plot?

1.

2.

3.

4.

7.1.2 Comments on the Construction Data

By looking at the time series plot (Figure 7.1) one can immediately detect an upward trend in the series. The seasonal pattern is also quite apparent, with peaks and valleys at time multiples of 12. This, of course, should not surprise you because we are dealing with monthly data, unadjusted for seasonal variations. Observe that fluctuations in the data remain relatively constant throughout the period. This suggests that a logarithmic transformation of the data, for example, is not absolutely necessary to stabilize the variance.

This brief analysis of the data should help us to select the type of differencing necessary to identify stationary models. What type of differencing would you recommend to make the series stationary?

1.

2.

3.

7.2 IDENTIFICATION

7.2.1 Autocorrelation and Partial Autocorrelation Functions

Table 7.1 contains the autocorrelations and partial autocorrelations of the series $\nabla_{12}z_t$, $\nabla\nabla_{12}z_t$, and $\nabla^2\nabla_{12}z_t$. Figures 7.2(a), 7.2(b), and 7.2(c) show the plots of these correlations. Study each of them carefully, and attempt to identify potential models. Space is provided in Section 7.2.2 for you to make comments on each of the series.

TABLE 7.1 Autocorrelations and Partial Autocorrelations

SERIES WITH D = 0 DS = 1 S = 12
MEAN =   75.29   SD =   252.2   (NOBS = 252)

AUTOCORRELATIONS

LAGS	ROW SE												
1–12	.06	.96	.88	.78	.67	.54	.40	.26	.11	−.03	−.16	−.27	−.36
13–24	.18	−.43	−.46	−.48	−.48	−.46	−.42	−.37	−.32	−.26	−.20	−.14	−.07
25–36	.21	−.02	.02	.04	.05	.05	.03	.00	−.03	−.06	−.10	−.14	−.18
37–48	.21	−.20	−.21	−.21	−.20	−.19	−.16	−.12	−.07	−.02	.03	.08	.13

PARTIAL AUTOCORRELATIONS   STANDARD ERROR =   .06

LAGS												
1–12	.96	−.57	−.05	−.13	−.17	−.22	.00	−.06	−.12	−.01	.01	.07
13–24	.17	−.09	−.02	.00	.00	−.04	−.08	−.04	−.04	.02	.05	.00
25–36	−.07	−.14	−.05	−.08	−.03	−.10	−.03	−.02	−.03	.02	−.04	.05
37–48	.05	−.01	−.08	.03	−.06	−.04	.05	−.01	−.02	−.06	.03	.00

SERIES WITH D = 1 DS = 1 S = 12
MEAN =   −0.2869   SD =   66.34   (NOBS = 251)

AUTOCORRELATIONS

LAGS	ROW SE												
1–12	.06	.62	.28	.17	.18	.20	.06	−.05	−.03	−.07	−.16	−.33	−.49
13–24	.11	−.35	−.24	−.23	−.25	−.24	−.13	−.06	−.06	−.08	−.07	.03	.14
25–36	.12	.20	.19	.16	.14	.16	.12	.06	.06	.05	.07	−.03	−.15
37–48	.13	−.18	−.13	−.11	−.09	−.13	−.15	−.08	−.04	.00	−.03	.00	.08

PARTIAL AUTOCORRELATIONS   STANDARD ERROR =   .06

LAGS												
1–12	.62	−.17	.13	.07	.08	−.19	.00	.03	−.13	−.14	−.22	−.28
13–24	.14	−.10	−.03	−.06	.04	−.02	−.02	.01	−.11	−.12	−.01	−.02
25–36	.14	−.05	−.02	−.07	.07	−.07	−.05	.01	−.11	−.02	−.13	−.08
37–48	.01	−.02	−.06	.07	−.05	−.09	.01	.00	−.02	−.06	−.01	−.06

SERIES WITH D = 2 DS = 1 S = 12
MEAN =   0.4640   SD =   57.75   (NOBS = 250)

AUTOCORRELATIONS

LAGS	ROW SE												
1–12	.06	−.03	−.32	−.15	−.03	.21	−.04	−.17	.05	.09	.10	.01	−.40
13–24	.08	.01	.14	.06	−.05	−.13	.06	.10	.03	−.03	−.13	−.02	.07
25–36	.09	.09	.04	−.03	−.04	.08	.02	−.08	−.01	−.03	.16	.04	−.13
37–48	.09	−.09	.02	−.01	.10	−.02	−.12	.05	.01	.08	−.08	−.07	.04

PARTIAL AUTOCORRELATIONS   STANDARD ERROR =   .06

LAGS												
1–12	−.03	−.32	−.19	−.18	.10	−.12	−.13	.02	.02	.07	.08	−.34
13–24	−.03	−.11	−.06	−.16	−.06	−.08	−.09	.02	.02	−.11	−.05	−.21
25–36	.01	−.02	.03	−.11	.05	.00	−.06	.04	−.04	.07	.01	−.08
37–48	−.03	.00	−.12	.03	.03	−.07	−.05	−.01	.01	−.03	.01	−.08

FIGURE 7.2(a)   Sample Autocorrelation Function and Partial Autocorrelation Function of $\nabla_{12}z_t$.

```
AUTOCORRELATION FUNCTION                    PARTIAL AUTOCORRELATION FUNCTION
                  WITH   D = 0 DS = 1 S = 12

 -0.8  -0.4   0.0   0.4   0.8             -0.8  -0.4   0.0   0.4   0.8
 * . . . * . . . * . . . * . . . *         * . . . * . . . * . . . * . . . *
   .                 ( . )         R   1 .                 ( . )             R
   .                ( . )        R   2 .          R        ( . )
   .               (   . )      R   3 .                    (R. )
   .               (   . )    R   4 .                    R . )
   .              (    . )  R   5 .                    R( . )
   .              (    . )R   6 .                    R( . )
   .              (    . R)   7 .                    ( R )
   .              ( . R . )   8 .                    (R. )
   .              ( R   . )   9 .                    R . )
   .             ( R .   )  10 .                    ( R )
   .            (R    .   ) 11 .                    ( R )
   .            R     .    ) 12 .                    ( .R)
   .          R(      .    ) 13 .                    ( . )R
   .         R (      .    ) 14 .                    (R. )
   .         R(       .    ) 15 .                    ( R )
   .         R(       .    ) 16 .                    ( R )
   .         R(       .    ) 17 .                    ( R )
   .          R       .    ) 18 .                    (R. )
   .          R       .    ) 19 .                    (R. )
   .          (R      .    ) 20 .                    (R. )
   .          ( R     .    ) 21 .                    (R. )
   .          (  R    .    ) 22 .                    ( R )
   .          (   R . )    ) 23 .                    ( .R)
   .          (    R.      ) 24 .                    ( R )
   .          (     R      ) 25 .                    (R. )
   .          (     R      ) 26 .                    R . )
   .          (      .R    ) 27 .                    (R. )
   .          (      .R    ) 28 .                    (R. )
   .          (      .R    ) 29 .                    ( R )
   .          (      R     ) 30 .                    (R. )
   .          (      R     ) 31 .                    ( R )
   .          (      R     ) 32 .                    ( R )
   .          (     R.     ) 33 .                    ( R )
   .          (    R .     ) 34 .                    ( R )
   .          (    R .     ) 35 .                    (R. )
   .          (   R  .     ) 36 .                    ( .R)
   .          (   R  .     ) 37 .                    ( .R)
   .          (   R  .     ) 38 .                    ( R )
   .          (   R  .     ) 39 .                    (R. )
   .          (   R  .     ) 40 .                    ( R )
   .          (   R  .     ) 41 .                    (R. )
   .          (    R .     ) 42 .                    (R. )
   .          (    R .     ) 43 .                    ( .R)
   .          (     R.     ) 44 .                    ( R )
   .          (      R     ) 45 .                    ( R )
   .          (      R     ) 46 .                    (R. )
   .          (      .R    ) 47 .                    ( R )
   .          (       . R  ) 48 .                    ( R )
```

FIGURE 7.2(b)   Sample Autocorrelation Function and Partial Autocorrelation Function of $\nabla\nabla_{12}Z_t$.

```
AUTOCORRELATION FUNCTION                    PARTIAL AUTOCORRELATION FUNCTION
                     WITH  D = 1 DS = 1 S = 12

  -0.8  -0.4   0.0   0.4   0.8              -0.8  -0.4   0.0   0.4   0.8
  * . * . * . * . * . * . * . * . *         * . * . * . * . * . * . * . * . *
  .                ( . )        R      1 .              ( . )        R
  .                ( . ) R      2 .            R( . )
  .                (  . R       3 .              ( . R
  .                (  . R       4 .              ( .R)
  .                (  . R       5 .              ( .R)
  .                ( .R )       6 .            R( . )
  .                ( R. )       7 .              ( R )
  .                (  R )       8 .              ( R )
  .                ( R. )       9 .            R . )
  .               (R . )       10 .            R . )
  .           R ( .  )         11 .            R( . )
  .          R   ( . )         12 .          R ( . )
  .           R ( .  )         13 .              ( . R
  .          R( .  )           14 .            (R. )
  .            .R .  )         15 .              ( R )
  .          R( .  )           16 .            (R. )
  .          R( . .  )         17 .              ( .R)
  .             ( R .    )     18 .              ( R )
  .             (  R.  )       19 .              ( R )
  .             (  R.  )       20 .              ( R )
  .             (  R.  )       21 .            R . )
  .             (  R. )        22 .            R . )
  .             (   R  )       23 .              ( R )
  .             (    . R )     24 .              ( R )
  .             (    . R)      25 .              ( . R
  .             (    . R)      26 .            (R. )
  .             (    . R )     27 .              ( R )
  .             (    . R )     28 .            (R. )
  .             (    . R )     29 .              ( .R)
  .             (    . R )     30 .            (R. )
  .             (   .R  )      31 .            (R. )
  .             (   .R  )      32 .              ( R )
  .             (   .R  )      33 .            R . )
  .             (   .R  )      34 .              ( R )
  .             (   R.  )      35 .            R . )
  .             ( R .  )       36 .            (R. )
  .             (R  .  )       37 .              ( R )
  .             ( R .  )       38 .              ( R )
  .             ( R .  )       39 .            (R. )
  .             (   R.  )      40 .              ( .R)
  .             (  R .  )      41 .            (R. )
  .             (  R .  )      42 .            (R. )
  .             (   R.. )      43 .              ( R )
  .             (   R.  )      44 .              ( R )
  .             (    R  )      45 .              ( R )
  .             (    R  )      46 .            (R. )
  .             (    R   )     47 .              ( R )
  .             (    .R  )     48 .            (R. )
```

FIGURE 7.2(c)   Sample Autocorrelation Function and Partial Autocorrelation Function of $\nabla^2\nabla_{12}z_t$.

```
AUTOCORRELATION FUNCTION                    PARTIAL AUTOCORRELATION FUNCTION
                     WITH   D = 2  DS = 1  S = 12

  -0.8  -0.4   0.0   0.4   0.8              -0.8  -0.4   0.0   0.4   0.8
  * . * . * . * . * . * . * . * . * . *     * . * . * . * . * . * . * . * . * . * . *
   .                 ( R )           1  .                 ( R )
   .           R   ( . )             2  .           R   ( . )
   .             R  .  )             3  .             R(  .  )
   .             ( R )               4  .             R(  .  )
   .             ( . )R              5  .             (  .R)
   .             (R. )               6  .             R  . )
   .           R( . )                7  .             R  . )
   .             ( .R)               8  .             ( R )
   .             ( .R)               9  .             ( R )
   .             ( . R              10  .             ( .R)
   .             ( R )              11  .             ( .R)
   .           R  ( . )             12  .           R ( . )
   .             ( R )              13  .             ( R )
   .             ( . R              14  .             R . )
   .             ( .R)              15  .             (R. )
   .           ( R. )              16  .             R . )
   .           (R . )              17  .             (R. )
   .           ( .R )              18  .             (R. )
   .           ( .R )              19  .             (R. )
   .           ( R )               20  .             ( R )
   .           ( R )               21  .             ( R )
   .           (R . )              22  .             R . )
   .           ( R )               23  .             (R. )
   .           ( .R )              24  .           R( . )
   .           ( .R )              25  .             ( R )
   .           ( .R )              26  .             ( R )
   .           ( R )               27  .             ( R )
   .           ( R. )              28  .             R . )
   .           ( .R )              29  .             ( .R)
   .           ( R )               30  .             ( R )
   .           ( R. )              31  .             (R. )
   .           ( R )               32  .             ( .R)
   .           ( R )               33  .             (R. )
   .           ( . R)              34  .             ( .R)
   .           ( .R )              35  .             ( R )
   .           (R . )              36  .             (R. )
   .           ( R. )              37  .             ( R )
   .           ( R )               38  .             ( R )
   .           ( R )               39  .             R . )
   .           ( .R )              40  .             ( R )
   .           ( R )               41  .             ( R )
   .           (R . )              42  .             (R. )
   .           ( .R )              43  .             (R. )
   .           ( R )               44  .             ( R )
   .           ( .R )              45  .             ( R )
   .           ( R. )              46  .             ( R )
   .           ( R. )              47  .             ( R )
   .           ( .R )              48  .             (R. )
```

7.2.2 Comments

A. Series $\nabla_{12}z_t$

1.

2.

3.

4.

Potential Models?

B. Series $\nabla\nabla_{12}z_t$

1.

2.

3.

4.

Potential Models?

C. Series $\nabla^2\nabla_{12}z_t$

1.

2.

3.

4.

Potential Models?

7.2.3 Series $\nabla_{12}z_t$

By looking at the plots in Figure 7.2(a), we observe that the autocorrelations of this series gradually decay to 0 in a damped oscillatory fashion, and that two prominent spikes appear in the partial autocorrelations at lags 1 and 2. We also see that several other values at higher lags of the partial autocorrelations lie outside the approximate 95% large-sample confidence limits. Furthermore, since the autocorrelations do not seem to die out quickly, and the value of the first autocorrelation is 0.96, we can safely conclude that this series is not stationary, and that additional differencing is required.

In the Appendix to this chapter however, we will analyze the following model:

$$(1 - \phi_1 B - \phi_2 B^2)\nabla_{12}z_t = \delta + a_t. \qquad \text{(Model A)}[2]$$

The purpose of this analysis is to show that by studying the residual patterns of an inappropriate formulation, one can make logical modifications to a misspecified model to arrive at a formulation which adequately depicts the behavior of the series.

7.2.4 Series $\nabla\nabla_{12}z_t$

The autocorrelations of the $\nabla\nabla_{12}z_t$ series, as Figure 7.2(b) shows, have prominent spikes at lags 1, 11, 12, and 13. For now, ignore the spike at lag 2. In the partial autocorrelations we observe a major spike at lag 1 and several minor ones at lags 2, 6, 11, 12, and 13. Let us try to filter out the more prominent spikes, and then see what additional information is contained in the residuals in order to formulate an appropriate model.

From theory, we can identify two moving average models with significant autocorrelations at lags 1, 11, 12, and 13, namely[3]

$$w_t = (1 - \theta B)(1 - \Theta B^{12})a_t, \qquad (7.1)$$

and

$$w_t = (1 - \theta_1 B - \theta_{12} B^{12} - \theta_{13} B^{13})a_t, \qquad (7.2)$$

where $w_t = \nabla\nabla_{12}z_t$. Indeed, in Section 4.4. we derived the *acf* of the multiplicative seasonal model (7.1) as

$$\rho_1 = -\theta/(1 + \theta^2)$$
$$\rho_{12} = -\Theta/(1 + \Theta^2)$$
$$\rho_{11} = \rho_{13} = \rho_1\rho_{12} = \frac{\theta\Theta}{(1 + \theta^2)(1 + \Theta^2)} \qquad (7.3)$$

---

[2] The constant δ has been incorporated in this model formulation because the mean of the series $\nabla_{12}z_t$ is 75.29 million dollars.

[3] In this chapter we have systematically omitted the subscript of the parameters where possible.

and all other autocorrelations are 0. The autocorrelations for (7.2) are

$$\rho_1 = (-\theta_1 + \theta_{12}\theta_{13})/\omega$$
$$\rho_{11} = \theta_1\theta_{12}/\omega$$
$$\rho_{12} = (-\theta_{12} + \theta_1\theta_{13})/\omega$$
$$\rho_{13} = -\theta_{13}/\omega \qquad\qquad (7.4)$$

with

$$\omega = (1 + \theta_1^2 + \theta_{12}^2 + \theta_{13}^2)$$

and all other autocorrelations are 0.

Because the *acf* of $\nabla\nabla_{12}z_t$ has also a spike at lag 2, we might be inclined to assume that the *acf* is dying out rather than showing a cutoff, and therefore we could prefer to include an AR(1) parameter.

However, because of the very strong evidence of symmetry about the autocorrelation at lag 12, evidence of both a regular MA(1) and a seasonal MA(1) parameter model, we propose to combine all evidence and estimate a model both containing the AR(1) parameter and the moving average parameters. During the diagnostic checking of the model we will then evaluate dropping the AR(1) and the MA(1) parameter.

Thus, we have identified two models which merit further investigation:

$$(1 - \phi B)\nabla\nabla_{12}z_t = (1 - \theta B)(1 - \Theta B^{12})a_t \qquad \text{(Model 1)}$$

and

$$(1 - \phi B)\nabla\nabla_{12}z_t = (1 - \theta_1 B - \theta_{12}B^{12} - \theta_{13}B^{13})a_t. \qquad \text{(Model 2)}$$

7.2.5 Series $\nabla^2\nabla_{12}z_t$

The autocorrelations in Figure 7.2(c) have two major spikes at lags 2 and 12. The graph of the partial autocorrelations shows one spike at lag 2 and decaying spikes at lags of multiples of 12. If we ignore the spikes at lag 2 in both the autocorrelations and partial autocorrelations, then the observed autocorrelations and partial autocorrelations closely resemble those of a model with a seasonal moving average parameter. Therefore, we propose also to use the model

$$\nabla^2\nabla_{12}z_t = (1 - \Theta B^{12})a_t \qquad \text{(Model 3)}$$

in the estimation process, and then to check the residuals. If, after going through the estimation phase of this model, we still observe a spike at lag 2 in either the residual autocorrelations or partial autocorrelations, we will then appropriately modify the model.

7.2.6 Summary

We have identified three models which merit further investigation. We would like to repeat that it is quite normal to tentatively entertain more than one model for further analysis. The objective of the identification phase is not to select the single correct model but to narrow down the choice of possible models that then will be subjected to further analysis. If you are puzzled by any of the logic in our analysis, please go back to the beginning of Section 7.2 and read it one more time. If you are wondering why we did not include a particular model which you perhaps thought of as being a likely candidate, please keep in mind that the evaluations of the three differenced series were based on *judgment*—that is, on what our experience has shown to be high payoff models.

For recapitulation and convenience, the three potential models are listed in Table 7.2. In the next section we will analyze these models in detail.

TABLE 7.2   Models to Be Estimated

Model 1	$(1 - \phi B)\nabla\nabla_{12}z_t = (1 - \theta B)(1 - \Theta B^{12})a_t$
Model 2	$(1 - \phi B)\nabla\nabla_{12}z_t = (1 - \theta_1 B - \theta_{12}B^{12} - \theta_{13}B^{13})a_t$
Model 3	$\nabla^2\nabla_{12}z_t = (1 - \Theta B^{12})a_t$

7.3 ESTIMATION

7.3.1 Model 1:
$$(1 - \phi B)\nabla\nabla_{12}z_t = (1 - \theta B)(1 - \Theta B^{12})a_t$$

Use the output below and in Figure 7.3 to evaluate Model 1 and suggest improvements.

```
MODEL WITH D = 1 DS = 1 S = 12
MEAN = -0.2869     SD =    66.34    (NOBS = 251)

LEAST SQUARES METHOD
PARAMETER ESTIMATES
====================
                EST      SE      EST/SE     95% CONF LIMITS
    AR( 1)     0.261    0.092     2.855     0.082     0.441
    MA( 1)    -0.489    0.082    -5.956    -0.650    -0.328
    SMA( 1)    0.684    0.047    14.625     0.592     0.776

EST.RES.SD =    4.1965E+01
EST.RES.SD(WITH BACK FORECAST) =      4.2855E+01

R SQR =   .605
DF = 248
NO F-STAT, NO INTERCEPT
```

```
CORRELATION MATRIX
          AR( 1)  MA( 1)
  MA( 1)   .747
  SMA( 1) −.129   −.102
```

```
AUTOCORRELATIONS OF RESIDUALS
  LAGS ROW SE
  1−12  .06  −.02   .04   .01  −.02   .15  −.04  −.04   .00   .01  −.11  −.06   .02
 13−24  .07  −.04   .00  −.04  −.14  −.10  −.04  −.03   .00   .03  −.12  −.03   .05
 25−36  .07   .08   .08  −.01  −.02   .05   .00  −.04   .03  −.04   .04  −.03  −.06
 37−48  .07  −.01   .03  −.07   .01  −.04  −.13   .02  −.02   .08  −.03  −.02   .02
```

```
CHI−SQUARE TEST              P−VALUE
Q(12) =  11.3       9 D.F.   .254
Q(24) =  26.0      21 D.F.   .207
Q(36) =  33.3      33 D.F.   .454
Q(48) =  43.6      45 D.F.   .530
```

```
AUTOCORRELATIONS OF FIRST DIFFERENCED RESIDUALS
  LAGS ROW SE
  1−12  .06  −.52   .04   .01  −.10   .18  −.09  −.02   .02   .06  −.08  −.01   .06
 13−24  .08  −.05   .04   .02  −.07  −.01   .02   .00   .02   .04  −.09   .01   .03
 25−36  .08   .01   .05  −.04  −.04   .06   .00  −.06   .07  −.07   .07  −.02  −.04
 37−48  .08   .01   .06  −.08   .07   .01  −.12   .10  −.07   .11  −.06  −.02  −.03
```

```
CHI−SQUARE TEST              P−VALUE
Q(12) =  86.7       9 D.F.   .000
Q(24) =  92.5      21 D.F.   .000
Q(36) =  101.      33 D.F.   .000
Q(48) =  119.      45 D.F.   .000
```

ANALYSIS OF MODEL 1: $(1 - \phi B)\nabla\nabla_{12}z_t = (1 - \theta B)(1 - \Theta B^{12})a_t$

Model 1 appears to be quite acceptable: The residuals are basically white noise, although the residual autocorrelation at lag 5 is relatively large, the residual autocorrelation Q statistics do not invalidate the hypothesis of model adequacy, and the values of the autoregressive and moving average parameters confirm that each parameter is significant. However, the correlation matrix suggests that the AR(1) and MA(1) parameters are not independent. Let us see what happens when we delete either the MA(1) or the AR(1) parameter. In other words, let us evaluate the following two alternative models:

$$(1 - \phi B\nabla\nabla_{12}z_t = (1 - \Theta B^{12})a_t \qquad \text{(Model 1.1)}$$

or

$$\nabla\nabla_{12}z_t = (1 - \theta B)(1 - \Theta B^{12})a_t. \qquad \text{(Model 1.2)}$$

FIGURE 7.3   Model 1: Autocorrelation Function of Residuals.

```
AUTOCORRELATION FUNCTION
       RESIDUALS

  -0.8  -0.4   0.0    0.4    0.8
 *  .  .  *  .  *  .  *  .  *  .  *  .  *  . .  *
          .                ( R  )          1 .
          .                ( .R)           2 .
          .                ( R )           3 .
          .                ( R )           4 .
          .                ( .  R          5 .
          .                (R.  )          6 .
          .                (R.  )          7 .
          .                ( R  )          8 .
          .                ( R  )          9 .
          .                R  . )         10 .
          .                (R.  )         11 .
          .                ( R  )         12 .
          .                (R.  )         13 .
          .                ( R  )         14 .
          .                (R.  )         15 .
          .                R  . )         16 .
          .                (R.  )         17 .
          .                (R.  )         18 .
          .                ( R  )         19 .
          .                ( R  )         20 .
          .                ( R  )         21 .
          .                R  . )         22 .
          .                ( R  )         23 .
          .                ( .R)          24 .
          .                ( .R)          25 .
          .                ( .R)          26 .
          .                ( R  )         27 .
          .                ( R  )         28 .
          .                ( .R)          29 .
          .                ( R  )         30 .
          .                (R.  )         31 .
          .                ( R  )         32 .
          .                (R.  )         33 .
          .                ( .R)          34 .
          .                ( R  )         35 .
          .                (R.  )         36 .
          .                ( R  )         37 .
          .                ( R  )         38 .
          .                (R.  )         39 .
          .                ( R  )         40 .
          .                (R.  )         41 .
          .                R  . )         42 .
          .                ( R  )         43 .
          .                ( R  )         44 .
          .                ( .R)          45 .
          .                (R.  )         46 .
          .                ( R  )         47 .
          .                ( R  )         48 .
```

7.3.2 Model 1.1:
$$(1 - \phi B)\nabla\nabla_{12}z_t = (1 - \Theta B^{12})a_t$$

Use the output below and in Figure 7.4 to evaluate Model 1.1. Suggest model improvements.

```
MODEL WITH D = 1 DS = 1 S = 12
MEAN = -0.2869    SD =   66.34    (NOBS = 251)

LEAST SQUARES METHOD
PARAMETER ESTIMATES
====================


            EST      SE      EST/SE     95% CONF LIMITS
  AR( 1)   0.573    0.052   11.004     0.471    0.675
  SMA( 1)  0.667    0.048   13.948     0.574    0.761


EST.RES.SD =    4.3519E+01
EST.RES.SD(WITH BACK FORECAST) =    4.4344E+01

R SQR =    .573
DF = 249
NO F-STAT, NO INTERCEPT

CORRELATION MATRIX
          AR( 1)
 SMA( 1) -.126

AUTOCORRELATIONS OF RESIDUALS
  LAGS ROW SE
   1-12   .06    .12 -.19 -.07 -.01  .15 -.02 -.08  .03  .02 -.11 -.07  .02
  13-24   .07    .00  .03 -.03 -.15 -.11 -.02  .01  .03 -.03 -.14 -.05  .06
  25-36   .07    .12  .09 -.03 -.04  .05  .00 -.04  .02  .00  .05 -.03 -.09
  37-48   .07    .00  .04 -.03  .02 -.05 -.13 -.01  .03  .08 -.03 -.07  .02

CHI-SQUARE TEST           P-VALUE
Q(12) =  26.2      10 D.F.   .004
Q(24) =  43.8      22 D.F.   .004
Q(36) =  55.0      34 D.F.   .013
Q(48) =  66.3      46 D.F.   .026

AUTOCORRELATIONS OF FIRST DIFFERENCED RESIDUALS
  LAGS ROW SE
   1-12   .06   -.32 -.25  .03 -.06  .19 -.06 -.10  .06  .07 -.09 -.03  .06
  13-24   .08   -.03  .06  .03 -.10 -.03  .04  .01  .05  .02 -.11 -.02  .04
  25-36   .08    .04  .06 -.06 -.06  .08 -.01 -.06  .04 -.04  .07 -.01 -.09
  37-48   .08    .03  .06 -.07  .08  .00 -.12  .06 -.01  .09 -.04 -.08  .00

CHI-SQUARE TEST           P-VALUE
Q(12) =  59.5      10 D.F.   .000
Q(24) =  68.7      22 D.F.   .000
Q(36) =  80.0      34 D.F.   .000
Q(48) =  95.0      46 D.F.   .000
```

FIGURE 7.4   Model 1.1: Autocorrelation Function of Residuals.

```
          AUTOCORRELATION FUNCTION
                RESIDUALS


    -0.8  -0.4   0.0   0.4   0.8
  *  .  * .  * ..  *  .  * ..  *  .  * ..  *
    .                    (  .  R           1 .
    .                  R(  .  )            2 .
    .                   (R.  )             3 .
    .                   ( R  )             4 .
    .                   (  .  R            5 .
    .                   ( R  )             6 .
    .                   (R.  )             7 .
    .                   ( R  )             8 .
    .                   ( R  )             9 .
    .                   R  .  )           10 .
    .                   (R.  )            11 .
    .                   ( R  )            12 .
    .                   ( R  )            13 .
    .                   (  .R)            14 .
    .                   ( R  )            15 .
    .                   R  .  )           16 .
    .                   R  .  )           17 .
    .                   ( R  )            18 .
    .                   ( R  )            19 .
    .                   ( R  )            20 .
    .                   (R.  )            21 .
    .                   R  .  )           22 .
    .                   (R.  )            23 .
    .                   (  .R)            24 .
    .                   (  .  R           25 .
    .                   (  .R)            26 .
    .                   ( R  )            27 .
    .                   (R.  )            28 .
    .                   (  .R)            29 .
    .                   ( R  )            30 .
    .                   (R.  )            31 .
    .                   ( R  )            32 .
    .                   ( R  )            33 .
    .                   (  .R)            34 .
    .                   ( R  )            35 .
    .                   (R.  )            36 .
    .                   ( R  )            37 .
    .                   (  .R)            38 .
    .                   (R.  )            39 .
    .                   ( R  )            40 .
    .                   (R.  )            41 .
    .                   R  .  )           42 .
    .                   ( R  )            43 .
    .                   ( R  )            44 .
    .                   (  .R)            45 .
    .                   ( R  )            46 .
    .                   (R.  )            47 .
    .                   ( R  )            48 .
```

ANALYSIS OF MODEL 1.1: $(1 - \phi B)\nabla\nabla_{12}z_t = (1 - \Theta B^{12})a_t$

The small spikes at lag 1 and lag 2 in the residual autocorrelations, $r_1 = 0.12$ and $r_2 = -0.19$, indicate that the MA(1) parameter cannot be deleted. Also, the omission of the MA(1) parameter causes the Q statistic to increase considerably. Based on this evidence, Model 1 appears to be a more adequate model than Model 1.1.

7.3.3 Model 1.2:
$$\nabla\nabla_{12}z_t = (1 - \theta B)(1 - \Theta B^{12})a_t$$

Use the output below and in Figure 7.5 to evaluate Model 1.2. Suggest improvements.

```
MODEL WITH D = 1 DS = 1 S = 12
MEAN = -0.2869     SD =    66.34     (NOBS = 251)

LEAST SQUARES METHOD
PARAMETER ESTIMATES
===================

            EST       SE      EST/SE      95% CONF LIMITS
 MA( 1)   -0.671    0.046    -14.491    -0.761    -0.580
 SMA( 1)   0.696    0.045     15.337     0.607     0.785

EST.RES.SD =    4.2306E+01
EST.RES.SD(WITH BACK FORECAST) =      4.3265E+01

R SQR =    .597
DF = 249
NO F-STAT, NO INTERCEPT

CORRELATION MATRIX
         MA( 1)
 SMA( 1)  -.013

AUTOCORRELATIONS OF RESIDUALS
  LAGS ROW SE
   1-12  .06    .07  .16 -.01  .03  .11 -.02 -.03 -.02 -.01 -.11 -.07  .00
  13-24  .07   -.04 -.02 -.07 -.15 -.12 -.07 -.04 -.03 -.05 -.12 -.03  .06
  25-36  .07    .09  .09  .01  .00  .03  .01 -.04  .03 -.04  .03 -.04 -.04
  37-48  .07   -.02  .02 -.06 -.02 -.06 -.14  .01 -.04  .08 -.04  .01  .02

CHI-SQUARE TEST              P-VALUE
Q(12) =  16.9      10 D.F.    .078
Q(24) =  37.8      22 D.F.    .019
Q(36) =  44.9      34 D.F.    .100
Q(48) =  56.2      46 D.F.    .143
```

```
AUTOCORRELATIONS OF FIRST DIFFERENCED RESIDUALS
  LAGS ROW SE
   1-12  .06  -.55  .14 -.12 -.02  .12 -.07 -.01  .00  .06 -.07 -.02  .06
  13-24  .08  -.03  .04  .02 -.06 -.01  .01  .01  .02  .03 -.09  .00  .04
  25-36  .08   .01  .05 -.04 -.03  .03  .01 -.06  .07 -.07  .08 -.04 -.01
  37-48  .09  -.01  .06 -.06  .05  .02 -.12  .10 -.09  .13 -.09  .02 -.06

CHI-SQUARE TEST            P-VALUE
Q(12) =  92.3     10 D.F.    .000
Q(24) =  97.0     22 D.F.    .000
Q(36) =  105.     34 D.F.    .000
Q(48) =  128.     46 D.F.    .000
```

ANALYSIS OF MODEL 1.2: $\nabla\nabla_{12}z_t = (1 - \theta B)(1 - \Theta B^{12})a_t$

The Q statistics of the residual autocorrelations are very high. The large residual autocorrelations are at lags 2 and 16, with $r_2 = 0.16$ and $r_{16} = -0.15$. These residual autocorrelations at lags 2 and 16 suggest that the residuals may not be purely white noise. Therefore, deleting the AR(1) parameter from Model 1 again does not seem to yield any improvements. On the contrary, the modified model is not able to filter out a large portion of the characteristic behavior of the transformed series.

In comparing Models 1.1 and 1.2, we notice that the seasonal parameter hardly changes in value. The deletions of the AR or MA parameter only resulted in an increase in the absolute value of the other nonseasonal parameter included in the formulation. This could have been predicted if we had analyzed the coefficient correlation matrix in Model 1 in more detail.

The two modifications made on Model 1 do, however, show that neither the AR(1) nor the MA(1) parameter can be deleted without significantly affecting model adequacy. Further modifications could be made to test the adequacy of Model 1, such as adding a second- or sixteenth-order moving average parameter to Model 1.2. But, more than likely, the results from these endeavors would only serve to confirm the adequacy of Model 1. Moreover, the addition of parameters could make a simple model complicated, and thus violate the principle of parsimony which cautions us to try to use the *smallest* possible number of parameters and still represent the model adequately.

At this point, we should be satisfied with the basic Model 1. We will now investigate whether Models 2 and 3 are better formulations than Model 1.

FIGURE 7.5   Model 1.2: Autocorrelation Function of Residuals.

```
          AUTOCORRELATION FUNCTION
                RESIDUALS

     -0.8  -0.4   0.0   0.4   0.8
   * .. *   *   * . *   *   * . *   *   * .. *
         .                ( .R)        1 .
         .                ( . R        2 .
         .                ( R )        3 .
         .                ( R )        4 .
         .                ( . R        5 .
         .                ( R )        6 .
         .                ( R )        7 .
         .                ( R )        8 .
         .                ( R )        9 .
         .                R . )       10 .
         .                (R. )       11 .
         .                ( R )       12 .
         .                (R. )       13 .
         .                ( R )       14 .
         .                (R. )       15 .
         .                R . )       16 .
         .                R . )       17 .
         .                (R. )       18 .
         .                (R. )       19 .
         .                ( R )       20 .
         .                (R. )       21 .
         .                R . )       22 .
         .                (R. )       23 .
         .                ( .R)       24 .
         .                ( .R)       25 .
         .                ( .R)       26 .
         .                ( R )       27 .
         .                ( R )       28 .
         .                ( .R)       29 .
         .                ( R )       30 .
         .                (R. )       31 .
         .                ( R )       32 .
         .                (R. )       33 .
         .                ( R )       34 .
         .                (R. )       35 .
         .                (R. )       36 .
         .                ( R )       37 .
         .                ( R )       38 .
         .                (R. )       39 .
         .                ( R )       40 .
         .                (R. )       41 .
         .                R . )       42 .
         .                ( R )       43 .
         .                (R. )       44 .
         .                ( .R)       45 .
         .                (R. )       46 .
         .                ( R )       47 .
         .                ( R )       48 .
```

7.3.4 Model 2:
$$(1 - \phi B)\nabla\nabla_{12}z_t = (1 - \theta_1 B - \theta_{12}B^{12} - \theta_{13}B^{13})a_t$$

Use the output below and in Figure 7.6 to evaluate Model 2. Suggest improvements.

```
MODEL WITH D = 1 DS = 1 S = 12
MEAN = -0.2869     SD =   66.34      (NOBS = 251)

LEAST SQUARES METHOD
PARAMETER ESTIMATES
====================
              EST       SE      EST/SE     95% CONF LIMITS
AR( 1)      0.267     0.092     2.898      0.086     0.448
MA( 1)     -0.479     0.083    -5.745     -0.643    -0.316
MA(12)      0.703     0.045    15.461      0.614     0.792
MA(13)      0.310     0.078     3.972      0.157     0.462

EST.RES.SD =    4.2007E+01
EST.RES.SD(WITH BACK FORECAST) =      4.2917E+01

R SQR =   .605
DF = 247
NO F-STAT, NO INTERCEPT

CORRELATION MATRIX
          AR( 1)  MA( 1)  MA(12)
  MA( 1)   .751
  MA(12)  -.144   -.129
  MA(13)  -.627   -.811    .379

ROOTS   REAL   IMAGINARY  SIZE
  AR    3.743    0.000    3.743
  MA   -2.270    0.000    2.270
  MA   -0.518    0.890    1.030
  MA   -0.518   -0.890    1.030
  MA    0.893    0.517    1.032
  MA    0.893   -0.517    1.032
  MA    0.514   -0.894    1.032
  MA    0.514    0.894    1.032
  MA   -0.891    0.510    1.027
  MA   -0.891   -0.510    1.027
  MA    1.032    0.000    1.032
  MA   -0.003    1.031    1.031
  MA   -0.003   -1.031    1.031
  MA   -1.023    0.000    1.023
```

AUTOCORRELATIONS OF RESIDUALS

LAGS	ROW SE											
1-12	.06	-.02	.03	.02	-.02	.15	-.04	-.04	.00	.02	-.11	-.06	.03
13-24	.07	-.06	.02	-.05	-.14	-.10	-.04	-.02	-.01	-.02	-.13	-.02	.06
25-36	.07	.06	.09	-.02	-.02	.06	.00	-.04	.03	-.03	.03	-.02	-.06
37-48	.07	-.01	.03	-.07	.02	-.04	-.13	.02	-.02	.08	-.04	-.02	.02

CHI-SQUARE TEST   P-VALUE

Q(12)	=	11.7	8 D.F.	.165
Q(24)	=	27.5	20 D.F.	.122
Q(36)	=	34.5	32 D.F.	.351
Q(48)	=	45.4	44 D.F.	.415

AUTOCORRELATIONS OF FIRST DIFFERENCED RESIDUALS

LAGS	ROW SE											
1-12	.06	-.52	.03	.01	-.11	.18	-.09	-.02	.01	.07	-.09	-.02	.09
13-24	.08	-.09	.07	.01	-.06	-.01	.02	.00	.01	.04	-.10	.01	.04
25-36	.08	-.01	.07	-.05	-.03	.06	-.01	-.05	.06	-.06	.06	-.01	-.04
37-48	.09	.00	.07	-.10	.08	.01	-.12	.09	-.07	.11	-.06	-.02	-.03

CHI-SQUARE TEST   P-VALUE

Q(12)	=	88.1	8 D.F.	.000
Q(24)	=	96.8	20 D.F.	.000
Q(36)	=	105.	32 D.F.	.000
Q(48)	=	124.	44 D.F.	.000

ANALYSIS OF MODEL 2: $(1 - \phi B)\nabla\nabla_{12}z_t = (1 - \theta_1 B - \theta_{12}B^{12} - \theta_{13}B^{13})a_t$

Although the residuals appear to be white noise, and the residual autocorrelation Q statistics assert the adequacy of the model, the correlation matrix suggests that there is a high degree of correlation between the parameter estimates, in particular between θ_1 and θ_{13}. In the identification stage of the series $\nabla\nabla_{12}z_t$ (see Table 7.1), we observed major spikes in the autocorrelations at lags 1, 11, 12, and 13. Theory tells us that both Models 1 and 2 possess these autocorrelation characteristics. However, because there is symmetry about r_{12}, with r_{11} almost identical to r_{13} and opposite in sign of the product of r_1 and r_{12}, we really have found a typical behavior of an autocorrelation function of a model with a first-order consecutive and a first-order seasonal moving average parameter.

Finally, Model 1 is a compact version of Model 2,

$$(1 - \phi B)\nabla\nabla_{12}z_t = (1 - \theta B)(1 - \Theta B^{12})a_t$$
$$= (1 - \theta B - \Theta B^{12} + \theta\Theta B^{13})a_t$$
$$= (1 - \theta_1 B - \theta_{12}B^{12} - \theta_{13}B^{13})a_t$$

FIGURE 7.6   Model 2: Autocorrelation Function of Residuals.

```
   AUTOCORRELATION FUNCTION
        RESIDUALS

-0.8  -0.4   0.0   0.4   0.8
*  .  *  *  .  *  .  *  .  *  .  *  .  *  ..  *
          .                ( R )          1  .
          .                ( R )          2  .
          .                ( R )          3  .
          .                ( R )          4  .
          .                ( . R          5  .
          .               (R. )           6  .
          .               (R. )           7  .
          .                ( R )          8  .
          .                ( R )          9  .
          .               R . )          10  .
          .               (R. )          11  .
          .                ( R )         12  .
          .               (R. )          13  .
          .                ( R )         14  .
          .               (R. )          15  .
          .               R . )          16  .
          .               (R. )          17  .
          .               (R. )          18  .
          .                ( R )         19  .
          .                ( R )         20  .
          .                ( R )         21  .
          .               R . )          22  .
          .                ( R )         23  .
          .                ( .R)         24  .
          .                ( .R)         25  .
          .                ( .R)         26  .
          .                ( R )         27  .
          .                ( R )         28  .
          .                ( .R)         29  .
          .                ( R )         30  .
          .               (R. )          31  .
          .                ( R )         32  .
          .                ( R )         33  .
          .                ( R )         34  .
          .                ( R )         35  .
          .               (R. )          36  .
          .                ( R )         37  .
          .                ( R )         38  .
          .               (R. )          39  .
          .                ( R )         40  .
          .               (R. )          41  .
          .               R . )          42  .
          .                ( R )         43  .
          .                ( R )         44  .
          .                ( .R)         45  .
          .               (R. )          46  .
          .                ( R )         47  .
          .                ( R )         48  .
```

TABLE 7.3   Parameter Estimates[a] for Models 1 and 2

	MODEL 1		MODEL 2
ϕ	0.261 (0.092)	ϕ	0.267 (0.092)
θ	−0.489 (0.082)	θ_1	−0.479 (0.083)
Θ	0.684 (0.047)	θ_{12}	0.703 (0.145)
		θ_{13}	0.310 (0.078)

[a] Numbers in parentheses are the large-sample standard errors. The implied θ_{13} parameter from Model 1 is $\theta_{13} = -\theta\Theta = 0.334$.

where $\theta_1 = \theta$, $\theta_{12} = \Theta$, and $\theta_{13} = -\theta\Theta$. Also, since the parameter values as shown in Table 7.3 are very similar for both models, rather than deleting parameters from Model 2 to remove the correlation between the moving average parameters, we will accept Model 1 as the best model which fits the series $\nabla\nabla_{12}z_t$.

For the Model 2 output we have also given the analysis of the roots of the polynomials. Before continuing, we must indicate that the root analysis may be difficult for some readers but can be omitted without affecting the further analysis in this Chapter.

As you will recall (see Section 5.3.1), the size of the roots of a stationary autoregressive polynomial must be larger than one. Similarly, the size of the roots of a moving average polynomial must also be larger than one for it to be invertible. For each of the models analyzed so far, the size of each root was larger than one. For the Model 2 we find that the AR(1) polynomial has a root with size equal to 3.743, certainly larger than one.[4] The interesting point shows up with the 13 roots of the moving average polynomials. One root has a size of 2.270, again larger than one. The remaining twelve roots are almost all equal in size.

Indeed, it is instructive to see what twelve roots of (approximately) equal size implies about a simpler model. Remember that for a first-order polynomial $(1 - cB)$, the root and size equal $1/c$. The second-order polynomial $(1 - cB^2)$ has two roots with size equal to $1/\sqrt{c}$. Therefore, the coefficient c can be calculated by $(1/\text{size})^2$. Similarly, if there are 12 roots with equal size, the coefficient in the polynomial can be recovered as $(1/\text{size})^{12}$. In our Model

[4] In principle we could generate tests for these roots as these are also estimates of underlying population roots. For a reference see Press (1982).

2, the size is approximately 1.03 and therefore the implied value of Φ in $(1 - \Phi B^{12})$ equals $\Phi = (1/1.03)^{12} = 0.701$. Notice that in Model 1, the estimate of $\Phi = 0.684$. All this indicates that a model which has a number of roots, equal to the span of the series, all of equal size, can be simplified by introducing a seasonal parameter.

7.3.5 Model 3:
$$\nabla^2 \nabla_{12} z_t = (1 - \Theta B^{12}) a_t$$

Use the output below and in Figure 7.7 to evaluate Model 3. Suggest improvements.

```
MODEL WITH D = 2 DS = 1 S = 12
MEAN =  0.4640      SD =    57.75     (NOBS = 250)

LEAST SQUARES METHOD
PARAMETER ESTIMATES
====================
              EST       SE      EST/SE    95% CONF LIMITS
SMA( 1)     0.649     0.049    13.241     0.553     0.745

EST.RES.SD =    4.9168E+01
EST.RES.SD(WITH BACK FORECAST) =      4.9923E+01

R SQR =   .278
DF = 249
NO F-STAT, NO INTERCEPT

AUTOCORRELATIONS OF RESIDUALS
  LAGS ROW SE
  1-12   .06  -.03 -.33 -.13 -.05  .16 -.04 -.10  .04  .05 -.08 -.04  .03
 13-24   .07   .02  .08  .02 -.12 -.07  .03  .05  .07 -.01 -.14 -.05  .06
 25-36   .08   .11  .09 -.06 -.06  .05 -.01 -.06  .01  .00  .07 -.01 -.09
 37-48   .08   .02  .06 -.03  .05 -.03 -.13  .02  .04  .09 -.04 -.10  .01

CHI-SQUARE TEST              P-VALUE
Q(12) =   45.3      11 D.F.   .000
Q(24) =   61.3      23 D.F.   .000
Q(36) =   74.7      35 D.F.   .000
Q(48) =   89.0      47 D.F.   .000

AUTOCORRELATIONS OF FIRST DIFFERENCED RESIDUALS
  LAGS ROW SE
  1-12   .06  -.35 -.24  .05 -.06  .19 -.06 -.10  .06  .07 -.09  .00  .04
 13-24   .08  -.04  .05  .04 -.09 -.03  .03  .00  .05  .03 -.11 -.01  .03
 25-36   .08   .02  .07 -.07 -.05  .08  .00 -.05  .03 -.05  .08  .01 -.10
 37-48   .08   .04  .07 -.10  .09  .01 -.13  .06 -.01  .09 -.03 -.10 -.01

CHI-SQUARE TEST              P-VALUE
Q(12) =   63.4      11 D.F.   .000
Q(24) =   72.4      23 D.F.   .000
Q(36) =   84.4      35 D.F.   .000
Q(48) =   104.      47 D.F.   .000
```

FIGURE 7.7   Model 3: Autocorrelation Function of Residuals.

```
     AUTOCORRELATION FUNCTION
             RESIDUALS

  -0.8  -0.4   0.0    0.4    0.8
  *   *   *   *   *   *   *   *   *   *   *   *
  . . . . . . . . . . . . . . . . . . .
  .                (R.  )           1 .
  .           R  ( . )              2 .
  .              R . )              3 .
  .              (R.  )             4 .
  .              ( .   R            5 .
  .              (R.  )             6 .
  .              (R.  )             7 .
  .              ( .R)              8 .
  .              ( .R)              9 .
  .              (R.  )            10 .
  .              (R.  )            11 .
  .              ( .R)             12 .
  .              ( R )             13 .
  .              ( .R)             14 .
  .              ( R )             15 .
  .              R . )             16 .
  .              (R.  )            17 .
  .              ( R )             18 .
  .              ( .R)             19 .
  .              ( .R)             20 .
  .              ( R )             21 .
  .              R . )             22 .
  .              (R.  )            23 .
  .              ( .R)             24 .
  .              ( .   R           25 .
  .              ( .R)             26 .
  .              (R.  )            27 .
  .              (R.  )            28 .
  .              ( .R)             29 .
  .              ( R )             30 .
  .              (R.  )            31 .
  .              ( R )             32 .
  .              ( R )             33 .
  .              ( .R)             34 .
  .              ( R )             35 .
  .              (R.  )            36 .
  .              ( R )             37 .
  .              ( .R)             38 .
  .              ( R )             39 .
  .              ( .R)             40 .
  .              ( R )             41 .
  .              R . )             42 .
  .              ( R )             43 .
  .              ( .R)             44 .
  .              ( .R)             45 .
  .              (R.  )            46 .
  .              (R.  )            47 .
  .              ( R )             48 .
```

ANALYSIS OF MODEL 3: $\nabla^2\nabla_{12}z_t = (1 - \Theta B^{12})a_t$

The high Q statistic for the residual autocorrelations and corresponding zero P-values indicate that Model 3 may be inadequate. The large spike at lag 2 in the residual autocorrelations, $r_2 = -0.33$, suggests that perhaps a moving average process of order 2 still needs to be filtered out. That is, the calculated residuals e_t can be represented as

$$e_t = (1 - \theta_2 B^2)a_t,$$

where a_t is white noise. If we rewrite Model 3 as

$$\nabla^2\nabla_{12}z_t = (1 - \Theta B^{12})e_t,$$

and substitute the above equation for e_t in this formulation, then we obtain Model 3.1,

$$\nabla^2\nabla_{12}z_t = (1 - \Theta B^{12})(1 - \theta_2 B^2)a_t. \qquad \text{(Model 3.1)}$$

7.3.6 Model 3.1:
$$\nabla^2\nabla_{12}z_t = (1 - \theta_2 B^2)(1 - \Theta B^{12})a_t$$

Use the output below and in Figure 7.8 to evaluate Model 3.1. Suggest improvements.

```
MODEL WITH D = 2 DS = 1 S = 12
MEAN =  0.4640     SD =    57.75     (NOBS = 250)

LEAST SQUARES METHOD
PARAMETER ESTIMATES
===================
              EST       SE      EST/SE    95% CONF LIMITS
  MA( 2)     0.495     0.055     8.965     0.387     0.603
  SMA( 1)    0.666     0.048    13.935     0.573     0.760

EST.RES.SD =     4.4833E+01
EST.RES.SD(WITH BACK FORECAST)  =     4.5805E+01

R SQR =   .402
DF = 248
NO F-STAT, NO INTERCEPT

CORRELATION MATRIX
          MA( 2)
  SMA( 1)    .094
```

```
AUTOCORRELATIONS OF RESIDUALS
  LAGS ROW SE
  1-12   .06  -.16   .07 -.16 -.04   .06 -.06 -.05 -.02   .01 -.09 -.03   .03
 13-24   .07   .00   .04 -.01 -.10 -.07 -.01   .02   .02   .00 -.11 -.02   .06
 25-36   .07   .09   .09 -.01 -.01   .03   .01 -.05   .04 -.05   .06 -.04 -.04
 37-48   .07  -.01   .05 -.04   .02 -.03 -.14   .05 -.04   .11 -.07   .00 -.03

CHI-SQUARE TEST              P-VALUE
Q(12) =  20.4      10 D.F.   .026
Q(24) =  29.1      22 D.F.   .142
Q(36) =  37.7      34 D.F.   .306
Q(48) =  51.4      46 D.F.   .271

AUTOCORRELATIONS OF FIRST DIFFERENCED RESIDUALS
  LAGS ROW SE
  1-12   .06  -.59   .20 -.14   .00   .10 -.05 -.01   .00   .05 -.07   .01   .03
 13-24   .09  -.02   .03   .03 -.06 -.01   .01   .01   .01   .04 -.09   .01   .02
 25-36   .09   .00   .05 -.04 -.01   .02   .02 -.07   .08 -.09   .08 -.04 -.01
 37-48   .09  -.01   .06 -.07   .05   .03 -.13   .12 -.10   .14 -.10   .03 -.08

CHI-SQUARE TEST              P-VALUE
Q(12) = 110.       10 D.F.   .000
Q(24) = 114.       22 D.F.   .000
Q(36) = 123.       34 D.F.   .000
Q(48) = 151.       46 D.F.   .000
```

ANALYSIS OF MODEL 3.1: $\nabla^2\nabla_{12}z_t = (1 - \theta_2B^2)(1 - \Theta B^{12})a_t$

The Q statistics for the residual autocorrelations are lower, and the corresponding P-values have improved but are still low, as compared to Model 3, suggesting that this model is still inadequate. The small spikes at lags 1 and 3 in the residual autocorrelations, both with values -0.16, suggest that perhaps more moving average parameters should be added. However, that would violate the principle of parsimony which, as stated earlier, tells us to use the *smallest possible* number of parameters to adequately represent the model. Thus, we can conclude that Model 1,

$$(1 - \phi B)\nabla\nabla_{12}z_t = (1 - \theta B)(1 - \Theta B^{12})a_t \qquad \text{(Model 1)}$$

is a better model for the residential construction series. The diagnostic checks which follow will only be made for Model 1.

FIGURE 7.8   Model 3.1: Autocorrelation Function of Residuals.

```
          AUTOCORRELATION FUNCTION
               RESIDUALS

     -0.8  -0.4   0.0   0.4   0.8
     *  .  *  *  *  .  *  *  *  .  *  *  *  .  *
         .                R  .  )           1  .
         .               (  .R)             2  .
         .                R  .  )           3  .
         .               (R.  )             4  .
         .               (  .R)             5  .
         .               (R.  )             6  .
         .               (R.  )             7  .
         .               (  R  )            8  .
         .               (  R  )            9  .
         .               (R.  )            10  .
         .               (  R  )           11  .
         .               (  R  )           12  .
         .               (  R  )           13  .
         .               (  .R)            14  .
         .               (  R  )           15  .
         .                R  .  )          16  .
         .               (R.  )            17  .
         .               (  R  )           18  .
         .               (  R  )           19  .
         .               (  R  )           20  .
         .               (  R  )           21  .
         .                R  .  )          22  .
         .               (  R  )           23  .
         .               (  .R)            24  .
         .               (  .R)            25  .
         .               (  .R)            26  .
         .               (  R  )           27  .
         .               (  R  )           28  .
         .               (  R  )           29  .
         .               (  R  )           30  .
         .               (R.  )            31  .
         .               (  .R)            32  .
         .               (R.  )            33  .
         .               (  .R)            34  .
         .               (R.  )            35  .
         .               (R.  )            36  .
         .               (  R  )           37  .
         .               (  .R)            38  .
         .               (R.  )            39  .
         .               (  R  )           40  .
         .               (  R  )           41  .
         .                R  .  )          42  .
         .               (  .R)            43  .
         .               (R.  )            44  .
         .               (  .  R            45  .
         .               (R.  )            46  .
         .               (  R  )           47  .
         .               (  R  )           48  .
```

7.3.7 Summary

Table 7.4 contains the models evaluated together with a brief characterization of them. You should study this table carefully. If the logic in any of the models confuses you, go back to the appropriate section and read it one more time.

TABLE 7.4 **Summary of All Estimated Models**

MODEL	REPRESENTATION	CHARACTERIZATION
1	$(1 - \phi B)\nabla\nabla_{12}z_t = (1 - \theta B)(1 - \Theta B^{12})a_t$	Residuals appear to be white noise; Q statistic adequate; possible correlation between AR(1) and MA(1)
1.1	$(1 - \phi B)\nabla\nabla_{12}z_t = (1 - \Theta B^{12})a_t$	Large spike at lag 1 in residual autocorrelations; high Q statistic; underspecification
1.2	$\nabla\nabla_{12}z_t = (1 - \theta B)(1 - \Theta B^{12})a_t$	Spikes at lags 2 and 16 in residual autocorrelations; high Q statistic; underspecification
2	$(1 - \phi B)\nabla\nabla_{12}z_t = (1 - \theta_1 B - \theta_{12}B^{12} - \theta_{13}B^{13})a_t$	High degree of correlation between the estimates of the MA parameters; overparameterization
3	$\nabla^2\nabla_{12}z_t = (1 - \Theta B^{12})a_t$	Major spike at lag 2 in residual autocorrelation; high Q statistic
3.1	$\nabla^2\nabla_{12}z_t = (1 - \theta_2 B^2)(1 - \Theta B^{12})a_t$	Spikes at lags 1 and 2 in residual autocorrelation; high Q statistic

7.4 DIAGNOSTIC CHECKS FOR MODEL 1:
$(1 - \phi B)\nabla\nabla_{12}z_t = (1 - \theta B)(1 - \Theta B^{12})a_t$

Three principal diagnostic checks will be applied to Model 1 to test its adequacy:

1. over- and underfitting of parameters;
2. analyses of residuals; and
3. evaluation of the cumulative periodogram of the series.

7.4.1 Overfitting and Underfitting

In Section 7.3 we looked at two underparameterized formulations of Model 1, namely, Models 1.1 and 1.2, and one overparameterized version, Model 2. In analyzing those models we concluded that the underparameterized models could not sufficiently filter out all the characteristic behavior of the transformed series, and that, because of the high degree of correlation between the estimates of the moving average parameters of Model 2, this formulation was also not an improvement over the formulation of Model 1. These conclusions serve to corroborate our presumption that Model 1 is the most adequate model we can find to fit the construction series.

We however suggest further evaluating Model 1 by examining two other overparameterized models:

$$(1 - \phi_1 B - \phi_2 B^2)\bigtriangledown\bigtriangledown_{12}z_t = (1 - \theta B)(1 - \Theta B^{12})a_t, \quad \text{(Model 1.3)}$$
$$(1 - \phi B)\bigtriangledown\bigtriangledown_{12}z_t = (1 - \theta_1 B - \theta_2 B^2)(1 - \Theta B^{12})a_t. \quad \text{(Model 1.4)}$$

In Model 1.3 (Figure 7.9) we added a second-order autoregressive parameter and in Model 1.4 (Figure 7.10) a second-order moving average parameter.

Using the estimation results given in Table 7.5 and contrasting them with the results of Model 1 (see Section 7.3.1), we see that in Model 1.3:

1. the AR(1) parameter has a much smaller point estimate and a confidence interval that includes 0;
2. the AR(2) parameter has a point estimate of 0.181, but its confidence interval also includes 0;
3. the AR(1) and AR(2) parameter estimates are highly correlated, an indication that the AR(2) parameter may be redundant; and
4. the Q statistics and the resulting P-values of the residual autocorrelations are very similar for Model 1 and Model 1.3,

For Model 1.4 we observe (see Table 7.6) that:

1. the MA parameter estimates are both insignificant, and are highly correlated not only with each other, but also with the AR(1) parameter estimate; and
2. the Q statistics and the corresponding P-values of the residual autocorrelations are again very similar for Model 1 and Model 1.4.

These results provide further evidence that the Model 1 selected to fit the residual construction data is quite appropriate.

TABLE 7.5   Model 1.3

```
MODEL WITH D = 1 DS = 1 S = 12
MEAN = -0.2869    SD =   66.34    (NOBS = 251)

LEAST SQUARES METHOD
PARAMETER ESTIMATES
===================
            EST      SE      EST/SE     95% CONF LIMITS
 AR( 1)    0.021    0.135    0.159     -0.242    0.285
 AR( 2)    0.181    0.108    1.668     -0.032    0.393
 MA( 1)   -0.720    0.112   -6.408     -0.940   -0.500
 SMA( 1)   0.682    0.047   14.438      0.589    0.774


EST.RES.SD =   4.1943E+01
EST.RES.SD(WITH BACK FORECAST) =    4.2846E+01

R SQR =  .607
DF = 247
NO F-STAT, NO INTERCEPT

CORRELATION MATRIX
          AR( 1)  AR( 2)  MA( 1)
 AR( 2)   -.729
 MA( 1)    .888   -.817
 SMA( 1)   .007   -.101    .037

AUTOCORRELATIONS OF RESIDUALS
  LAGS ROW SE
  1-12   .06    .00   .02 -.07   .02   .12 -.02 -.05   .00   .01 -.10 -.06   .02
 13-24   .07   -.02   .01 -.04  -.14  -.10 -.03 -.01   .00  -.03 -.12 -.04   .06
 25-36   .07    .09   .08 -.01  -.02   .04  .00 -.04   .03  -.03  .04 -.04  -.05
 37-48   .07   -.01   .03 -.05   .01  -.04 -.14  .02  -.02   .09 -.05 -.02   .00

CHI-SQUARE TEST          P-VALUE
Q(12) =  9.65     8 D.F.   .290
Q(24) = 24.2     20 D.F.   .233
Q(36) = 31.9     32 D.F.   .470
Q(48) = 42.9     44 D.F.   .519

AUTOCORRELATIONS OF FIRST DIFFERENCED RESIDUALS
  LAGS ROW SE
  1-12   .06   -.51   .05 -.09  -.01   .12 -.05 -.04   .03   .06 -.08 -.01   .05
 13-24   .08   -.03   .04  .02  -.07  -.01  .02  .01   .02   .03 -.09  .00   .03
 25-36   .08    .02   .05 -.05  -.03   .05  .01 -.06   .07  -.07  .08 -.03  -.03
 37-48   .08    .00   .06 -.07   .05   .02 -.12  .10  -.07   .13 -.09  .00  -.05

CHI-SQUARE TEST          P-VALUE
Q(12) = 76.0      8 D.F.   .000
Q(24) = 81.3     20 D.F.   .000
Q(36) = 89.5     32 D.F.   .000
Q(48) = 110.     44 D.F.   .000
```

FIGURE 7.9   Model 1.3: Autocorrelation Function of Residuals.

```
              AUTOCORRELATION FUNCTION
                    RESIDUALS

        -0.8  -0.4   0.0    0.4    0.8
        *  .  .  *  .  .  *  .  *  .  *  ..  *  .  *  .  *  ..  *
                           ( R )          1  .
                           ( R )          2  .
                           (R. )          3  .
                           ( R )          4  .
                           ( . R          5
                           ( R )          6  .
                           (R. )          7  .
                           ( R )          8  .
                           ( R )          9  .
                           R . )         10  .
                           (R. )         11  .
                           ( R )         12  .
                           ( R )         13  .
                           ( R )         14  .
                           (R. )         15  .
                           R . )         16  .
                           R . )         17  .
                           (R. )         18  .
                           ( R )         19  .
                           ( R )         20  .
                           ( R )         21  .
                           R . )         22  .
                           (R. )         23  .
                           ( .R)         24  .
                           ( .R)         25  .
                           ( .R)         26  .
                           ( R )         27  .
                           ( R )         28  .
                           ( .R)         29  .
                           ( R )         30  .
                           (R. )         31  .
                           ( R )         32  .
                           ( R )         33  .
                           ( .R)         34  .
                           (R. )         35  .
                           (R. )         36  .
                           ( R )         37  .
                           ( R )         38  .
                           (R. )         39  .
                           ( R )         40  .
                           (R. )         41  .
                           R . )         42  .
                           ( R )         43  .
                           ( R )         44  .
                           ( .R)         45  .
                           (R. )         46  .
                           ( R )         47  .
                           ( R )         48  .
```

TABLE 7.6   Model 1.4

```
MODEL WITH D = 1 DS = 1 S = 12
MEAN = -0.2869    SD =   66.34    (NOBS = 251)

LEAST SQUARES METHOD
PARAMETER ESTIMATES
===================
           EST       SE      EST/SE     95% CONF LIMITS
AR( 1)    0.493    0.309     1.599     -0.112    1.098
MA( 1)   -0.254    0.323    -0.785     -0.887    0.380
MA( 2)    0.176    0.230     0.767     -0.274    0.627
SMA( 1)   0.682    0.047    14.489      0.590    0.774


EST.RES.SD =   4.2006E+01
EST.RES.SD(WITH BACK FORECAST) =   4.2907E+01


R SQR =  .605
DF = 247
NO F-STAT, NO INTERCEPT

CORRELATION MATRIX
         AR( 1)  MA( 1)  MA( 2)
  MA( 1)   .981
  MA( 2)   .962    .961
  SMA( 1) -.023   -.011    .012

AUTOCORRELATIONS OF RESIDUALS
  LAGS ROW SE
  1-12   .06  -.01  .03 -.03 -.02  .14 -.04 -.04  .00  .01 -.11 -.06  .02
  13-24  .07  -.03  .01 -.04 -.14 -.10 -.04 -.02  .00 -.03 -.12 -.03  .06
  25-36  .07   .08  .08 -.01 -.02  .04  .00 -.04  .03 -.04  .04 -.03 -.05
  37-48  .07  -.01  .03 -.06  .01 -.04 -.13  .02 -.02  .08 -.04 -.02  .01

CHI-SQUARE TEST          P-VALUE
Q(12) =  10.3      8 D.F.   .245
Q(24) =  24.6     20 D.F.   .219
Q(36) =  32.0     32 D.F.   .464
Q(48) =  42.5     44 D.F.   .536

AUTOCORRELATIONS OF FIRST DIFFERENCED RESIDUALS
  LAGS ROW SE
  1-12   .06  -.52  .05 -.04 -.07  .16 -.09 -.02  .01  .06 -.08 -.01  .06
  13-24  .08  -.04  .04  .02 -.07 -.01  .02  .00  .02  .03 -.09  .00  .03
  25-36  .08   .01  .05 -.04 -.04  .05  .00 -.06  .07 -.07  .07 -.02 -.03
  37-48  .08   .00  .06 -.08  .06  .01 -.12  .10 -.07  .11 -.07 -.01 -.04

CHI-SQUARE TEST          P-VALUE
Q(12) =  83.4      8 D.F.   .000
Q(24) =  88.9     20 D.F.   .000
Q(36) =  97.0     32 D.F.   .000
Q(48) =  116.     44 D.F.   .000
```

FIGURE 7.10 Model 1.4: Autocorrelation Function of Residuals.

```
          AUTOCORRELATION FUNCTION
               RESIDUALS

   -0.8   -0.4    0.0    0.4    0.8
   *  .  .  .  *  .  .  *  .  .  *  .  .  *  .  .  *
        .                ( R )              1 .
        .                ( R )              2 .
        .                ( R )              3 .
        .                ( R )              4 .
        .                ( . R              5 .
        .               (R. )               6 .
        .               (R. )               7 .
        .                ( R )              8 .
        .                ( R )              9 .
        .               R . )              10 .
        .               (R. )              11 .
        .                ( R )             12 .
        .                ( R )             13 .
        .                ( R )             14 .
        .               (R. )              15 .
        .               R . )              16 .
        .               (R. )              17 .
        .               (R. )              18 .
        .                ( R )             19 .
        .                ( R )             20 .
        .                ( R )             21 .
        .               R . )              22 .
        .                ( R )             23 .
        .                ( .R)             24 .
        .                ( .R)             25 .
        .                ( .R)             26 .
        .                ( R )             27 .
        .                ( R )             28 .
        .                ( .R)             29 .
        .                ( R )             30 .
        .               (R. )              31 .
        .                ( R )             32 .
        .               (R. )              33 .
        .                ( .R)             34 .
        .               (R. )              35 .
        .               (R. )              36 .
        .                ( R )             37 .
        .                ( R )             38 .
        .               (R. )              39 .
        .                ( R )             40 .
        .               (R. )              41 .
        .               R . )              42 .
        .                ( R )             43 .
        .                ( R )             44 .
        .                ( .R)             45 .
        .               (R. )              46 .
        .                ( R )             47 .
        .                ( R )             48 .
```

7.4.2 Residual Analysis

Autocorrelation checks If a model is adequately specified, then its estimated residuals should be white noise. In Section 7.3.1 and Figure 7.3 we saw that the estimated autocorrelations of the residuals of Model 1 fell well within the 95% large-sample confidence limits. This is a good indication that the residuals are white noise.

Chi-square test The results of the "portmanteau" lack-of-fit test applied to the first 48 autocorrelations of the residuals of Model 1 show that the Q statistic, which is approximately χ^2 distributed, has a value of 43.6. With 45 degrees of freedom, the P-value is 0.530. A P-value of 0.530 means that there is a 53 percent chance that the residuals come from a white noise series. The Q statistic based on the first 24 autocorrelations is 26.0 with 21 degrees of freedom and a P-value of only 0.207. The value obtained here does not substantiate the hypothesis of model adequacy nor does it show that the residuals are not white noise. Therefore, we should rely on results of other tests for more definite conclusions, such as those presented in Section 7.4.1 and those discussed below.

First differences Next, as discussed in Chapter 5, we should also evaluate if the first differences of the residuals are governed by an MA(1) model with moving average parameter $\theta = 1$. Such a model implies a $\rho_1 = -0.5$. The large spike at lag 1, $r_1 = -0.52$, in the autocorrelations of the first differenced residuals of Model 1 (see Figure 7.11) shows that the underlying process governing this series is very likely an MA(1) process with $\theta = 1$.

FIGURE 7.11   Model 1: Autocorrelation Function of First Differenced Residuals.

```
               AUTOCORRELATION FUNCTION
               FIRST DIFFERENCED RESIDUALS

         -0.8  -0.4   0.0    0.4    0.8
          * . . * . * . * . * . * . * . * . * . . * .
     1  .           R      ( .  )
     2  .                  ( .R)
     3  .                  ( R  )
     4  .                  R .  )
     5  .                  ( .  )R
     6  .                  (R.  )
     7  .                  ( R  )
     8  .                  ( R  )
     9  .                  ( .R)
    10  .                  (R.  )
    11  .                  ( R  )
    12  .                  ( .R)
    13  .                  (R.  )
    14  .                  ( .R)
    15  .                  ( R  )
    16  .                  (R.  )
    17  .                  ( R  )
    18  .                  ( R  )
    19  .                  ( R  )
    20  .                  ( R  )
    21  .                  ( .R)
    22  .                  (R.  )
    23  .                  ( R  )
    24  .                  ( R  )
    25  .                  ( R  )
    26  .                  ( .R)
    27  .                  (R.  )
    28  .                  (R.  )
    29  .                  ( .R)
    30  .                  ( R  )
    31  .                  (R.  )
    32  .                  ( .R)
    33  .                  (R.  )
    34  .                  ( .R)
    35  .                  ( R  )
    36  .                  (R.  )
    37  .                  ( R  )
    38  .                  ( .R)
    39  .                  (R.  )
    40  .                  ( .R)
    41  .                  (  R  )
    42  .                  (R .  )
    43  .                  (  .R )
    44  .                  ( R.  )
    45  .                  (  .R )
    46  .                  ( R.  )
    47  .                  (  R  )
    48  .                  ( R.  )
```

7.4.3 Cumulative Periodogram

As discussed in Chapter 5, we can detect periodic nonrandomness by calculating the normalized cumulative periodogram of the residuals. For a white noise series, the plot of the normalized cumulative periodogram against the frequency should be scattered about a straight line joining the points (0, 0) and (0.5, 1).

Deviations from the straight line can be statistically assessed by the use of the Kolmogorov–Smirnov confidence limits. If the residual series are white noise then the points on the cumulative periodogram should deviate from the straight line and cross the Kolmogorov–Smirnov confidence limits only with the stated probability. Figure 7.12 contains the plot of the normalized cumulative periodogram for the residuals of Model 1. There is not a single point that exceeds the approximate 90% Kolmogorov–Smirnov confidence limits, thus indicating again that Model 1 is adequate.

FIGURE 7.12   Cumulative Periodogram of the Residuals of Model 1.

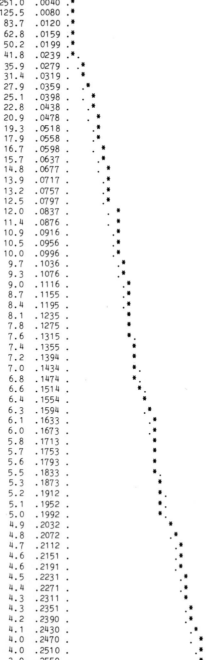

```
NORMALIZED CUMULATIVE PERIODOGRAM (CP) OF RESIDUALS
EXPECTED CP PLOTTED WITH (.); ACTUAL CP PLOTTED WITH (*)
IF ACTUAL EXCEEDS 90% KOLMOGOROV-SMIRNOV LIMITS, PLOTTED WITH (+)

PERIOD FREQ.     .1   .2   .3   .4   .5   .6   .7   .8   .9  1.0
                 ....*....*....*....*....*....*....*....*....*....*.
251.0  .0040 .*
125.5  .0080 .*
 83.7  .0120 .*
 62.8  .0159 .*
 50.2  .0199 .*
 41.8  .0239 .*.
 35.9  .0279 . .*
 31.4  .0319 .  *
 27.9  .0359 . .*
 25.1  .0398 .  . *
 22.8  .0438 .  .*
 20.9  .0478 .  . *
 19.3  .0518 .  .*
 17.9  .0558 .  .*
 16.7  .0598 .  . *
 15.7  .0637 .  .*
 14.8  .0677 .  . *
 13.9  .0717 .  .*
 13.2  .0757 .  .*
 12.5  .0797 .  .*
 12.0  .0837 .  . *
 11.4  .0876 .  . *
 10.9  .0916 .  . *
 10.5  .0956 .  .*
 10.0  .0996 .  .*
  9.7  .1036 .   .*
  9.3  .1076 .   .*
  9.0  .1116 .   .*
  8.7  .1155 .   .*
  8.4  .1195 .   .*
  8.1  .1235 .   *
  7.8  .1275 .   *
  7.6  .1315 .   *.
  7.4  .1355 .   *
  7.2  .1394 .    *
  7.0  .1434 .    *.
  6.8  .1474 .    *.
  6.6  .1514 .    *.
  6.4  .1554 .    *
  6.3  .1594 .    .*
  6.1  .1633 .     .*
  6.0  .1673 .     .*
  5.8  .1713 .     *
  5.7  .1753 .     *
  5.6  .1793 .     *
  5.5  .1833 .     *.
  5.3  .1873 .     *
  5.2  .1912 .     *.
  5.1  .1952 .     *.
  5.0  .1992 .     *.
  4.9  .2032 .      *
  4.8  .2072 .      .*
  4.7  .2112 .      .*
  4.6  .2151 .      .*
  4.6  .2191 .      .*
  4.5  .2231 .       .*
  4.4  .2271 .       .*
  4.3  .2311 .       *
  4.3  .2351 .       .*
  4.2  .2390 .       .*
  4.1  .2430 .        .*
  4.0  .2470 .        . *
  4.0  .2510 .        .*
  3.9  .2550 .        .*
```

```
3.9  .2590 .                                              . *
3.8  .2629 .                                               *
3.7  .2669 .                                               *
3.7  .2709 .                                               .*
3.6  .2749 .                                               .*
3.6  .2789 .                                               .*
3.5  .2829 .                                               .*
3.5  .2869 .                                               *
3.4  .2908 .                                              *
3.4  .2948 .                                              *
3.3  .2988 .                                              *
3.3  .3028 .                                              *.
3.3  .3068 .                                              *.
3.2  .3108 .                                              * .
3.2  .3147 .                                              * .
3.1  .3187 .                                              * .
3.1  .3227 .                                               * .
3.1  .3267 .                                               * .
3.0  .3307 .                                                *  .
3.0  .3347 .                                                *  .
3.0  .3386 .                                                 * .
2.9  .3426 .                                                 *  .
2.9  .3466 .                                                  *  .
2.9  .3506 .                                                   * .
2.8  .3546 .                                                   * .
2.8  .3586 .                                                   * .
2.8  .3625 .                                                    *.
2.7  .3665 .                                                    .*
2.7  .3705 .                                                     *
2.7  .3745 .                                                     .*
2.6  .3785 .                                                      *
2.6  .3825 .                                                      .
2.6  .3865 .                                                      .*
2.6  .3904 .                                                       *
2.5  .3944 .                                                       *
2.5  .3984 .                                                       *
2.5  .4024 .                                                        *
2.5  .4064 .                                                        *
2.4  .4104 .                                                         *
2.4  .4143 .                                                         *
2.4  .4183 .                                                          *
2.4  .4223 .                                                          .*
2.3  .4263 .                                                           .*
2.3  .4303 .                                                           *
2.3  .4343 .                                                           .*
2.3  .4382 .                                                            *
2.3  .4422 .                                                           *
2.2  .4462 .                                                            .
2.2  .4502 .                                                            *.
2.2  .4542 .                                                            * .
2.2  .4582 .                                                           *  .
2.2  .4622 .                                                          *   .
2.1  .4661 .                                                          .
2.1  .4701 .                                                         *
2.1  .4741 .                                                         .*
2.1  .4781 .                                                         *
2.1  .4821 .                                                         *
2.1  .4861 .                                                        .*
2.0  .4900 .                                                        .*
2.0  .4940 .                                                         *.
2.0  .4980 .                                                         *.
           ...........................................................
```

7.5 FORECASTING

The forecasts of the model $(1 - \phi B)\nabla\nabla_{12} z_t = (1 - \theta B)(1 - \Theta B^{12})a_t$ are excellent. The percent deviation of the one step ahead forecasts from the actual figures are in general very small. Tables 7.7 and 7.8 contain the one step ahead forecasts and 95% confidence limits. Figures 7.13 and 7.14 contain the plots of these forecasts. All of the one step ahead forecasts in Table 7.7 were generated using estimates based on the first 264 observations (January 1947–December 1969), while the forecasts in Table 7.8 were made by updating the parameter estimates each time a new observation became available. For example, after the forecasts were obtained using 264 data points, the parameters of the same model specification were re-estimated using 265 data points, and with this new set of estimates, new forecasts were generated.

TABLE 7.7 One Step Ahead Forecast for Model 1
$(1 - 0.261B)\nabla\nabla_{12} z_t = (1 + 0.489B)(1 - 0.684B^{12})a_t$

	MILLIONS OF DOLLARS				
			95% CONF. LIMIT		DEVIATION FROM
MONTH	FORECAST	ACTUAL	LOWER	UPPER	ACTUAL (%)
Jan 1969	2170	2133	2088	2253	−1.75
Feb	1948	1940	1866	2030	−0.42
Mar	2119	2195	2037	2202	+3.44
Apr	2480	2540	2398	2563	+2.35
May	2770	2810	2687	2852	+1.44
Jun	3009	2962	2927	3091	−1.59
Jul	3073	2974	2990	3155	−3.32*
Aug	2998	2880	2916	3080	−4.10*
Sep	2791	2763	2709	2873	−1.01
Oct	2641	2648	2558	2723	+0.28
Nov	2558	2482	2476	2640	−3.06
Dec	2309	2288	2226	2391	−0.91
Jan 1970	1949	1961	1867	2032	+0.59
Feb	1766	1765	1684	1849	−0.08
Mar	2019	1986	1937	2101	−1.68
Apr	2322	2297	2240	2404	−1.09
May	2558	2485	2476	2640	−2.94
Jun	2616	2592	2533	2698	−0.91
Jul	2592	2650	2510	2674	+2.18
Aug	2568	2707	2486	2650	+5.13*
Sep	2629	2721	2547	2712	+3.36*
Oct	2640	2747	2558	2723	+3.89*
Nov	2618	2735	2536	2700	+4.28*
Dec	2581	2627	2499	2664	+1.74

* Actual data outside the 95% approximate large-sample confidence limit.

TABLE 7.8 One Step Ahead Forecast with Updating[a] for Model 1

$$(1 - \phi B)\nabla\nabla_{12}z_t = (1 - \theta B)(1 - \Theta B^{12})a_t$$

MILLIONS OF DOLLARS

MONTH	ϕ	95% CONF. LIMIT LOWER	UPPER	θ	95% CONF. LIMIT LOWER	UPPER	Θ	95% CONF. LIMIT LOWER	UPPER	FORE-CAST	ACTUAL	95% CONF. LIMIT LOWER	UPPER	% DEVIA-TION FROM ACTUAL
Jan 1969	0.261	0.082	0.441	−0.489	−0.650	−0.328	0.684	0.592	0.776	2170	2133	2088	2253	−1.75
Feb	0.257	0.077	0.436	−0.494	−0.653	−0.334	0.682	0.589	0.774	1930	1940	1848	2012	+0.52
Mar	0.256	0.077	0.435	−0.493	−0.653	−0.334	0.681	0.589	0.773	2125	2195	2043	2207	+3.21
Apr	0.250	0.071	0.429	−0.502	−0.660	−0.343	0.678	0.585	0.771	2516	2540	2433	2598	+0.96
May	0.254	0.077	0.432	−0.501	−0.658	−0.343	0.674	0.581	0.767	2782	2810	2699	2864	+1.01
Jun	0.262	0.085	0.438	−0.494	−0.652	−0.337	0.673	0.580	0.766	3022	2962	2940	3104	−2.03
Jul	0.258	0.081	0.436	−0.493	−0.652	−0.334	0.667	0.572	0.761	3045	2974	2963	3128	−2.40
Aug	0.251	0.075	0.428	−0.509	−0.664	−0.355	0.682	0.592	0.772	2965	2880	2882	3047	−2.93*
Sep	0.279	0.105	0.453	−0.493	−0.648	−0.337	0.680	0.589	0.771	2797	2763	2664	2831	+0.56
Oct	0.275	0.102	0.448	−0.494	−0.650	−0.339	0.680	0.589	0.771	2647	2648	2564	2730	+0.04
Nov	0.274	0.102	0.446	−0.495	−0.649	−0.340	0.680	0.590	0.770	2558	2482	2476	2641	−3.07
Dec	0.267	0.095	0.439	−0.504	−0.657	−0.351	0.685	0.596	0.774	2270	2288	2187	2353	+0.79
Jan 1970	0.264	0.092	0.436	−0.504	−0.667	−0.351	0.685	0.596	0.774	1984	1961	1901	2067	−1.16
Feb	0.270	0.098	0.441	−0.497	−0.651	−0.344	0.682	0.593	0.771	1761	1765	1678	1844	+0.24
Mar	0.270	0.099	0.441	−0.497	−0.650	−0.344	0.682	0.593	0.771	1970	1986	1887	2052	+0.82
Apr	0.269	0.098	0.439	−0.499	−0.651	−0.347	0.680	0.591	0.769	2290	2297	2207	2373	+0.30
May	0.269	0.099	0.439	−0.498	−0.650	−0.346	0.679	0.591	0.768	2534	2485	2452	2617	−1.99
Jun	0.267	0.097	0.437	−0.499	−0.651	−0.348	0.681	0.593	0.769	2623	2592	2540	2705	−1.18
Jul	0.269	0.100	0.439	−0.501	−0.652	−0.350	0.677	0.589	0.765	2644	2650	2562	2727	+0.22
Aug	0.268	0.099	0.437	−0.502	−0.652	−0.351	0.677	0.589	0.765	2651	2707	2569	2733	+2.07
Sep	0.263	0.094	0.432	−0.506	−0.656	−0.356	0.683	0.597	0.770	2677	2721	2594	2759	+1.63
Oct	0.267	0.099	0.435	−0.507	−0.656	−0.359	0.682	0.595	0.768	2658	2747	2575	2740	+3.24*
Nov	0.286	0.118	0.453	−0.496	−0.646	−0.347	0.686	0.600	0.772	2717	2735	2634	2800	+0.67
Dec	0.289	0.124	0.455	−0.496	−0.645	−0.347	0.687	0.601	0.772	2607	2627	2524	2690	+0.76

* Actual data outside the 95% forecast approximate large-sample confidence limit.
[a] The estimates reported for any particular month have been obtained using data up to, but not including, that month.

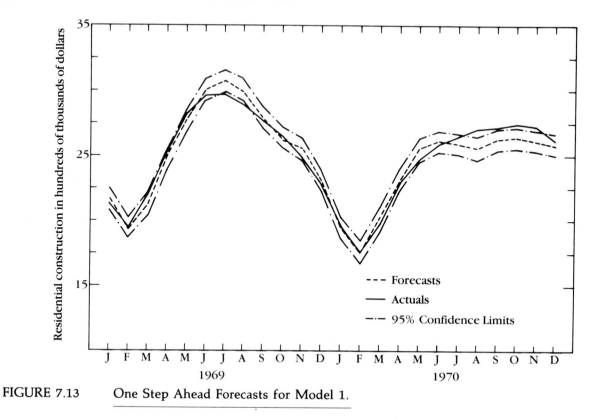

FIGURE 7.13   One Step Ahead Forecasts for Model 1.

This process was repeated 23 times. Table 7.8 also contains the point estimates of the parameters after each updating.

The percentage of the one step ahead forecasts which fall within a set confidence interval must correspond to the confidence coefficient,[5] for example, 95% of the forecasts should fall within the lower and upper limits of the 95% confidence interval. As can be seen, six of the 24 months for which one step ahead forecasts were generated using the fixed parameter model fell outside the range of the 95% confidence interval, while only two were outside the range using the updated parameter model. Theoretically, only 1.2 points should have fallen outside the forecast's 95% confidence interval. Since for a minimum mean squared error forecast the one step ahead forecast errors are uncorrelated (see Chapter 6), the probability of having a number of points lie outside a confidence interval can be calculated using the binomial distribution. The probability of x or less points falling outside the 95% confidence interval, given that the number of forecast periods is 24 and the probability of having a point fall outside the band is 0.05, is given in Table 7.9.

---

[5] Strictly speaking this applies to the case where the parameters are known.

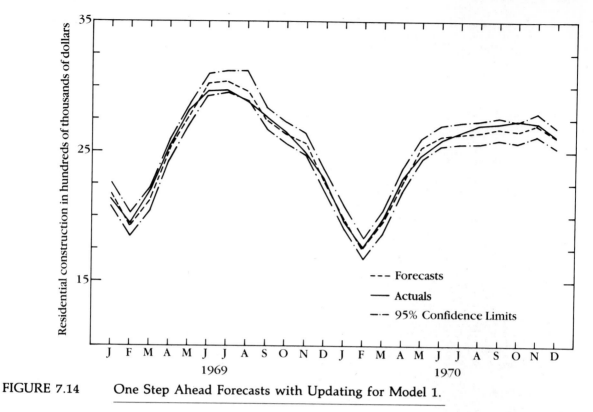

FIGURE 7.14   One Step Ahead Forecasts with Updating for Model 1.

TABLE 7.9   Probability of X or Less Points Falling Outside the 95% Confidence Interval

X	PROBABILITY
0	0.2920
1	0.6608
2	0.8841
3	0.9703
4	0.9940

From Table 7.9 we observe that if the one step ahead forecast errors are white noise, then the probability of having two or less points outside the 95% confidence interval is 0.8841. If this probability was only 0.10, then we would conclude that the forecasts were not so great. However, with this probability as high as 0.88, we also have to conclude that, given independent

TABLE 7.10   Twelve-Month Forecasts for Model 1[a]
$(1 - 0.296B)\nabla\nabla_{12}z_t = (1 + 0.490B)(1 - 0.686B^{12})a_t$

| | | MILLIONS OF DOLLARS | |
| | | 95% CONF. LIMIT | |
MONTH	FORECAST	LOWER	UPPER
Jan 1971	2344	2261	2426
Feb	2165	1995	2334
Mar	2378	2140	2615
Apr	2676	2382	2970
May	2892	2550	3233
Jun	3044	2661	3428
Jul	3121	2699	3543
Aug	3143	2686	3599
Sep	3098	2609	3587
Oct	3032	2513	3551
Nov	2951	2403	3499
Dec	2804	2228	3379

[a] *Base:* December 1970

FIGURE 7.15   Twelve-Month Forecast for Model 1.

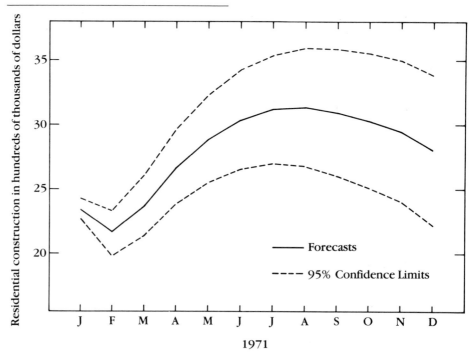

forecast errors, the chance of having more than two points falling outside is very small.

Nevertheless, it is rather remarkable that even with all the changes and fluctuations taking place in the economy, a relatively simple model was able to generate so many accurate forecasts for 24 consecutive months.

In Table 7.10 and Figure 7.15 we have given forecast results for the year 1971 after re-estimating the same model but using data up to December 1970.

7.6 CONCLUSIONS

The model $(1 - \phi B)\nabla\nabla_{12}z_t = (1 - \theta B)(1 - \Theta B^{12})a_t$ fits the United States residential construction data covering the period from January 1947 to December 1970 remarkably well. One step ahead forecasts generated by this model using updated parameters are extremely accurate.

APPENDIX: MODEL A

In this appendix we show that by studying the residual patterns of an inappropriate formulation we can make logical modifications to a misspecified model and arrive at a formulation which adequately depicts the behavior of the series.

In Section 7.2 we identified Model A,

$$(1 - \phi_1 B - \phi_2 B^2)\nabla_{12}z_t = \delta + a_t \qquad \text{(Model A)}$$

as a nonstationary model which perhaps with additional differencing could be used to depict the behavior of the construction data. Because the mean of series $\nabla_{12}z_t$ is 75.29 millions of current dollars, we included a constant, δ, in Model A. In this appendix we will analyze this model.

7.A.1 Model A:
$$(1 - \phi_1 B - \phi_2 B^2)\nabla_{12}z_t = \delta + a_t$$

Evaluate this model using the estimation output below and in Figure 7.A.1. Suggest possible model improvements.

```
MODEL WITH D = 0 DS = 1 S = 12
MEAN =   75.29     SD =    252.2    (NOBS = 252)

LEAST SQUARES METHOD
PARAMETER ESTIMATES
====================
                EST       SE      EST/SE      95% CONF LIMITS
    AR( 1)     1.599    0.044     36.551      1.513     1.685
    AR( 2)    -0.654    0.044    -14.842     -0.740    -0.568
    CONST      4.861    3.282      1.481     -1.572    11.294

EST.RES.SD =     5.0210E+01
EST.RES.SD(WITH BACK FORECAST) =      5.0366E+01

R SQR =  .961
DF = 249
F =     3054.     (2,249 DF) P-VALUE =    .000

CORRELATION MATRIX
          AR( 1)  AR( 2)
   AR( 2)  -.961
   CON( 3) -.094    .006
```

```
AUTOCORRELATIONS OF RESIDUALS
  LAGS ROW SE
  1-12  .06   .07  -.20  -.06   .03   .24   .00  -.13   .06   .08   .06  -.03  -.40
 13-24  .08  -.03   .08   .00  -.09  -.16   .02   .05  -.01  -.05  -.13  -.03   .06
 25-36  .08   .09   .04  -.02  -.03   .07   .02  -.07  -.01  -.02   .13   .01  -.13
 37-48  .09  -.10   .00  -.03   .06  -.04  -.12   .03   .00   .07  -.06  -.04   .05

CHI-SQUARE TEST            P-VALUE
Q(12) =  77.7      10 D.F.   .000
Q(24) =  96.5      22 D.F.   .000
Q(36) = 113.       34 D.F.   .000
Q(48) = 127.       46 D.F.   .000

AUTOCORRELATIONS OF FIRST DIFFERENCED RESIDUALS
  LAGS ROW SE
  1-12  .06  -.35  -.22   .02  -.06   .24  -.06  -.18   .09   .02   .05   .15  -.39
 13-24  .09   .13   .10   .01  -.02  -.13   .08   .05   .00   .02  -.10   .01   .03
 25-36  .09   .04   .01  -.03  -.07   .09   .02  -.08   .04  -.09   .15   .02  -.09
 37-48  .09  -.04   .06  -.06   .11  -.01  -.12   .09  -.05   .11  -.08  -.05   .03

CHI-SQUARE TEST            P-VALUE
Q(12) = 118.       10 D.F.   .000
Q(24) = 135.       22 D.F.   .000
Q(36) = 153.       34 D.F.   .000
Q(48) = 174.       46 D.F.   .000
```

ANALYSIS OF MODEL A

Based only on the coefficient estimates we conclude that both AR coefficients are highly significant. However, the correlation matrix suggests that the AR(1) and AR(2) parameter estimates are highly correlated. Therefore we propose to evaluate if the second-order autoregressive parameter can be deleted from the model without significantly affecting the model performance.

The spikes at lags 2, 5, and 12 in the residual autocorrelations are further evidence of model inadequacy. The accompanying Q statistics are rather inflated and consequently the P-values are low, which indicate that Model A is not adequate.

Therefore, we suggest deleting the AR(2) parameter and repeating the analysis procedure on the following AR(1) model:

$$(1 - \phi B)\nabla_{12}z_t = \delta + a_t. \qquad \text{(Model A.1)}$$

FIGURE 7.A.1   Model A: Autocorrelation Function of Residuals.

```
           AUTOCORRELATION FUNCTION
                  RESIDUALS

   -0.8  -0.4    0.0    0.4    0.8
   *  *   *   *   *  *   *   *  *  *   *  *   *  *   *
   .                  ( .R)              1 .
   .               R( . )               2 .
   .                (R. )               3 .
   .                ( .R)               4 .
   .                ( . )  R            5 .
   .                ( R )               6 .
   .               R . )               7 .
   .                ( .R)               8 .
   .                ( .R)               9 .
   .                ( .R)              10 .
   .                (R. )              11 .
   .          R     ( . )             12 .
   .                ( R )              13 .
   .                ( .R)              14 .
   .                ( R )              15 .
   .                (R. )              16 .
   .               R . )              17 .
   .                ( R )              18 .
   .                ( .R)              19 .
   .                ( R )              20 .
   .                (R. )              21 .
   .               R . )              22 .
   .                ( R )              23 .
   .                ( .R)              24 .
   .                ( .R)              25 .
   .                ( .R)              26 .
   .                ( R )              27 .
   .                ( R )              28 .
   .                ( .R)              29 .
   .                ( R )              30 .
   .                (R. )              31 .
   .                ( R )              32 .
   .                ( R )              33 .
   .                ( . R              34 .
   .                ( R )              35 .
   .               R . )              36 .
   .                (R .  )            37 .
   .                (  R  )            38 .
   .                (  R  )            39 .
   .                (  .R )            40 .
   .                ( R.  )            41 .
   .                (R .  )            42 .
   .                (  R  )            43 .
   .                (  R  )            44 .
   .                (  .R )            45 .
   .                ( R.  )            46 .
   .                ( R.  )            47 .
   .                (  .R )            48 .
```

7.A.2 Model A.1:
$$(1 - \phi B)\nabla_{12}z_t = \delta + a_t$$

Use the output below in Figure 7.A.2 to evaluate Model A.1. Suggest improvements.

```
MODEL WITH D = 0  DS = 1  S = 12
MEAN =   75.29     SD =   252.2    (NOBS = 252)

LEAST SQUARES METHOD
PARAMETER ESTIMATES
====================
              EST      SE      EST/SE     95% CONF LIMITS
AR( 1)       0.969    0.015    62.729     0.939     0.999
CONST        3.638    4.114     0.884    -4.425    11.701

EST.RES.SD =    6.5895E+01
EST.RES.SD(WITH BACK FORECAST) =    6.5977E+01

R SQR =  .932
DF = 250
F =    3440.    (1,250 DF) P-VALUE =   .000

CORRELATION MATRIX
          AR( 1)
  CON( 2)  -.426

AUTOCORRELATIONS OF RESIDUALS
   LAGS ROW SE
   1-12   .06    .62   .29   .18   .18   .21   .07  -.04  -.03  -.06  -.16  -.32  -.48
  13-24   .11   -.35  -.24  -.23  -.25  -.24  -.13  -.06  -.06  -.08  -.08   .02   .13
  25-36   .12    .20   .19   .15   .14   .16   .11   .06   .05   .05   .06  -.04  -.16
  37-48   .13   -.18  -.14  -.12  -.09  -.13  -.15  -.08  -.04   .00  -.02   .01   .08

CHI-SQUARE TEST            P-VALUE
Q(12) =   248.       11 D.F.    .000
Q(24) =   359.       23 D.F.    .000
Q(36) =   414.       35 D.F.    .000
Q(48) =   452.       47 D.F.    .000

AUTOCORRELATIONS OF FIRST DIFFERENCED RESIDUALS
   LAGS ROW SE
   1-12   .06   -.05  -.31  -.14  -.03   .22  -.04  -.17   .06   .09   .09   .00  -.39
  13-24   .08    .02   .14   .05  -.05  -.13   .06   .09   .02  -.03  -.12  -.02   .07
  25-36   .09    .10   .03  -.03  -.04   .08   .02  -.07   .00  -.02   .15   .03  -.12
  37-48   .09   -.09   .03   .00   .09  -.02  -.11   .04   .00   .08  -.07  -.05   .03

CHI-SQUARE TEST            P-VALUE
Q(12) =   94.9      11 D.F.    .000
Q(24) =   115.      23 D.F.    .000
Q(36) =   133.      35 D.F.    .000
Q(48) =   148.      47 D.F.    .000
```

FIGURE 7.A.2   Model A.1: Autocorrelation Function of Residuals.

```
            AUTOCORRELATION FUNCTION
                RESIDUALS

     -0.8  -0.4   0.0   0.4   0.8
     *  .  *  *  *  .  *  *  *  .  .  *  .  *  .  .  *
     .                   (   .   )         R           1  .
     .                   (   .   )  R                  2  .
     .                   (   .   R                     3  .
     .                   (   .   R                     4  .
     .                   (   .   R                     5  .
     .                   (  .R  )                      6  .
     .                   ( R.   )                      7  .
     .                   (  R   )                      8  .
     .                   ( R.   )                      9  .
     .                   (R  .   )                     10 .
     .               R   (   .   )                     11 .
     .             R     (   .   )                     12 .
     .               R   (   .   )                     13 .
     .              R(   .   )                         14 .
     .               R   .   )                         15 .
     .              R(   .   )                         16 .
     .              R(   .   )                         17 .
     .                   ( R   .   )                   18 .
     .                   (  R.   )                     19 .
     .                   (  R.   )                     20 .
     .                   (  R.   )                     21 .
     .                   (  R.   )                     22 .
     .                   (   R   )                     23 .
     .                   (   .   R )                   24 .
     .                   (   .    R)                   25 .
     .                   (   .    R)                   26 .
     .                   (   .  R )                    27 .
     .                   (   .  R )                    28 .
     .                   (   .  R )                    29 .
     .                   (   .  R )                    30 .
     .                   (   .R  )                     31 .
     .                   (   .R  )                     32 .
     .                   (   .R  )                     33 .
     .                   (   .R  )                     34 .
     .                   (  R.   )                     35 .
     .                   ( R .   )                     36 .
     .                   (R  .   )                     37 .
     .                   ( R .   )                     38 .
     .                   ( R .   )                     39 .
     .                   (  R.   )                     40 .
     .                   ( R .   )                     41 .
     .                   ( R .   )                     42 .
     .                   (  R.   )                     43 .
     .                   (  R.   )                     44 .
     .                   (   R   )                     45 .
     .                   (   R   )                     46 .
     .                   (   R   )                     47 .
     .                   (   .R  )                     48 .
```

ANALYSIS OF MODEL A.1

Again the coefficient estimates are highly significant. However, the residual autocorrelations indicate major departures from white noise. The major spikes in the residual autocorrelations at lags 1, 11, 12, and 13 suggest that the residual errors can be modeled by either one of the following two processes:

$$e_t = (1 - \theta B)(1 - \Theta B^{12}) a_t \qquad\qquad (A.1)$$

or

$$e_t = (1 - \theta_1 B - \theta_{12} B^{12} - \theta_{13} B^{13}) a_t, \qquad\qquad (A.2)$$

although the symmetry in the *acf* would indicate that (A.1) is the preferred choice.

If we rewrite Model A.1 as

$$(1 - \phi B) \nabla_{12} z_t = \delta + e_t$$

and substitute either (A.1) or (A.2) in the above expression, we then obtain the following two models:

$$(1 - \phi B) \nabla_{12} z_t = \delta + (1 - \theta B)(1 - \Theta B^{12}) a_t \qquad \text{(Model A.1.1)}$$

and

$$(1 - \phi B) \nabla_{12} z_t = \delta + (1 - \theta_1 B - \theta_{12} B^{12} - \theta_{13} B^{13}) a_t. \qquad \text{(Model A.1.2)}$$

The analysis of these two models follows.

7.A.3 Model A.1.1:
$(1 - \phi B) \nabla_{12} z_t = \delta + (1 - \theta B)(1 - \Theta B^{12}) a_t$

Use the output below and in Figure 7.A.3 to evaluate Model A.1.1. Suggest improvements.

```
MODEL WITH D = 0 DS = 1 S = 12
MEAN =    75.29    SD =    252.2    (NOBS = 252)

LEAST SQUARES METHOD
PARAMETER ESTIMATES
===================

            EST       SE      EST/SE     95% CONF LIMITS
AR( 1)    0.948    0.020    46.634     0.908    0.988
MA( 1)   -0.679    0.046   -14.671    -0.770   -0.589
SMA( 1)   0.703    0.045    15.623     0.615    0.791
CONST     3.905    2.032     1.922    -0.078    7.888

EST.RES.SD =    4.1899E+01
EST.RES.SD(WITH BACK FORECAST) =    4.2832E+01
```

```
R SQR =  .973
DF = 248
F =    2960.    (3,248 DF) P-VALUE =  .000

CORRELATION MATRIX
          AR( 1)  MA( 1) SMA( 1)
  MA( 1)   .144
  SMA( 1)  .104    .003
  CON( 4) -.724   -.102   -.090

ROOTS    REAL    IMAGINARY   SIZE
  AR    1.0546   0.0000    1.0546
  MA   -1.4718   0.0000    1.4718
  SMA   1.4228   0.0000    1.4228

AUTOCORRELATIONS OF RESIDUALS
  LAGS ROW SE
   1-12   .06    .08   .17   .00   .04   .12  -.01  -.02  -.01   .00  -.10  -.07   .01
  13-24   .07   -.04  -.02  -.07  -.16  -.12  -.08  -.05  -.04  -.06  -.13  -.05   .04
  25-36   .07    .07   .07  -.01  -.02   .01  -.01  -.06   .02  -.04   .03  -.05  -.05
  37-48   .07   -.03   .01  -.05  -.01  -.05  -.13   .01  -.04   .08  -.04   .02   .03

CHI-SQUARE TEST              P-VALUE
Q(12) =  17.5       9 D.F.   .041
Q(24) =  40.1      21 D.F.   .007
Q(36) =  46.2      33 D.F.   .063
Q(48) =  56.6      45 D.F.   .115

AUTOCORRELATIONS OF FIRST DIFFERENCED RESIDUALS
  LAGS ROW SE
   1-12   .06   -.55   .15  -.12  -.02   .11  -.07  -.01   .00   .06  -.07  -.02   .06
  13-24   .08   -.04   .04   .02  -.06  -.01   .01   .01   .01   .03  -.09   .00   .04
  25-36   .08    .01   .05  -.04  -.02   .03   .01  -.06   .07  -.07   .08  -.04  -.01
  37-48   .09   -.01   .06  -.06   .05   .02  -.12   .11  -.09   .13  -.10   .02  -.06

CHI-SQUARE TEST              P-VALUE
Q(12) =  93.2       9 D.F.   .000
Q(24) =  98.1      21 D.F.   .000
Q(36) =  106.      33 D.F.   .000
Q(48) =  129.      45 D.F.   .000
```

FIGURE 7.A.3   Model A.1.1: Autocorrelation Function of Residuals.

```
AUTOCORRELATION FUNCTION
       RESIDUALS

-0.8  -0.4   0.0   0.4   0.8
*  .  . .  .  .  .  .  .  .  .  .  .  . .  .  .  *
               .                  ( .R)        1  .
               .                  ( . )R       2  .
               .                  ( R )        3  .
               .                  ( .R)        4  .
               .                  ( . R        5  .
               .                  ( R )        6  .
               .                  ( R )        7  .
               .                  ( R )        8  .
               .                  ( R )        9  .
               .                  (R. )       10  .
               .                  (R. )       11  .
               .                  ( R )       12  .
               .                  (R. )       13  .
               .                  ( R )       14  .
               .                  (R. )       15  .
               .                  R  . )      16  .
               .                  R  . )      17  .
               .                  (R. )       18  .
               .                  (R. )       19  .
               .                  (R. )       20  .
               .                  (R. )       21  .
               .                  R  . )      22  .
               .                  (R. )       23  .
               .                  ( .R)       24  .
               .                  ( .R)       25  .
               .                  ( .R)       26  .
               .                  ( R )       27  .
               .                  ( R )       28  .
               .                  ( R )       29  .
               .                  ( R )       30  .
               .                  (R. )       31  .
               .                  ( R )       32  .
               .                  (R. )       33  .
               .                  ( R )       34  .
               .                  (R. )       35  .
               .                  (R. )       36  .
               .                  ( R )       37  .
               .                  ( R )       38  .
               .                  (R. )       39  .
               .                  ( R )       40  .
               .                  (R. )       41  .
               .                  R  . )      42  .
               .                  ( R )       43  .
               .                  (R. )       44  .
               .                  ( .R)       45  .
               .                  (R. )       46  .
               .                  ( R )       47  .
               .                  ( R )       48  .
```

ANALYSIS OF MODEL A.1.1

Although the model performance has improved, the inflated residual autocorrelation Q statistics and the resulting low P-values still indicate that the model does not adequately represent the stochastic process.

Since the value for the autoregressive parameter is very large, (0.948), it is highly probable that a consecutive differencing operation might be required. We also included the roots of the polynomials from which we clearly see that the autoregressive root is very close to unity. Recall that incorporation of a consecutive difference operation is like having an autoregressive parameter whose value is one. Thus, the analysis suggests that we should investigate the following model:

$$\nabla\nabla_{12}z_t = \delta + (1 - \theta B)(1 - \Theta B^{12})a_t. \qquad \text{(Model 1.2)}$$

Notice that this is Model 1.2, analyzed in Section 7.3.3. We therefore immediately propose to add an autoregressive parameter and evaluate model A.1.1.1:

$$(1 - \phi B)\nabla\nabla_{12}z_t = \delta + (1 - \theta B)(1 - \Theta B^{12})a_t. \quad \text{(Model A.1.1.1)}$$

7.A.4 Model A.1.1.1: $(1 - \phi B)\nabla\nabla_{12}z_t = \delta + (1 - \theta B)(1 - \Theta B^{12})a_t$

Use the output below and in Figure 7.A.4 to evaluate this model. Suggest improvements.

```
MODEL WITH D = 1 DS = 1 S = 12
MEAN = -0.2869     SD =    66.34    (NOBS = 251)

LEAST SQUARES METHOD
PARAMETER ESTIMATES
====================
            EST        SE      EST/SE     95% CONF LIMITS
 AR( 1)    0.261     0.092      2.849      0.082     0.441
 MA( 1)   -0.489     0.082     -5.944     -0.650    -0.328
 SMA( 1)   0.684     0.047     14.572      0.592     0.776
 CONST    -0.098     1.427     -0.069     -2.895     2.699

EST.RES.SD =    4.2051E+01
EST.RES.SD(WITH BACK FORECAST) =     4.2941E+01

R SQR =   .605
DF = 247
F =     125.9    (3,247 DF) P-VALUE =    .000
```

```
CORRELATION MATRIX
         AR( 1)  MA( 1) SMA( 1)
  MA( 1)   .747
 SMA( 1)  -.130   -.102
 CON( 4)   .019    .015   -.049

ROOTS   REAL    IMAGINARY   SIZE
  AR    3.8241    0.0000    3.8241
  MA   -2.0461    0.0000    2.0461
 SMA    1.4615    0.0000    1.4615

AUTOCORRELATIONS OF RESIDUALS
  LAGS ROW SE
  1-12  .06  -.02  .04  .01 -.02  .15 -.04 -.04  .00  .01 -.11 -.06  .02
 13-24  .07  -.04  .00 -.04 -.14 -.10 -.04 -.03  .00 -.03 -.12 -.03  .05
 25-36  .07   .08  .08 -.01 -.02  .05  .00 -.04  .03 -.04  .04 -.03 -.06
 37-48  .07  -.01  .03 -.07  .01 -.04 -.13  .02 -.02  .08 -.03 -.02  .02

CHI-SQUARE TEST           P-VALUE
Q(12) =  11.3       9 D.F.   .254
Q(24) =  26.0      21 D.F.   .207
Q(36) =  33.3      33 D.F.   .454
Q(48) =  43.6      45 D.F.   .530

AUTOCORRELATIONS OF FIRST DIFFERENCED RESIDUALS
  LAGS ROW SE
  1-12  .06  -.52  .04  .00 -.10  .18 -.09 -.02  .02  .06 -.08 -.01  .06
 13-24  .08  -.05  .04  .02 -.07 -.01  .02  .00  .02  .04 -.09  .01  .03
 25-36  .08   .01  .05 -.04 -.04  .06  .00 -.06  .07 -.07  .07 -.02 -.04
 37-48  .08   .01  .06 -.08  .07  .01 -.12  .10 -.07  .11 -.06 -.02 -.03

CHI-SQUARE TEST           P-VALUE
Q(12) =  86.7       9 D.F.   .000
Q(24) =  92.6      21 D.F.   .000
Q(36) =  101.      33 D.F.   .000
Q(48) =  119.      45 D.F.   .000
```

ANALYSIS OF MODEL A.1.1.1

Note that this model is similar to Model 1 (discussed in Section 7.3.1) except for the constant term. However, as this constant term is insignificant, we should expect that the estimation results are very similar. Indeed, the parameter estimates are identical.

What do you expect the results will be for Model A.1.2?

$$(1 - \phi B)\nabla_{12}z_t = \delta + (1 - \theta_1 B - \theta_{12} B^{12} - \theta_{13} B^{13})a_t \quad \text{(Model A.1.2)}$$

FIGURE 7.A.4   Model A.1.1.1: Autocorrelation Function of Residuals.

```
AUTOCORRELATION FUNCTION
    RESIDUALS

-0.8  -0.4   0.0   0.4   0.8
*  .  *  .  *  .  *  .  *  .  *  .  *  .  *  .  *
           .                ( R )              1 .
           .                ( .R)              2 .
           .                ( R )              3 .
           .                ( R )              4 .
           .                ( .  R             5 .
           .                (R. )              6 .
           .                (R. )              7 .
           .                ( R )              8 .
           .                ( R )              9 .
           .                R . )             10 .
           .                (R. )             11 .
           .                ( R )             12 .
           .                (R. )             13 .
           .                ( R )             14 .
           .                (R. )             15 .
           .                R . )             16 .
           .                (R. )             17 .
           .                (R. )             18 .
           .                ( R )             19 .
           .                ( R )             20 .
           .                ( R )             21 .
           .                R . )             22 .
           .                ( R )             23 .
           .                ( .R)             24 .
           .                ( .R)             25 .
           .                ( .R)             26 .
           .                ( R )             27 .
           .                ( R )             28 .
           .                ( .R)             29 .
           .                ( R )             30 .
           .                (R. )             31 .
           .                ( R )             32 .
           .                (R. )             33 .
           .                ( .R)             34 .
           .                ( R )             35 .
           .                (R. )             36 .
           .                ( R )             37 .
           .                ( R )             38 .
           .                (R. )             39 .
           .                ( R )             40 .
           .                (R. )             41 .
           .                R . )             42 .
           .                ( R )             43 .
           .                ( R )             44 .
           .                ( .R)             45 .
           .                (R. )             46 .
           .                ( R )             47 .
           .                ( R )             48 .
```

7.A.5 Model A.1.2:
$$(1 - \phi B)\nabla_{12} z_t = \delta + (1 - \theta_1 B - \theta_{12} B^{12} - \theta_{13} B^{13}) a_t$$

Use the output below and in Figure 7.A.5 to evaluate this model. Suggest improvements.

```
MODEL WITH D = 0 DS = 1 S = 12
MEAN =    75.29     SD =    252.2      (NOBS =  252)


LEAST SQUARES METHOD
PARAMETER ESTIMATES
====================
            EST       SE      EST/SE      95% CONF LIMITS
AR( 1)     0.949    0.020    46.559      0.909      0.989
MA( 1)    -0.681    0.046   -14.662     -0.772     -0.590
MA(12)     0.720    0.043    16.650      0.635      0.804
MA(13)     0.467    0.057     8.196      0.355      0.578
CONST      3.876    2.029     1.910     -0.102      7.853


EST.RES.SD =    4.1950E+01
EST.RES.SD(WITH BACK FORECAST) =    4.2892E+01


R SQR =  .973
DF = 247
F =    2215.    (4,247 DF) P-VALUE =  .000


CORRELATION MATRIX
          AR( 1)  MA( 1)  MA(12)  MA(13)
  MA( 1)    .138
  MA(12)    .102   -.007
  MA(13)   -.005   -.622    .546
  CON( 5)  -.728   -.098   -.089   -.009


ROOTS   REAL   IMAGINARY   SIZE
  AR    1.054    0.000     1.054
  MA   -1.543    0.000     1.543
  MA   -0.517    0.889     1.029
  MA   -0.517   -0.889     1.029
  MA    0.891    0.515     1.030
  MA    0.891   -0.515     1.030
  MA    0.514    0.892     1.029
  MA    0.514   -0.892     1.029
  MA   -0.891    0.509     1.026
  MA   -0.891   -0.509     1.026
  MA    1.030    0.000     1.030
  MA   -0.002    1.029     1.029
  MA   -0.002   -1.029     1.029
  MA   -1.019    0.000     1.019
```

```
AUTOCORRELATIONS OF RESIDUALS
  LAGS ROW SE
  1-12   .06    .08    .17    .00    .04    .12  -.01  -.01  -.01    .00  -.10  -.07    .02
  13-24  .07  -.06    .00  -.08  -.15  -.13  -.08  -.05  -.05  -.05  -.14  -.04    .05
  25-36  .07    .05    .09  -.02  -.01    .01  -.01  -.05    .01  -.03    .02  -.04  -.05
  37-48  .07  -.03    .02  -.06    .00  -.06  -.13    .01  -.04    .08  -.04    .02    .03
```

```
CHI-SQUARE TEST              P-VALUE
Q(12) =  17.6        8 D.F.    .024
Q(24) =  40.8       20 D.F.    .004
Q(36) =  46.6       32 D.F.    .046
Q(48) =  57.6       44 D.F.    .082
```

```
AUTOCORRELATIONS OF FIRST DIFFERENCED RESIDUALS
  LAGS ROW SE
  1-12   .06  -.55    .15  -.12  -.02    .12  -.07    .00  -.01    .06  -.07  -.03    .09
  13-24  .08  -.07    .07  -.01  -.04  -.02    .01    .02    .01    .04  -.09    .00    .05
  25-36  .08  -.02    .07  -.06  -.01    .03    .01  -.06    .06  -.06    .06  -.03  -.01
  37-48  .09  -.02    .07  -.08    .06    .01  -.12    .10  -.09    .13  -.10    .03  -.06
```

```
CHI-SQUARE TEST              P-VALUE
Q(12) =  95.7        8 D.F.    .000
Q(24) =  103.       20 D.F.    .000
Q(36) =  110.       32 D.F.    .000
Q(48) =  135.       44 D.F.    .000
```

ANALYSIS OF MODEL A.1.2

The results of this model are very similar to those of Model A.1.1.: residual autocorrelations show spikes at lags 2 and possibly 16; inflated Q statistics; high value for the AR(1) parameter estimate; and so on. Note also the high degree of correlation between the estimates of the moving average parameters. The discussion presented in Section 7.3.4 about the roots of the moving average polynomial is also applicable here. A total of 12 roots of approximately equal size indicates that the model could be simplified. The MA(13) could be represented as the product of an MA(1) and a seasonal MA(1).

We would favor Model A.1.1 over A.1.2 because of parsimony. At the same time, the analysis of Model A.1.1 led to the alternative Model A.1.1.1, and this model suggested that Model 1 of Section 7.3.1 was to be preferred.

We thus have shown that by analyzing the residual patterns of an inappropriate formulation, one can make logical alterations to the misspecified model to arrive at a formulation which adequately depicts the behavior of the series.

FIGURE 7.A.5   Model A.1.2: Autocorrelation Function of Residuals.

```
        AUTOCORRELATION FUNCTION
               RESIDUALS

    -0.8  -0.4   0.0   0.4   0.8
  *  .  *  .  .  *  .  *  .  .  *  .  *  .  *  .  .  *
              .                   ( .R)        1 .
              .                   ( . )R       2 .
              .                   ( R )        3 .
              .                   ( .R)        4 .
              .                   ( . R        5 .
              .                   ( R )        6 .
              .                   ( R )        7 .
              .                   ( R )        8 .
              .                   ( R )        9 .
              .                   (R. )       10 .
              .                   (R. )       11 .
              .                   ( R )       12 .
              .                   (R. )       13 .
              .                   ( R )       14 .
              .                   (R. )       15 .
              .                   R . )       16 .
              .                   R . )       17 .
              .                   (R. )       18 .
              .                   (R. )       19 .
              .                   (R. )       20 .
              .                   (R. )       21 .
              .                   R . )       22 .
              .                   (R. )       23 .
              .                   ( .R)       24 .
              .                   ( .R)       25 .
              .                   ( .R)       26 .
              .                   ( R )       27 .
              .                   ( R )       28 .
              .                   ( R )       29 .
              .                   ( R )       30 .
              .                   (R. )       31 .
              .                   ( R )       32 .
              .                   (R. )       33 .
              .                   ( R )       34 .
              .                   (R. )       35 .
              .                   (R. )       36 .
              .                   ( R )       37 .
              .                   ( R )       38 .
              .                   (R. )       39 .
              .                   ( R )       40 .
              .                   (R. )       41 .
              .                   R . )       42 .
              .                   ( R )       43 .
              .                   (R. )       44 .
              .                   ( .R)       45 .
              .                   (R. )       46 .
              .                   ( R )       47 .
              .                   ( R )       48 .
```

CHAPTER 8

UNITED STATES UNEMPLOYMENT

Note: This chapter was written jointly with Sergio Koreisha.

8.1 OBJECTIVE

The objective of the analysis in this chapter is to examine the effectiveness of the Box–Jenkins technique as a predictive test for determining the monthly number of unemployed persons in the United States.

Multiplicative seasonal ARIMA models will be constructed for unadjusted monthly data covering the period January 1948–February 1975 (a total of 326 observations) and will then be used to generate monthly one step ahead forecasts of the number of unemployed persons for the period March 1975–February 1977. The forecasted figures will be compared with the actual ones. If confidence in the predictability of the models is to be established, then the percentage of the forecasts falling within a set confidence interval must be the same as the confidence coefficient; for example, 95% of the forecasts should fall within the lower and upper limits of the 95% confidence interval.

8.2 UNEMPLOYMENT DATA

The U.S. Department of Commerce monthly publication, *Survey of Current Business*, was the source of all unemployment data used in the study.[1] *Unemployment* in this study refers to the unemployed labor force, persons 16 years of age and over. Appendix A contains the unadjusted monthly unemployment figures for the period January 1948–October 1977. Only the first 326 observations, covering the period January 1948–February 1975, have been used in the model building phase. The remaining observations found in Appendix A are used for model validation. Figure 8.1 contains a graph of the data.

A look at this graph reveals that the pattern of unemployment was sharply altered at the onset of the energy crisis in the last months of 1973. After declining for almost one and one-half years and reaching the low figure of 3.8 million in October 1973, the number of unemployed persons began to increase thereafter. At first, the rise was gradual, but in the closing months of 1974 the number of unemployed persons turned into a steep climb as the growth in the economy began to slow down. By February 1975, the figure had reached a record high of 8.3 million unemployed.

Some of the factors which contributed in 1974 to this unprecedented post-depression unemployment level include:

1. a general weakness in retail sales highlighted by a very sharp drop in the demand for automobiles which eventually took its toll on related industries;

---

[1] Data for January 1975 and later can be found in various issues of the U.S. Department of Commerce, Office of Business Economics, *Survey of Current Business*. Data before January 1974 can be found in the U.S. Department of Commerce, Office of Business Economics, *1975 Business Statistics*, The Biennial Supplement to the *Survey of Current Business* and earlier biennial editions of this publication.

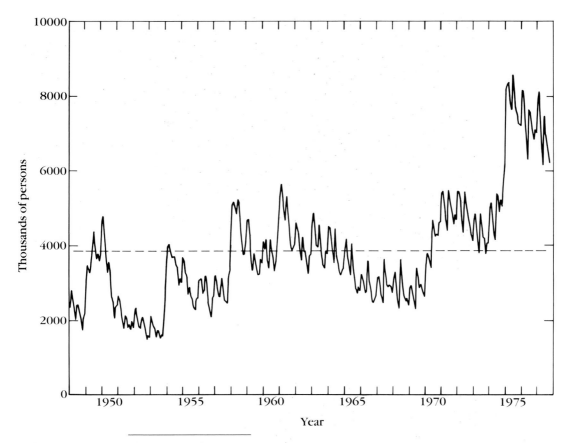

FIGURE 8.1   U.S. Unemployment.

2. the combination of high interest rates and a shortage of mortgage funds resulted in the most severe housing crisis since World War II; this situation was aggravated later in 1974 by a slackening in nonresidential construction;

3. a decline in industrial orders; and

4. the coal miners' strike of 1974.

8.3 MODEL IDENTIFICATION

A look at the plot of the unemployment data (Figure 8.1) suggests that a logarithmic or any other variance stabilizing transformation is not needed.[2] However, the existence of a yearly seasonality component (span = 12) is

---

[2] Attempts to identify potential models by taking logs did not provide fruitful results.

TABLE 8.1   Sample Autocorrelations of the Observed Series and of the First and Second Differences

```
SERIES WITH D = 0 DS = 0 S = 0
MEAN =    3504.        SD =     1081.      (NOBS = 326)

AUTOCORRELATIONS
  LAGS ROW SE
   1-12  .06    .90  .79  .72  .69  .67  .62  .60  .53  .48  .48  .51  .52
  13-24  .18    .43  .35  .30  .29  .31  .29  .30  .26  .24  .28  .32  .37
  25-36  .20    .31  .26  .24  .24  .26  .24  .25  .21  .19  .22  .26  .30
  37-48  .21    .23  .17  .14  .13  .15  .13  .13  .10  .08  .11  .16  .19

CHI-SQUARE TEST            P-VALUE
Q(12) =  .163E+04  12 D.F.   .000
Q(24) =  .204E+04  24 D.F.   .000
Q(36) =  .231E+04  36 D.F.   .000
Q(48) =  .242E+04  48 D.F.   .000


SERIES WITH D = 1 DS = 0 S = 0
MEAN =    18.33       SD =     412.5      (NOBS = 325)

AUTOCORRELATIONS
  LAGS ROW SE
   1-12  .06    .05 -.11 -.22 -.08  .23 -.12  .23 -.11 -.26 -.17  .01  .72
  13-24  .09   -.03 -.18 -.25 -.12  .18 -.16  .22 -.11 -.25 -.15 -.01  .69
  25-36  .11   -.02 -.17 -.21 -.10  .19 -.14  .21 -.12 -.23 -.12 -.01  .66
  37-48  .13   -.02 -.14 -.20 -.10  .16 -.13  .19 -.10 -.23 -.12  .01  .64

CHI-SQUARE TEST            P-VALUE
Q(12) =  276.      12 D.F.   .000
Q(24) =  549.      24 D.F.   .000
Q(36) =  804.      36 D.F.   .000
Q(48) =  .105E+04  48 D.F.   .000


SERIES WITH D = 2 DS = 0 S = 0
MEAN =    -1.009      SD =     569.1      (NOBS = 324)

AUTOCORRELATIONS
  LAGS ROW SE
   1-12  .06   -.42 -.02 -.14 -.09  .35 -.37  .37 -.10 -.13 -.05 -.28  .76
  13-24  .11   -.31 -.04 -.11 -.09  .33 -.37  .37 -.10 -.13 -.02 -.29  .74
  25-36  .14   -.29 -.06 -.08 -.09  .32 -.35  .35 -.11 -.12  .00 -.29  .71
  37-48  .16   -.29 -.03 -.08 -.09  .30 -.32  .32 -.09 -.12 -.01 -.26  .66

CHI-SQUARE TEST            P-VALUE
Q(12) =  432.      12 D.F.   .000
Q(24) =  837.      24 D.F.   .000
Q(36) =  .123E+04  36 D.F.   .000
Q(48) =  .157E+04  48 D.F.   .000
```

readily recognized. Further verification of this seasonal pattern is obtained by examining the autocorrelation of the series consecutively differenced. Table 8.1 contains the sample autocorrelations (r_k) of the observed series z_t as well as the sample autocorrelations of the first and second differences of the series ∇z_t and $\nabla^2 z_t$. As expected, the correlations at lags of multiples of 12 are very large. Figure 8.2 contains the first 48 sample autocorrelations of the second differences of the series. This figure clearly shows the very large autocorrelations at lags 12, 24, 36, and 48 which do not die out. Clearly the series is still not stationary.

In order to identify potential stationary models showing the relationships (a) between observations for successive months in a year, and (b) between observations for the same month in successive years, autocorrelations were estimated for the following additional set of differenced series:

1. First difference of the yearly component
    of the series. $\qquad\qquad\qquad\qquad\qquad$ $\nabla_{12} z_t$
2. First difference of the monthly component
    and first difference of the yearly component $\qquad$ $\nabla\nabla_{12} z_t$
    of the series.
3. Second difference of the monthly component
    and first difference of the yearly component $\qquad$ $\nabla^2\nabla_{12} z_t$.
    of the series.

Table 8.2 contains the first 48 sample autocorrelations for these three differences series and the partial autocorrelations of $\nabla^2\nabla_{12} z_t$. Figure 8.3 contains the autocorrelation and partial autocorrelation function of $\nabla^2\nabla_{12} z_t$.

Examining the first seasonal differences of the data, $\nabla_{12} z_t$, it is clear that this differencing is not sufficient to make the new series stationary. The values of the autocorrelation do not die out rapidly. They oscillate in sign very gradually: persistently positive at first, then persistently negative, and for the lags 44–48, again positive. The differencing $\nabla\nabla_{12}$ markedly reduces the correlation throughout, but does not provide a clear indication of a suitable model to represent the data. Possibly we could include a second-order autoregressive polynomial as well as a seasonal first-order moving average term. However, an examination of the difference $\nabla^2\nabla_{12} z_t$ reveals that correlations at lags 1, 11, 12, and 13 appear significant and the r_{11} and r_{13} are of the same sign (see Figure 8.3). Therefore we tentatively identify the following two models, namely:[3]

---

[3] We have again omitted the subscripts of the parameters where possible. Also, there is no need to include a constant term in Models A or B as the mean of $\nabla^2\nabla_{12} z_t$ equals -0.08×10^3 persons, and can therefore be put equal to 0.

FIGURE 8.2   Sample Autocorrelation Function of the Second Differences.

```
            AUTOCORRELATION FUNCTION
            WITH D = 2 DS = 0 S = 0

      -0.8  -0.4   0.0   0.4   0.8
      * .. . . *  . * . * . * .. . . * . *. * . *
       .            R    ( .  )              1 .
       .                 ( R  )              2 .
       .                 R  . )              3 .
       .                 (R. )               4 .
       .                 ( . )    R          5 .
       .            R    ( . )               6 .
       .                 ( . )    R          7 .
       .                 R  . )              8 .
       .                 R  . )              9 .
       .                 (R. )               10 .
       .           R (   . )                 11 .
       .                 (   . )       R     12 .
       .          R (    . )                 13 .
       .                 ( R. )              14 .
       .                 (R . )              15 .
       .                 ( R. )              16 .
       .                 (   . ) R           17 .
       .          R (    . )                 18 .
       .                 (   . ) R           19 .
       .                 ( R. )              20 .
       .                 ( R . )             21 .
       .                 (   R )             22 .
       .                 R    . )            23 .
       .                 (    . )      R     24 .
       .                 R    . )            25 .
       .                 (  R. )             26 .
       .                 (  R. )             27 .
       .                 (  R. )             28 .
       .                 (    . )R           29 .
       .                 R(    . )           30 .
       .                 (    . )R           31 .
       .                 ( R  . )            32 .
       .                 ( R . )             33 .
       .                 (   R )             34 .
       .                 R    . )            35 .
       .                 (    . )      R     36 .
       .                 (R    . )           37 .
       .                 (     R )           38 .
       .                 (   R. )            39 .
       .                 (   R. )            40 .
       .                 (    . R)           41 .
       .                 R    . )            42 .
       .                 (    . R            43 .
       .                 ( R. )              44 .
       .                 ( R  . )            45 .
       .                 (   R )             46 .
       .                 (R   . )            47 .
       .                 (    . )   R        48 .
```

TABLE 8.2   Sample Autocorrelations of $\nabla_{12}z_t$, $\nabla\nabla_{12}z_t$, and $\nabla^2\nabla_{12}z_t$

```
SERIES WITH D = 0 DS = 1 S = 12
MEAN =     127.1      SD =     922.2     (NOBS = 314)

AUTOCORRELATIONS
  LAGS ROW SE
  1-12   .06    .94   .86   .77   .67   .56   .44   .32   .20   .09  -.02  -.12  -.21
  13-24  .16   -.26  -.31  -.33  -.34  -.34  -.33  -.30  -.28  -.26  -.23  -.21  -.18
  25-36  .18   -.15  -.12  -.10  -.08  -.07  -.06  -.06  -.06  -.06  -.06  -.07  -.07
  37-48  .18   -.07  -.07  -.06  -.05  -.03  -.01   .01   .04   .07   .11   .15   .19

CHI-SQUARE TEST          P-VALUE
Q(12) = .107E+04   12 D.F.   .000
Q(24) = .140E+04   24 D.F.   .000
Q(36) = .143E+04   36 D.F.   .000
Q(48) = .147E+04   48 D.F.   .000

SERIES WITH D = 1 DS = 1 S = 12
MEAN =     8.067      SD =     269.5     (NOBS = 313)

AUTOCORRELATIONS
  LAGS ROW SE
  1-12   .06    .23   .30   .15   .12   .15   .07  -.05   .00  -.05  -.14  -.03  -.46
  13-24  .08   -.13  -.15  -.12  -.11  -.12  -.10   .01   .01  -.02  -.02  -.05   .03
  25-36  .08    .03  -.01   .02   .09   .06   .05   .02  -.05   .02   .04  -.03  -.02
  37-48  .08   -.09  -.04  -.02  -.08  -.10  -.03  -.16  -.01  -.07  -.02   .04   .07

CHI-SQUARE TEST          P-VALUE
Q(12) = 143.       12 D.F.   .000
Q(24) = 174.       24 D.F.   .000
Q(36) = 182.       36 D.F.   .000
Q(48) = 207.       48 D.F.   .000

SERIES WITH D = 2 DS = 1 S = 12
MEAN =  -0.8333E-01  SD =     335.7     (NOBS = 312)

AUTOCORRELATIONS
  LAGS ROW SE
  1-12   .06   -.55   .14  -.07  -.05   .08   .02  -.11   .06   .03  -.14   .35  -.49
  13-24  .09    .22  -.03   .01   .01  -.02  -.06   .07   .02  -.02   .02  -.07   .05
  25-36  .09    .02  -.04  -.03   .06   .00   .00   .03  -.09   .03   .06  -.04   .04
  37-48  .09   -.07   .02   .05  -.03  -.06   .13  -.18   .14  -.07  -.01   .02  -.02

CHI-SQUARE TEST          P-VALUE
Q(12) = 235.       12 D.F.   .000
Q(24) = 257.       24 D.F.   .000
Q(36) = 265.       36 D.F.   .000
Q(48) = 297.       48 D.F.   .000

PARTIAL AUTOCORRELATIONS        STANDARD ERROR = .06
  LAGS
  1-12    -.55  -.22  -.15  -.20  -.09   .05  -.10  -.08   .05  -.14   .30  -.21
  13-24   -.24  -.08  -.05  -.12  -.09  -.08  -.14  -.02   .08  -.11   .10  -.10
  25-36   -.10  -.06  -.14  -.08   .00  -.06  -.01  -.03  -.02   .01   .08  -.02
  37-48   -.06  -.03  -.07  -.01  -.08   .08  -.07  -.05  -.10  -.08  -.07  -.10
```

FIGURE 8.3 Autocorrelation and Partial Autocorrelation Function of $\nabla^2\nabla_{12}z_t$.

```
AUTOCORRELATION FUNCTION              PARTIAL AUTOCORRELATION FUNCTION
                   WITH D = 2 DS = 1 S = 12

 -0.8  -0.4   0.0   0.4   0.8           -0.8  -0.4   0.0   0.4   0.8
 . . . . . . . . . . . . . . . .         . . . . . . . . . . . . . . . .
      .      R      ( . )            1  .      R      ( . )
      .             ( . R            2  .           R( . )
      .             (R. )            3  .            R . )
      .             (R. )            4  .           R( . )
      .             ( .R)            5  .            (R. )
      .             ( R )            6  .            ( .R)
      .            R . )             7  .            (R. )
      .             ( .R)            8  .            (R. )
      .             ( R )            9  .            ( .R)
      .            R . )            10  .            R . )
      .             ( . )  R        11  .            ( . )  R
      .      R      ( . )           12  .           R( . )
      .             ( . )R          13  .      R    ( . )
      .             ( R  )          14  .            (R. )
      .             ( R  )          15  .            (R. )
      .             ( R  )          16  .            R . )
      .             ( R  )          17  .            (R. )
      .             ( R. )          18  .            (R. )
      .             ( .R )          19  .            R . )
      .             ( R  )          20  .            ( R )
      .             ( R  )          21  .            ( .R)
      .             ( R  )          22  .            R . )
      .             ( R. )          23  .            ( . R
      .             ( .R )          24  .            R . )
      .             ( R  )          25  .            (R. )
      .             ( R. )          26  .            (R. )
      .             ( R  )          27  .            R . )
      .             ( .R )          28  .            (R. )
      .             ( R  )          29  .            ( R )
      .             ( R  )          30  .            (R. )
      .             ( R  )          31  .            ( R )
      .             ( R. )          32  .            (R. )
      .             ( R  )          33  .            ( R )
      .             ( .R )          34  .            ( R )
      .             ( R. )          35  .            ( .R)
      .             ( .R )          36  .            ( R )
      .             ( R. )          37  .            (R. )
      .             ( R  )          38  .            ( R )
      .             ( .R )          39  .            (R. )
      .             ( R  )          40  .            ( R )
      .             ( R. )          41  .            (R. )
      .             ( . R)          42  .            ( .R)
      .            R . )            43  .            (R. )
      .             ( . R)          44  .            (R. )
      .             ( R. )          45  .            R . )
      .             ( R  )          46  .            (R. )
      .             ( R  )          47  .            (R. )
      .             ( R  )          48  .            R . )
```

Model A: $(0,2,1) \times (0,1,1)_{12}$ or $\nabla^2 \nabla_{12} z_t = (1 - \theta B)(1 - \Theta B^{12}) a_t$ (8.1)

Model B:[4] $(0,2,3) \times (0,1,0)_{12}$ or $\nabla^2 \nabla_{12} z_t = (1 - \theta_1 B - \theta_{12} B^{12} - \theta_{13} B^{13}) a_t$.   (8.2)

Given the above analysis we could now proceed along two roads. We could analyze the series $\nabla \nabla_{12} z_t$ using the ARIMA$(2,1,0) \times (0,1,1)_{12}$ model and then check on the adequacy of this specification. Alternatively, we could rely directly on the $\nabla^2 \nabla_{12} z_t$ transformation and evaluate Model A and Model B. We feel that the nature of the sample autocorrelations is somewhat clearer for the latter transformation and therefore we will basically evaluate $\nabla^2 \nabla_{12} z_t$.

The population autocorrelations for Models A and B are nonzero for lags 1, 11, 12, and 13. In fact, the population autocorrelations for the multiplicative seasonal model A, $(0,2,1) \times (0,1,1)_{12}$, have been calculated in Section 4.4 [see also equation (7.3)] as

$$\rho_1 = \frac{-\theta}{(1 + \theta^2)}$$

$$\rho_{12} = \frac{-\Theta}{(1 + \Theta^2)}$$

$$\rho_{11} = \rho_{13} = \rho_1 \rho_{12} = \frac{\theta \Theta}{(1 + \theta^2)(1 + \Theta^2)} \qquad (8.3)$$

and all other autocorrelations are 0. The autocorrelations for Model B, $(0,2,3) \times (0,1,0)_{12}$, have also been presented in Chapter 7, equation (7.4), as

$$\rho_1 = (-\theta_1 + \theta_{12}\theta_{13})/\omega$$

$$\rho_{11} = \theta_1 \theta_{12}/\omega$$

$$\rho_{12} = (-\theta_{12} + \theta_1 \theta_{13})/\omega$$

$$\rho_{13} = -\theta_{13}/\omega$$

with

$$\omega = (1 + \theta_1^2 + \theta_{12}^2 + \theta_{13}^2), \qquad (8.4)$$

and all other autocorrelations are 0. For these results, see also Table 4.3, Models 1 and 2. Note that in Model B, ρ_{13} is in general not equal to ρ_{11}, as ρ_{13} is not restricted to be equal to $\rho_1 \rho_{12}$.

Assuming that the correct model is of either form (8.1) or (8.2), the approximate variances for the estimated autocorrelations at lags *higher than 13* can be approximated by Bartlett's formula [see, e.g., (4.48)]. For Models A and B the approximate variances are

---

[4] The notation $(0,2,3) \times (0,1,0)_{12}$ is not completely clear. In particular, the 3 for the moving average polynomial refers to the number of parameters included (here, three) and not the order of the polynomial (here, 13). When there are omitted terms in the polynomial the number of parameters in the polynomial does not agree with the order.

$$\text{Var}(r_k) = \frac{1 + 2(\rho_1^2 + \rho_{11}^2 + \rho_{12}^2 + \rho_{13}^2)}{n}, \qquad k > 13, \qquad (8.5)$$

where $n = N - d - sD$ is the number of differences $\nabla^2 \nabla_{12} z_t$.

The estimate of the approximate variance is obtained from (8.5) by replacing the population autocorrelations by the sample autocorrelations, for example,

$$\widehat{\text{Var}}(r_k) = \frac{1 + 2(r_1^2 + r_{11}^2 + r_{12}^2 + r_{13}^2)}{n}, \qquad k > 13. \qquad (8.6)$$

Finally the large standard error, $\text{SE}(r_k)$, is obtained by taking the square root of $\widehat{\text{Var}}(r_k)$. For the differenced data $\nabla^2 \nabla_{12} z_t$, with $n = 326 - 2 - 12 = 312$, this estimate becomes

$$\text{SE}(r_k) = \left\{ \frac{1 + 2[(-0.55)^2 + (0.35)^2 + (-0.49)^2 + (0.22)^2]}{312} \right\}^{1/2}$$
$$= 0.088. \qquad (8.7)$$

From equation (4.49), Section 4.5, we might have inferred that we should use all r_k, $k \leq 13$, and not just r_1, r_{11}, r_{12}, and r_{13}. Of course, in equation (8.5), the use of the population autocorrelations ρ_2, ρ_3, . . . , ρ_{10} would not make any difference, as these correlations are 0. Under correct model specification, the r_2, r_3, . . . , r_{10} will be close to 0 and the inclusion or exclusion of these correlations from (8.6) should not make any difference. In theory, we have to use the estimator appropriate for the model evaluated. And, for Models A and B, this implies using equation (8.5) as stated above. Inclusion of all the correlations will produce a somewhat inflated standard error. For the unemployment data, including these additional correlations produces a $\text{SE}(r_k)$ of 0.091 for $k > 13$, which for all practical applications is equal to the value obtained in (8.7).

8.4 ESTIMATION

Table 8.3 contains the parameter point estimates,[5] the estimate of the large-sample standard error (SE) together with the 95% confidence limits, and

---

[5] The least squares estimates of the parameters of the model which minimize the sum of the squares residuals were determined by using program ESTIMA. This program requires as input guess values of the parameters for each model.

Guess values for the parameters of each model were calculated by equating the sample autocorrelations (r_1, r_{11}, r_{12}, and r_{13}) with their population values [see equations (8.3) and (8.4)] and solving for its unknown parameters.

The guess values used were $\theta = 0.8$ and $\Theta = 0.7$ for Model A, and $\theta_1 = 0.8$, $\theta_{12} = 0.7$, and $\theta_{13} = -0.5$ for Model B.

TABLE 8.3   Parameter Estimates for Model A and Model B

MODEL	ESTIMATES	SE	95% CONFIDENCE LIMITS LOWER	UPPER	RSD
A: $(0,2,1)\times(0,1,1)_{12}$	$\hat{\theta}=0.796$	0.035	0.728	0.864	215
	$\hat{\Theta}=0.700$	0.044	0.613	0.768	
B: $(0,2,3)\times(0,1,0)_{12}$	$\hat{\theta}_1=0.803$	0.026	0.751	0.854	214
	$\hat{\theta}_{12}=0.696$	0.030	0.637	0.755	
	$\hat{\theta}_{13}=-0.519$	0.038	-0.594	-0.444	

TABLE 8.4   Correlation Matrices of the Parameter Estimates

		θ	Θ	
Model A	θ	1.0		
$(0,2,1)\times(0,1,1)_{12}$	Θ	0.047	1.0	
		θ_1	θ_{12}	θ_{13}
Model B	θ_1	1.0		
$(0,2,3)\times(0,1,0)_{12}$	θ_{12}	-0.831	1.0	
	θ_{13}	-0.110	-0.410	1.0

the estimate of the residual standard deviation (RSD) for Models A and B.

Table 8.4 contains the estimates of the sampling correlation matrices of the parameter estimates of the two models. As we can see, the parameter estimates of the Model A, $(0,2,1)\times(0,1,1)_{12}$, seem to be independent, while the parameter estimates of the Model B, $(0,2,3)\times(0,1,0)_{12}$, are negatively correlated, particularly the parameter estimates of θ_1 and θ_{12}. The high degree of correlation between the two parameter estimates suggests that perhaps one of the parameters could be eliminated from that model without affecting the model adequacy.

8.5 DIAGNOSTIC CHECKS

Three principal diagnostic checks have been applied to test the validity of the two models.

FIGURE 8.4   Model A: Autocorrelation Function of the Residuals.

```
          AUTOCORRELATION FUNCTION
                OF RESIDUALS

     -0.8   -0.4    0.0    0.4    0.8
      *  .  .  *  .  *  .  *  .  *  .  *  .  *  .  *
               .              ( R )              1  .
               .              ( .  R             2  .
               .              (R. )              3  .
               .              ( R )              4  .
               .              ( R )              5  .
               .              ( R )              6  .
               .              (R. )              7  .
               .              ( R )              8  .
               .              (R. )              9  .
               .              (R. )             10  .
               .              ( R )             11  .
               .              (R. )             12  .
               .              (R. )             13  .
               .              (R. )             14  .
               .              (R. )             15  .
               .              ( R )             16  .
               .              ( R )             17  .
               .              (R. )             18  .
               .              ( .R)             19  .
               .              ( .R)             20  .
               .              ( R )             21  .
               .              ( R )             22  .
               .              (R. )             23  .
               .              ( .R)             24  .
               .              ( R )             25  .
               .              (R. )             26  .
               .              ( R )             27  .
               .              ( .R)             28  .
               .              ( R )             29  .
               .              ( R )             30  .
               .              ( R )             31  .
               .              (R. )             32  .
               .              (R. )             33  .
               .              ( .R)             34  .
               .              (R. )             35  .
               .              ( .R)             36  .
               .              (R. )             37  .
               .              ( R )             38  .
               .              ( R )             39  .
               .              ( R )             40  .
               .              (R. )             41  .
               .              ( R )             42  .
               .             R  . )             43  .
               .              ( R )             44  .
               .              (R. )             45  .
               .              ( R )             46  .
               .              ( R )             47  .
               .              ( .R)             48  .
```

8.5.1 Overfitting

Notice that Model B is an overparameterized model (in comparison to Model A), as it contains one additional parameter. Indeed, Model B can be logically derived from Model A with the following parameter relations: $\theta_1 = \theta$, $\theta_{12} = \Theta$, and $\theta_{13} = -\theta\Theta$. In estimating Model B, we did not impose the constraint on the θ_{13} parameter. It is therefore remarkable that the estimates for Model A presented in Table 8.3 nevertheless do satisfy this relationship and in particular $-\hat{\theta}\hat{\Theta} = -0.557$, a value close to the point estimate $\hat{\theta}_{13} = -0.519$.

Furthermore, the correlation matrix of the three parameter estimates in Model B suggests that there exists a significant correlation between $\hat{\theta}_1$ and $\hat{\theta}_{12}$ which also indicates that the two-parameter Model A might be more appropriate.

Finally, since the sample autocorrelations of $\nabla^2\nabla_{12}z_t$ indicated that all the population autocorrelation are nonzero except ρ_1, ρ_{11}, ρ_{12}, and ρ_{13}, and based on experience we conclude that there is no reason to fit more elaborate models than the ones considered up to now. We therefore propose to tentatively hold off estimating any other models and to first analyze the residuals of Model A and Model B to check for model adequacies.

8.5.2 Residual Analysis

If the estimated models are adequate, their residuals should be approximately white noise. Figures 8.4 and 8.5 show the sample autocorrelation function of the residuals with their 95% confidence limits for Models A and B, respectively. In both cases, the residuals are well within the white noise confidence band. Furthermore, as expected, the sample autocorrelations of the first differences of the residuals for both models strongly suggest a moving average process of order 1 (see Figures 8.6 and 8.7).

Q STATISTIC

The overall adequacy of the models can be tested by the Ljung–Box residual portmanteau test of model adequacy, Q statistic (see Section 5.3.2). This test is designed to evaluate jointly the first K values of the residual autocorrelations, $r_k(\hat{a})$, instead of each autocorrelation individually.

Table 8.5 contains for the two models the values of the Q statistic, the degrees of freedom (D.F.) and the P-values based on the first 36 autocorrelations. The P-values for both models are high and differ by only 3% in favor of Model B. The P-values indicate that, based on evidence present in the residuals, the hypothesis of white noise errors is almost certainly—more than 70% certain—to hold. Therefore, the Q statistics do not provide any startling evidence of model inadequacy.

FIGURE 8.5   Model B: Autocorrelation Function of the Residuals.

```
     AUTOCORRELATION FUNCTION
         OF RESIDUALS

  -0.8   -0.4    0.0    0.4    0.8
 *  .  *  .  *  .  *  .  *  .  *  .  *
                  ( R )            1  .
           .      (  . R           2  .
           .      (R.  )           3  .
           .      ( R )            4  .
           .      ( R )            5  .
           .      ( R )            6  .
           .      (R.  )           7  .
           .      ( R )            8  .
           .      (R.  )           9  .
           .      (R.  )          10  .
           .      ( R )           11  .
           .      (R.  )          12  .
           .      (R.  )          13  .
           .      (R.  )          14  .
           .      (R.  )          15  .
           .      ( R )           16  .
           .      ( R )           17  .
           .      (R.  )          18  .
           .      (  .R)          19  .
           .      (  .R)          20  .
           .      ( R )           21  .
           .      ( R )           22  .
           .      (R.  )          23  .
           .      (  .R)          24  .
           .      ( R )           25  .
           .      (R.  )          26  .
           .      ( R )           27  .
           .      (  .R)          28  .
           .      ( R )           29  .
           .      ( R )           30  .
           .      (R.  )          31  .
           .      (R.  )          32  .
           .      (R.  )          33  .
           .      (  .R)          34  .
           .      (R.  )          35  .
           .      (  .R)          36  .
           .      (R.  )          37  .
           .      ( R )           38  .
           .      ( R )           39  .
           .      ( R )           40  .
           .      (R.  )          41  .
           .      ( R )           42  .
           .      R  .  )         43  .
           .      (R.  )          44  .
           .      R  .  )         45  .
           .      ( R )           46  .
           .      ( R )           47  .
           .      (  .R)          48  .
```

FIGURE 8.6   Model A: Autocorrelation Function of the First Differences of Residuals.

```
            AUTOCORRELATION FUNCTION
            FIRST DIFFERENCED RESIDUALS

          -0.8   -0.4    0.0    0.4    0.8
          *  .   *  .  *  . *  . *  . *  . *  . *
     1 .         R       (  . )
     2 .                 (  .  R
     3 .                 R .  )
     4 .                 (  R  )
     5 .                 (  .R)
     6 .                 (  R  )
     7 .                 (R.  )
     8 .                 (  .R)
     9 .                 (  R  )
    10 .                 (R.  )
    11 .                 (  .   R
    12 .                 (  R  )
    13 .                 (  R  )
    14 .                 (  R  )
    15 .                 (  R  )
    16 .                 (  .R)
    17 .                 (  R  )
    18 .                 (R.  )
    19 .                 (  .R)
    20 .                 (  R  )
    21 .                 (  R  )
    22 .                 (  R  )
    23 .                 (R.  )
    24 .                 (  .R)
    25 .                 (  R  )
    26 .                 (R.  )
    27 .                 (  R  )
    28 .                 (  .R)
    29 .                 (  R  )
    30 .                 (  R  )
    31 .                 (  R  )
    32 .                 ( .R  )
    33 .                 (R.  )
    34 .                 (  .   R
    35 .                 R .  )
    36 .                 (  .   R
    37 .                 (R.  )
    38 .                 (  .R)
    39 .                 (  R  )
    40 .                 (  R  )
    41 .                 (R.  )
    42 .                 (  .R)
    43 .                 R .  )
    44 .                 (  .R)
    45 .                 (R.  )
    46 .                 (  .R)
    47 .                 (R.  )
    48 .                 (  .R)
```

FIGURE 8.7   Model B: Autocorrelation Function of the First Differences of Residuals.

```
                AUTOCORRELATION FUNCTION
                FIRST DIFFERENCED RESIDUALS

           -0.8   -0.40   0.0    0.4    0.8
            * . * . * . * . * . * . * . * . *
      1  .        R       ( . )
      2  .                ( . R
      3  .                R . )
      4  .                ( R )
      5  .                ( .R)
      6  .                ( R )
      7  .                (R. )
      8  .                ( .R)
      9  .                ( R )
     10  .                (R. )
     11  .                ( . R
     12  .                (R. )
     13  .                ( R )
     14  .                ( R )
     15  .                ( R )
     16  .                ( R )
     17  .                ( R )
     18  .                (R. )
     19  .                ( .R)
     20  .                ( R )
     21  .                ( R )
     22  .                ( R )
     23  .                (R. )
     24  .                ( .R)
     25  .                ( R )
     26  .                (R. )
     27  .                ( R )
     28  .                ( .R)
     29  .                ( R )
     30  .                ( R )
     31  .                ( R )
     32  .                ( R )
     33  .                (R. )
     34  .                ( . R
     35  .              R . )
     36  .                ( . R
     37  .                (R. )
     38  .                ( .R)
     39  .                ( R )
     40  .                ( R )
     41  .                (R. )
     42  .                ( . R
     43  .                R . )
     44  .                ( .R)
     45  .                (R. )
     46  .                ( .R)
     47  .                (R. )
     48  .                ( .R)
```

TABLE 8.5   Portmanteau Test Applied to the Residuals

MODELS	Q STATISTIC	χ^2D.F.	P-VALUE
A: $\nabla^2\nabla_{12}z_t = (1 - \theta B)(1 - \Theta B^{12})a_t$	28.7	34	0.724
B: $\nabla^2\nabla_{12}z_t = (1 - \theta_1 B - \theta_{12}B^{12} - \theta_{13}B^{13})a_t$	27.1	33	0.756

8.5.3. Cumulative Periodogram

As discussed in Chapter 5, periodic nonrandomness can be evaluated by calculating the normalized cumulative periodogram $C(f_j)$ of the residuals. For a white noise series, the plot of $C(f_j)$ against the frequency f_j or the period $1/f_j$ should be scattered about a straight line joining points $(0, 0)$ and $(0.5, 1)$. Deviations from the straight line can be assessed by the use of the large-sample Kolmogorov–Smirnov confidence limits.

Figures 8.8 and 8.9 contain the plots of the cumulative periodograms for Models A and B respectively. The points on the cumulative periodogram for Model A oscillate about the expected straight line with very small ampli-

FIGURE 8.8   Model A: Cumulative Periodogram.

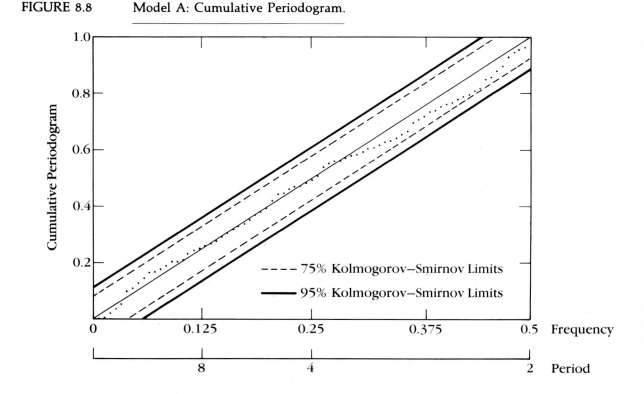

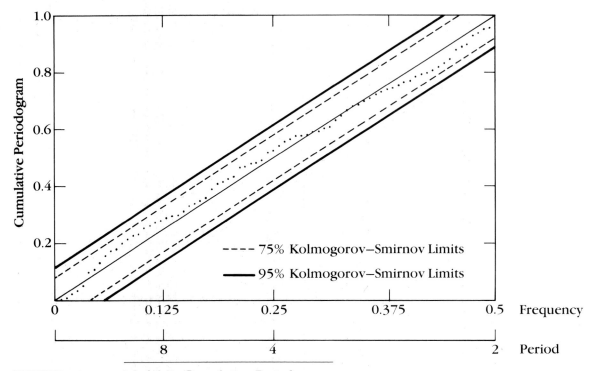

FIGURE 8.9   Model B: Cumulative Periodogram.

tude, while the points of the cumulative periodogram for the Model B, although still clustered about the expected straight line, tend to be persistently on one side of the line and then on the other. The oscillation amplitudes of both models are well within the 95% large-sample Kolgomorov–Smirnov confidence limits. So, based on results of the cumulative periodogram, we again slightly prefer Model A to Model B.

The diagnostic checks indicate that both models fit the data quite well. However, the correlation matrix of Model B suggests very high correlation among the parameter estimates and therefore we may conclude that the two-parameter Model A parsimoniously depicts the behavior of the data. Therefore, the forecasts will be based solely on Model A.

8.6 FORECASTING

The forecasts generated by the model $\nabla^2\nabla_{12}z_t = (1 - \theta B)(1 - \Theta B^{12})a_t$ were very accurate. The percent deviations of the one step ahead forecasts from the actual figures were, in general, very small. Tables 8.6 and 8.7 contain the one step ahead point forecasts together with the 95% large-sample confidence limits. All the one step ahead forecasts in Table 8.6 were generated using the values for the parameter estimates which best fit the first 326 observations (January 1948–February 1975), while the forecasts in Table 8.7 were made by updating the parameter estimates each time a new observation become available. Figures 8.10 and 8.11 contain the plots of the one step ahead forecasts for the fixed and updated models, respectively.

If confidence in the predictability of the model is to be established, then the percentage of the forecasts falling within a set confidence inter-

TABLE 8.6   One Step Ahead Forecasts
$\nabla^2 \nabla_{12} z_t = (1 - 0.796B)(1 - 0.700B^{12})a_t$

THOUSANDS OF PERSONS

MONTH	FORECAST	ACTUAL	95% CONF. LIMIT LOWER	95% CONF. LIMIT UPPER	DEVIATION FROM ACTUAL (%)
Mar 1975	8403	8359	7981	8824	− 0.52
Apr	8298	7820	7877	8720	− 6.12
May	7426	7623	7005	7848	+ 2.78
Jun	8808	8569	8387	9230	− 2.79
Jul	8158	8209	7736	8578	+ 0.62
Aug	7744	7696	7323	8166	− 0.63
Sep	7534	7522	7113	7956	− 0.16
Oct	7127	7244	6705	7548	+ 1.62
Nov	7450	7231	7029	7872	− 3.03
Dec	7019	7195	6597	7440	+ 2.45
Jan 1976	8212	8174	7790	8633	− 0.46
Feb	8136	8033	7714	8557	− 1.28
Mar	7558	7525	7136	7979	− 0.43
Apr	6428	6890	6007	6849	+ 6.71*
May	6597	6304	6176	7018	− 4.65
Jun	6861	7655	6440	7282	−10.37*
Jul	7700	7577	7279	8121	− 1.62
Aug	7346	7323	6925	7767	− 0.31
Sep	7408	7026	6987	7829	− 5.44
Oct	6625	6833	6204	7046	+ 3.04
Nov	6905	7095	6484	7326	+ 2.68
Dec	7334	7022	6913	7755	− 4.44
Jan 1977	7964	7848	7543	8385	− 1.48
Feb	7554	8109	7133	7975	+ 6.84*

* Actual data outside the 95% approximate large-sample confidence limit.

TABLE 8.7 One Step Ahead Forecasts with Updating[a]

$$\nabla^2 \nabla_{12} z_t = (1 - \theta B)(1 - \Theta B^{12}) a_t$$

THOUSANDS OF PERSONS

MONTH	θ	95% CONF. LIMIT LOWER	95% CONF. LIMIT UPPER	Θ	95% CONF. LIMIT LOWER	95% CONF. LIMIT UPPER	FORE-CAST	ACTUAL	95% CONF. LIMIT LOWER	95% CONF. LIMIT UPPER	DEVIATION FROM ACTUAL (%)
Mar 1975	0.796	0.728	0.864	0.700	0.613	0.786	8404	8359	7981	8824	−0.52
Apr	0.798	0.731	0.865	0.700	0.613	0.786	8332	7820	7911	8753	−6.55*
May	0.812	0.748	0.876	0.699	0.612	0.786	7833	7623	7410	8257	−2.76
Jun	0.814	0.750	0.878	0.699	0.612	0.786	8979	8569	8555	9403	−4.79
Jul	0.814	0.750	0.879	0.705	0.619	0.791	8491	8209	8065	8916	−3.43
Aug	0.806	0.740	0.872	0.707	0.621	0.792	7971	7696	7545	8397	−3.57
Sep	0.798	0.731	0.865	0.703	0.617	0.789	7746	7522	7320	8172	−2.98
Oct	0.786	0.718	0.855	0.705	0.620	0.790	7299	7244	6872	7725	−0.75
Nov	0.783	0.714	0.852	0.704	0.619	0.789	7487	7231	7061	7913	−3.54
Dec	0.773	0.703	0.843	0.706	0.623	0.790	7211	7195	6785	7637	−0.22
Jan 1976	0.772	0.703	0.842	0.706	0.623	0.790	8216	8174	7791	8641	−0.51
Feb	0.771	0.702	0.841	0.708	0.628	0.789	8179	8033	7755	8604	−1.82
Mar	0.766	0.696	0.837	0.705	0.624	0.785	7696	7525	7271	8120	−2.27
Apr	0.761	0.691	0.832	0.702	0.621	0.783	6891	6890	6467	7315	−0.02
May	0.761	0.691	0.831	0.702	0.622	0.782	6460	6304	6036	6883	−2.47
Jun	0.758	0.687	0.828	0.700	0.620	0.780	7145	7655	6722	7568	+6.67*
Jul	0.771	0.703	0.839	0.705	0.625	0.785	7276	7577	6851	7702	+3.97
Aug	0.770	0.701	0.839	0.706	0.626	0.786	7150	7323	6724	7577	+2.36
Sep	0.767	0.698	0.836	0.707	0.628	0.787	7282	7026	6856	7708	−3.65
Oct	0.774	0.707	0.842	0.710	0.631	0.789	6743	6833	6316	7169	+1.32
Nov	0.774	0.706	0.841	0.711	0.632	0.789	6985	7095	6559	7411	+1.55
Dec	0.773	0.705	0.840	0.710	0.632	0.789	7125	7022	6699	7550	−1.46
Jan 1977	0.774	0.707	0.842	0.711	0.633	0.789	8061	7848	7636	8486	−2.71
Feb	0.778	0.711	0.844	0.718	0.643	0.794	7809	8109	7384	8233	+3.71
Mar	0.781	0.715	0.847	0.724	0.649	0.798	7814	7556	7388	8239	−3.41

* Actual data outside the 95% approximate large-sample confidence limit.
• The estimates reported for any particular month have been obtained using data up to, but not including, that month.

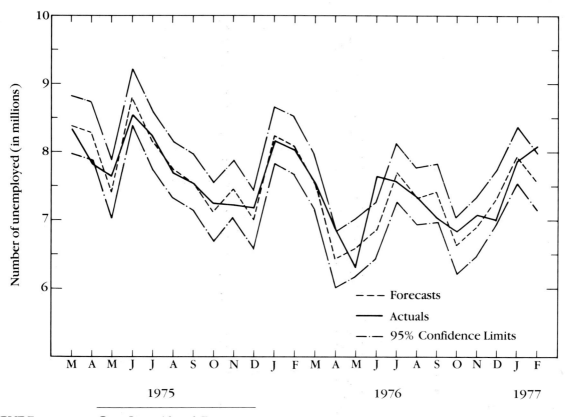

FIGURE 8.10   One Step Ahead Forecasts.

val must be the same as the range of the confidence interval itself. For example, 95% of the forecasts should fall within the lower and upper limits of the 95% confidence interval. As can be seen, values for four of the 24 months for which forecasts were generated using the fixed parameter model fell outside the range of the 95% confidence interval, while only two were outside the range using the updated parameter model. Theoretically, only 1.2 points should have fallen outside the forecasts' 95% confidence band.

Nevertheless, it is rather remarkable that even with all the changes and fluctuations taking place in the economy, a relatively simple model was able to generate so many accurate forecasts for 24 consecutive months.

As a last forecast evaluation, we made unemployment projections for the next 12 months based on parameter estimates incorporating data up to February 1977. Table 8.8 contains these forecasts and Figure 8.12 contains a plot of these projections.

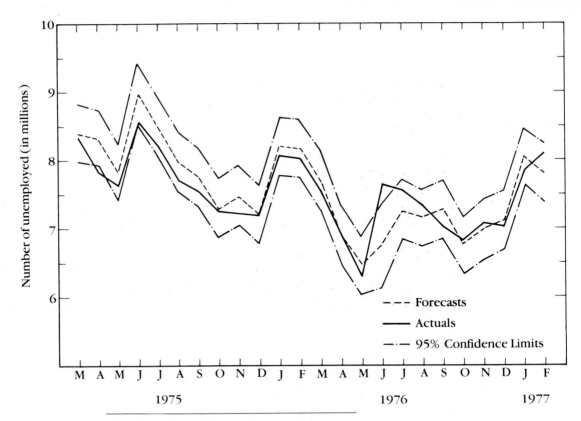

FIGURE 8.11   One Step Ahead Forecasts with Updating.

TABLE 8.8   12-Month Forecast[a]

Model: $\nabla^2\nabla_{12}z_t = (1 - 0.781B)(1 - 0.724B^{12})a_t$

| | | THOUSANDS OF PERSONS | |
| | | 95% CONF. LIMIT | |
MONTH	FORECAST	LOWER	UPPER
Mar 1977	7814	7388	8239
Apr	7307	6636	7978
May	6957	6049	7865
Jun	8090	6941	9239
Jul	7856	6458	9255
Aug	7494	5836	9152
Sep	7363	5435	9291
Oct	7134	4929	9342
Nov	7328	4829	9827
Dec	7321	4521	10120
Jan 1978	8303	5192	11410
Feb	8391	4958	11820

[a] *Base:* February 1977

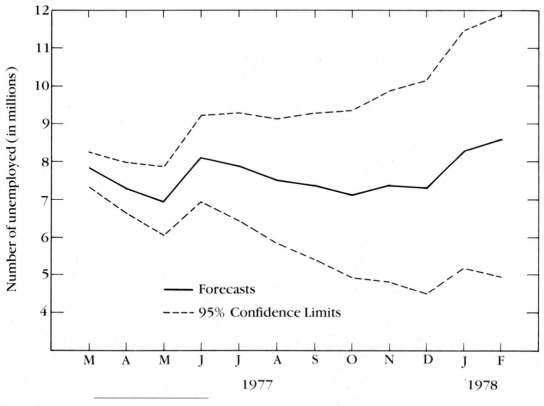

FIGURE 8.12   12-Month Forecast.

8.7 CONCLUSIONS

The model $\nabla^2 \nabla_{12} z_t = (1 - \theta B)(1 - \Theta B^{12})a_t$ fits the unadjusted unemployment data covering the period from January 1948–February 1977 remarkably well. Similarly, one step ahead forecasts generated by this model using updated parameters were very reliable. Only two forecasts during the 24-month testing period lay outside the 95% large-sample confidence limits.

CHAPTER 9

THE CLOROX COMPANY CASE

The case you are about to analyze is based on an actual company setting. The analyses of the problems found in this case require not only skill in formulating Box–Jenkins ARIMA models to forecast shipments of pint-size bleach, but also require that you put to use your managerial talents.

Within the case setting, suppose you are Sergio de Silva. When presenting your report to Martha Meyer, your supervisor, state clearly what assumptions you had to make to formulate your recommendations. This report should include recommendations for how much pint-size bleach the Philadelphia plant will have to produce in advance in case of a strike, for where and when shipments should be made, as well as an estimate of how much the strike will cost.

Note: This chapter was written by Sergio Koreisha.

THE CLOROX COMPANY[1]

In late August of 1977 Mr. Robert Pearson, general manufacturing manager of the Household Products Division of The Clorox Company, asked Ms. Martha Meyer, manager of the Operations Planning Department, to prepare a strike contingency plan for the Philadelphia plant. (See Figure 9.1 for the organizational chart.) The plan, Pearson explained, should include the strategy as well as the cost for supplying the plant's sales territory with all Clorox products from the nearby plants. Also to be included in the report were forecasts of the demand for pint-size bleach for the other four plants that received their pint supply from the Philadelphia plant.

The labor contract was to expire at midnight September 30. Management estimated that if a strike were to occur it would last for approximately four to five weeks, that is, for the entire month of October. The alternatives to be evaluated for the strike contingency plan were to be based on demand figures for just the month of October. Extensions to the plan were to be made only if a settlement were not in sight after three weeks of negotiations.

The Philadelphia Plant

The Philadelphia plant is the largest of the 12 Clorox plants located in the United States. Like all other plants, Philadelphia manufactures only a portion of the products sold in its territory. The remainder of the products comes from various other plants. The matrix in Figure 9.2 contains the products manufactured at each plant location as well as the normal pattern of interplant shipment of products not manufactured at a particular location.

Four filling lines[2] are used for the four products made in Philadelphia:

Line #1: Bleach, pint size, and all three sizes of Formula 409®
Line #2: Bleach, quarts and half gallons, and both sizes of Liquid Plumr®
Line #3: Bleach, gallons and 1-½ gallons
Line #4: Clorox 2®

However, only three lines are in use at any one time. This is because there is not sufficient volume (demand) to justify having more than three line crews per shift.[3] In fact, the plant's average monthly production for pint-size bleach and all sizes of Formula 409® can be produced in 1-½ weeks using two shifts.

Adjacent to the bottling area of the plant, but still part of the Philadelphia

---

[1] This case was prepared as the basis for class discussion rather than to illustrate either effective or ineffective handling of an administrative situation.

[2] Lines #1, #2, and #3 are liquid lines; line #4 is a dry line.

[3] Philadelphia operates five days a week with two shifts per day.

FIGURE 9.1   Simplified Organization Chart of The Clorox Company.

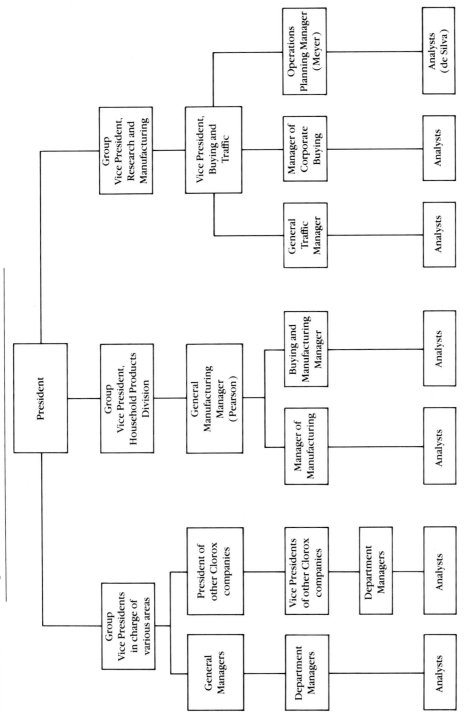

FIGURE 9.2 Matrix of Interplant Shipments.

SOURCE	BOS	BAL	TOL	RAL	PHI	ATL	POR	LA	DAL	ST.L	CHI	OMA
							DESTINATION					
BOS	Bleach*											
BAL		Bleach* LP										
TOL			Bleach* CL2 LP									
RAL					Bleach*							
PHI	Pint 409 CL2 LP FP	Pint 409 CL2 FP	Pint	Pint	Bleach 409 CL2 LP							
ATL				409 CL2 LP FP		Bleach* 409 CL2 LP						
POR							Bleach* LP					
LA**							Pint 409 CL2 FP	Bleach 409 CL2 LP FP	409 CL2 FP			
DAL									Bleach* LP	LP		
ST.L						Pint			Pint	Bleach	Pint	Pint
CHI**			409 FP		FP	FP				CL2 409 FP	Bleach* CL2 LP 409 FP	409 CL2 LP FP
OMA	Litter***	Litter	Litter	Litter	Litter	Litter	Litter	Litter	Litter	Litter	Litter	Bleach* Litter

* All bottle sizes except pints.
** Food products are manufactured at several non-Clorox plant locations and then shipped to Chicago and Los Angeles for distribution to other plants.
*** Cat litter.

Cities		*Products*
BOS = Boston, MA	POR = Portland, OR	LP = Liquid Plumr®
BAL = Baltimore, MD	LA = Los Angeles, CA	CL2 = Clorox 2®
TOL = Toledo, OH	DAL = Dallas, TX	409 = Formula 490®
RAL = Raleigh, NC	ST.L = St. Louis, MO	FP = Food Products
PHI = Philadelphia, PA	CHI = Chicago, IL	Pint = Pints of bleach
ATL = Atlanta, GA	OMA = Omaha, NE	

TABLE 9.1 Average Monthly Shipments and Freight Rates of All Clorox Products to Main Cities in the United States Serviced by the Philadelphia Plant

CITIES	WEIGHT (IN THOUSANDS OF POUNDS)	FREIGHT RATES ($/CWT) FROM					
		PHI*	BOS*	BAL*	TOL*	ATL*	RAL*
Connecticut							
Cheshire	379.1	0.64	0.75	0.83	1.29	1.82	1.46
North Haven	870.4	0.61	0.72	0.81	1.29	1.82	1.46
Sharon	123.1	0.64	0.74	0.81	1.29	1.82	1.46
Suffield	206.5	0.74	0.69	0.87	1.33	1.82	1.46
Total	1579.1						
New Jersey							
Carlstadt	1037.7	0.57	0.83	0.76	1.26	1.79	1.42
Edison Township	1734.9	0.52	0.75	0.71	1.26	1.79	1.42
Elizabeth	2668.3	0.49	0.75	0.69	1.18	1.79	1.42
Hawthorne	390.0	0.40	0.80	0.75	1.23	1.79	1.42
Kearny	995.8	0.49	0.75	0.69	1.19	1.79	1.42
Linden	642.5	0.57	0.75	0.69	1.28	1.79	1.42
Malaga	313.4	0.49	0.70	0.69	1.18	1.79	1.42
Woodbridge	2054.4	0.51	0.72	0.70	1.21	1.79	1.42
Total	9837.0						
New York							
Brentwood	390.7	0.81	1.25	1.17	1.41	1.82	1.57
Bronx	1759.8	0.54	1.25	1.17	1.41	1.70	1.45
Brooklyn	1191.9	0.61	1.25	1.17	1.41	1.70	1.45
Central Islip	734.1	0.81	1.35	1.21	1.41	1.82	1.57
East Farmingdale	1354.6	0.90	1.40	1.25	1.41	2.00	1.75
East Syracuse	434.1	0.92	1.45	1.37	1.35	2.19	1.96
Garden City	544.7	0.43	1.49	0.58	1.41	1.69	1.45
Mount Kisco	475.8	0.54	1.28	1.19	1.41	1.70	1.45
Schenectady	528.4	0.64	1.18	1.19	1.38	2.01	1.76
Syracuse	708.8	0.92	1.45	1.37	1.35	2.22	2.00
Waterford	482.7	0.64	1.25	1.19	1.38	2.02	1.57
Westbury	563.0	0.85	1.29	1.20	1.50	1.90	1.65
Total	9168.6						
Pennsylvania							
King of Prussia	686.7	0.52	1.30	0.90	1.47	1.72	1.46
Oaks	385.8	0.52	1.30	0.90	1.47	1.72	1.46
Philadelphia	3570.1	0.32	1.28	0.89	1.35	1.70	1.45
Yeadon	683.8	0.52	1.30	0.90	1.47	1.72	1.46
Total	5326.4						

* *Key:* PHI = Philadelphia, BOS = Boston, BAL = Baltimore, TOL = Toledo, ATL = Atlanta, RAL = Raleigh

TABLE 9.2   Product Distribution by State (Percent of Lbs.)

STATE	BLEACH	LIQUID PLUMR®	CLOROX 2®	LITTER	FORMULA 409®	FOOD PRODUCTS
Connecticut	84.00	1.00	7.00	3.00	1.00	4.00
New Jersery	81.00	2.00	11.00	3.00	0.25	2.75
New York	85.00	2.00	6.00	2.00	2.00	3.00
Pennsylvania	79.00	3.00	8.00	5.00	2.00	3.00

plant, is a large 300,000 sq. ft. warehouse facility. At present slightly less than ⅔ of it is being utilized.

The plant's sales territory includes the states of New York, New Jersey, Connecticut, and parts of Pennsylvania. Approximately 90% of the demand volume for all Clorox products in this territory is generated in the cities found in Table 9.1. Table 9.2 contains the product mix distribution for each state supplied by Philadelphia.

Managing the Situation

Immediately after the meeting with Pearson, Meyer called her assistant, Sergio de Silva, into her office to discuss the Philadelphia problem. De Silva had recently graduated from the University of California, Berkeley with a Master's Degree in Operations Research. He was the department's forecasting expert.

After explaining some of the details of the conversation she had with Pearson, she added:

"We need to give Pearson plausible numbers next week. The earlier the better. Do you think you could come up with the necessary forecasts within this time period?"

"Well," replied de Silva, "it shouldn't be too difficult to develop Box–Jenkins ARIMA models for the pint demand for all the plants that receive their supply from Philadelphia. I've just finished gathering the pint-size bleach data for all the plants, and the Box–Jenkins program is running beautifully." (See Table 9.3.) "However, I don't have the data for the other products readily available. Remember how much hassle we went through to get the bleach numbers?"

Meyer pondered for a minute or so and then said:

"I've an idea: We need really accurate numbers for the pints because we need to start October's production in two weeks or so. Let's develop ARIMA forecasting models for the pints. For the others we can just use monthly shipment averages. I'll call up the plant managers of the plants that will be involved in supplying Philadelphia's territory to see how much slack production capacity they have available, so that we can come up with a contingency plan." (See Table 9.4.) "We can get the necessary freight rates from Traffic, and the manufacturing costs we can get from Report E-31." (See Tables 9.1, 9.5, and 9.6.) "We can generate the cost figures based on aggregate data." (See Tables 9.7 and 9.8.)

TABLE 9.3   Shipments of Pint-Size Bleach from Various Plants (November 1966–August 1977)[a,b]

	JAN	FEB	MAR	APR	MAY	JUN	JUL	AUG	SEP	OCT	NOV	DEC
Baltimore												
1966											18.11	6.77
1967	7.43	8.97	10.78	10.83	11.65	10.73	11.50	9.40	10.47	8.73	9.18	8.88
1968	9.51	8.30	9.52	11.34	10.78	11.73	11.21	14.18	6.75	7.06	9.80	7.70
1969	9.14	10.34	11.35	11.24	13.53	14.36	13.09	10.05	9.25	16.04	16.13	7.50
1970	9.44	10.12	12.68	28.39	12.85	11.59	14.59	10.62	9.74	9.83	10.26	8.77
1971	9.77	9.78	10.04	27.96	11.85	13.75	12.36	12.84	10.23	12.15	10.66	10.30
1972	12.46	12.32	12.92	15.19	32.33	16.98	14.11	13.10	11.82	15.25	14.05	12.38
1973	16.71	12.86	16.11	18.06	18.72	16.86	13.42	15.69	18.37	16.89	28.23	12.74
1974	15.75	18.32	20.18	23.24	20.18	22.43	23.10	21.15	20.29	25.54	29.13	17.02
1975	15.95	19.12	23.18	25.01	23.48	27.25	23.54	23.71	22.16	23.49	20.11	24.47
1976	19.80	24.44	27.90	23.05	27.35	25.92	19.69	22.72	19.77	22.69	20.22	20.70
1977	22.30	22.96	25.43	29.77	28.69	29.43	26.29	25.95				
Boston												
1966											19.61	16.63
1967	19.18	20.48	21.47	20.66	21.95	23.01	22.33	18.12	17.50	18.81	20.56	14.11
1968	19.09	19.24	20.16	22.78	20.49	25.02	20.52	18.78	16.05	17.90	19.89	16.86
1969	18.59	20.29	21.23	21.62	21.15	24.28	20.34	19.70	17.21	17.39	21.69	17.27
1970	19.02	19.10	22.44	21.32	23.77	22.73	22.87	18.46	17.25	17.71	18.45	15.66
1971	16.70	19.31	19.82	23.71	22.73	25.63	17.76	20.52	16.23	17.49	18.83	20.60
1972	18.40	17.23	17.88	23.09	23.36	21.27	24.42	17.89	16.21	19.12	20.71	14.58
1973	21.18	18.53	19.16	24.57	20.64	17.96	22.00	19.33	20.70	17.66	23.01	19.40
1974	20.13	22.13	22.60	23.70	22.89	23.38	22.69	22.45	21.68	16.33	22.17	20.73
1975	19.53	21.52	24.45	24.46	25.39	25.63	21.66	23.97	19.79	23.83	18.89	20.51
1976	17.86	24.00	28.83	19.03	25.70	33.29	16.05	21.76	19.82	23.03	21.00	24.89
1977	19.46	27.51	23.84	26.82	26.71	27.75	26.38	25.95				
Philadelphia												
1966											34.34	32.40
1967	35.45	36.34	57.80	32.39	51.66	58.83	38.85	35.30	33.47	31.78	39.21	29.25
1968	58.46	35.65	51.86	35.06	39.35	45.26	39.50	36.72	31.33	30.27	38.01	32.88
1969	38.20	40.69	45.69	45.64	49.97	51.97	47.27	61.79	30.88	37.84	39.05	46.29
1970	60.55	40.54	50.00	49.37	59.06	52.77	57.17	43.88	73.03	29.42	38.46	34.72
1971	54.72	29.52	39.92	50.28	54.78	55.33	46.97	68.79	39.71	53.77	38.67	40.10
1972	43.12	44.51	55.55	64.90	55.36	62.24	59.95	44.72	44.20	52.86	60.07	40.45
1973	47.96	54.26	65.72	72.43	50.19	49.14	61.71	60.45	68.10	55.71	60.47	59.16
1974	67.33	74.85	67.50	76.82	62.39	103.54	60.56	61.33	68.74	54.64	67.56	70.66
1975	66.72	92.43	76.34	88.19	74.04	100.92	60.89	85.35	87.41	68.82	62.50	94.37
1976	72.41	91.81	83.41	73.18	96.50	79.57	64.97	116.77	60.43	64.03	70.92	101.22
1977	60.00	73.75	96.72	102.80	90.82	98.20	97.70	107.10				

[a] August 1977 data are preliminary and are based on shipments of the first three weeks of August.
[b] Figures are thousands of cases.

TABLE 9.3 (*continued*)

	JAN	FEB	MAR	APR	MAY	JUN	JUL	AUG	SEP	OCT	NOV	DEC
Raleigh												
1966											48.40	42.52
1967	45.44	48.88	49.57	48.41	55.31	52.03	50.27	43.32	41.13	39.73	50.16	34.60
1968	42.08	43.30	44.22	52.32	46.86	52.98	45.31	40.58	34.80	36.33	44.41	37.26
1969	43.72	41.28	48.20	46.12	46.54	50.93	42.53	40.31	39.39	40.51	45.80	41.65
1970	39.43	47.49	47.08	46.94	54.82	52.61	49.13	40.20	37.01	41.91	44.98	30.52
1971	34.51	39.35	44.28	50.48	52.28	49.23	42.11	44.93	35.39	40.04	39.60	39.61
1972	40.04	36.40	40.30	48.18	49.95	47.48	49.64	36.82	36.99	39.54	41.85	33.71
1973	40.02	36.85	36.91	48.19	43.22	34.50	40.81	43.34	43.58	37.58	43.41	43.18
1974	41.23	41.82	46.54	48.45	40.70	44.70	45.77	45.69	42.56	34.15	44.57	40.41
1975	38.54	40.71	44.60	49.52	44.11	47.71	39.09	44.28	37.22	40.63	37.29	40.01
1976	32.45	40.20	47.52	37.75	44.82	46.52	34.08	41.59	35.29	37.74	39.93	38.80
1977	34.24	40.69	43.35	47.38	45.97	47.40	40.75	43.73				
Toledo												
1966											89.34	80.53
1967	91.77	89.73	93.60	89.33	109.35	99.43	106.69	116.06	77.33	88.87	119.76	63.94
1968	83.76	84.84	89.07	101.12	116.14	121.94	85.61	79.90	73.62	75.27	91.71	74.87
1969	92.39	89.90	107.20	110.03	153.72	113.60	112.46	71.59	86.24	85.41	102.39	101.08
1970	86.21	106.44	122.57	104.54	155.63	123.44	184.62	79.95	65.27	118.33	113.16	57.36
1971	82.14	104.56	89.32	116.13	142.41	131.94	71.38	104.49	107.95	74.35	90.06	116.81
1972	91.86	77.65	101.02	107.53	126.48	107.53	147.43	102.31	78.26	88.21	85.50	99.08
1973	99.13	115.30	86.06	115.79	132.70	83.43	145.32	107.33	106.07	103.72	100.11	147.93
1974	103.84	154.11	91.87	122.43	156.50	100.47	144.61	136.82	95.65	94.67	124.16	166.46
1975	88.91	116.86	131.72	148.98	103.54	162.10	101.66	138.95	116.47	112.41	86.65	162.46
1976	96.11	104.48	99.55	102.76	174.79	115.96	77.70	107.88	102.24	111.44	86.71	136.37
1977	102.76	81.51	94.56	119.32	134.30	180.91	75.95	94.98				
Total of All Plants												
1966											209.80	178.85
1967	199.26	204.41	233.22	201.62	249.92	244.04	229.64	222.20	179.89	187.93	238.87	150.78
1968	212.90	190.84	214.84	222.62	233.62	256.92	202.15	190.16	162.54	166.82	203.83	169.52
1969	202.03	202.50	233.67	234.64	284.91	255.15	235.69	203.45	182.97	197.20	225.07	213.79
1970	214.65	223.69	254.76	250.56	306.13	263.15	328.38	193.11	202.29	217.20	225.30	147.03
1971	197.83	202.51	203.35	268.55	284.05	275.86	190.58	251.57	209.50	197.80	197.82	227.44
1972	205.89	188.11	227.67	258.88	287.48	255.51	295.56	214.84	187.48	214.98	228.18	200.20
1973	234.00	237.80	223.96	279.03	265.42	201.89	283.26	246.15	256.83	231.56	255.23	282.42
1974	248.37	311.22	248.69	294.64	302.66	294.51	296.73	287.45	248.92	225.33	287.59	315.27
1975	229.64	290.64	300.29	336.16	270.57	363.61	246.84	316.09	283.05	269.17	225.44	341.82
1976	238.64	284.91	287.21	255.77	369.16	301.27	212.49	310.71	237.54	258.92	238.78	321.97
1977	238.75	246.42	283.91	326.09	326.48	383.69	266.53	297.70				

TABLE 9.4   Monthly Slack Capacity[a]

PRODUCT	BOS	BAL	TOL	RAL	ATL
Bleach	3000	6000	10000	3000	8000
Clorox 2®	——	——	2000	——	3000
Liquid Plumr®	——	250	500	——	1000
Formula 409®	——	——	——	——	1000

[a] All other cities do not have any slack capacity.

TABLE 9.5   Some Interplant Freight Rates ($/cwt)[a]

	PHI	CHI	BAL	TOL	BOS	ATL	RAL
Philadelphia	0.25						
Chicago	1.50	0.26					
Baltimore	0.83	2.33	0.22				
Toledo	1.24	0.85	1.65	0.21			
Boston	0.75	2.25	1.75	1.89	0.28		
Atlanta	1.82	1.80	1.80	1.75	2.25	0.23	
Raleigh	1.69	3.00	1.60	2.80	2.30	1.35	0.22

[a] Entries along the diagonal of this table are transportation costs incurred to ship goods to outside warehouses located in the same cities as the plants.

TABLE 9.6   Manufacturing Costs for Some Plants ($/case)

PRODUCT	PHI	BOS	BAL	TOL	RAL	ATL
Bleach	6.18	6.36	6.68	6.10	5.84	5.79
Clorox 2®	11.08	——	——	11.17	——	10.84
Liquid Plumr®	10.00	——	10.06	10.48	——	9.18
Formula 409®	15.57	——	——	——	——	10.39

TABLE 9.7   Average Monthly Demand for the Products Shipped to the Territories of the Boston and Baltimore Plants

PLANT	TOTAL (THOUSANDS OF LBS.)	BLEACH	DISTRIBUTION (PERCENT)				
			LIQUID PLUMR®	CLOROX 2®	LITTER	FORMULA 409®	FOOD PRODUCTS
Boston	7889.6	84	3	7	1	1	4
Baltimore	9051.5	81	1	5	6	2	5

TABLE 9.8   Average Distribution of Clorox Products by Size for All Plants

PRODUCT	SIZE	PERCENT OF LBS.
Bleach	Pints	10
	Quarts	12
	Halves	34
	Gallons	29
	Kings	15
Clorox 2®	24–24 oz.[a]	12
	12–40 oz.	26
	8–61 oz.	38
	6–100 oz.	24
Formula 409®	22 oz.	12
	32 oz.	47
	64 oz.	41
Liquid Plumr®	Quarts	50
	Halves	50
Food Products	Small	35
	Large	65

[a] 24–24 oz. = 24 cartons per case of 24 ounces each; 12–40 oz. = 12 cartons per case of 40 ounces each.

TABLE 9.9   Pallet Statistics for the Products

PRODUCT	SIZE	LBS/CASE	CASES/PALLET
Bleach	Pints	30.50	56
	Quarts	45.00	48
	Halves	39.50	60
	Gallons	38.75	35
	Kings	58.75	45
Clorox 2®	24–24 oz	40.75	60
	12–40 oz.	33.00	84
	8–61 oz.	33.75	70
	6–100 oz.	41.25	75
Formula 409®	22 oz.	21.00	75
	32 oz.	28.00	55
	64 oz.	39.00	40
Liquid Plumr®	Quarts	47.75	50
	Halves	31.75	60
Food Products	Small	40.25	144
	Large	48.75	78

TABLE 9.10   Monthly Outside Warehouse Costs
(Includes Loading and Unloading)

CITY	COST ($/PALLET)
Baltimore	3.00
Boston	2.75
Philadelphia	3.25
Raleigh	2.50
Toledo	3.15

"Great!" explained de Silva. "I think that by the end of the week we can get together to write up the report. I'll get the data on pallet configuration and call up a few outside warehouses to get some quotes on storage rates." (See Tables 9.9 and 9.10.) "We have to assume that plants that are to receive shipments of pints from Philadelphia in September to satisfy October's demand won't have sufficient storage space to accommodate all that volume."

"Right!" Meyer continued, "We won't, however, have to incur storage costs for the products which will be supplied from the other plants to the Philadelphia plant shipment area because they can probably load the trucks as the products come off the line."

After discussing a few more details, de Silva returned to his office to start on the phone calls and on the formulation of the ARIMA models.

CHAPTER 10

___

TRANSFER FUNCTION MODELS

In previous chapters we analyzed the forecasting of a single time series without explicitly using information contained in other related time series. In many forecasting situations, other events will systematically influence the series to be forecasted (the dependent variables) and therefore there is a need to go beyond a univariate forecasting model. We must build a forecasting model that incorporates more than one time series and introduces explicitly the dynamic characteristics of the system. Such a model is called a *multiple time series model,* or *transfer function model.* These time series included in the model to explain the behavior of dependent variables are sometimes called *leading indicator variables.* Therefore, a transfer function model is sometimes also called a *leading indicator model.* We will not use this last term because, in principle, a transfer function model could incorporate feedback. In this book we will only pay lip service to causality and a causal transfer function model. A complete treatment of this topic would needlessly increase the level of mathematical sophistication. For a discussion of this and related topics see Granger and Newbold (1973), Zellner and Palm (1974, 1975), Pierce and Haugh (1977), Wallis (1977), Zellner (1979), Geweke (1983), Granger (1980), and Palm and Zellner (1980).

Typical examples of transfer function models are forecasting sales using advertising expenditures [Helmer and Johansson (1977), Kyle (1978)], aggregate consumption and advertising [Ashley et al. (1980)], and the consumption of alcoholic beverages in Finland [Leskinen and Teräsvirta (1976)].

Special cases of transfer function models are *intervention models.* These models are typically constructed as a means of assessing the impact of a discrete intervention or event on a stochastic process. Because these intervention models require a rather simple theory of impact not all interventions can be modeled with such an intervention model. Nevertheless, there are many situations where such simple theories of impact are justified. Examples of such models include assessment of the impact of marketing mix manipulations [effect of Procter and Gamble's promotion of the American Dental Association endorsement of Crest on the market shares of Crest and Colgate dentifrices, Wichern and Jones (1977)], the effect of public policy on photochemical smog data in Los Angeles [Box and Tiao (1975), Tiao et al. (1975a and b)], the impact of new traffic laws [Campbell and Ross (1968), Glass (1968), Glass et al. (1975), Ross et al. (1970)], and the impact of gun control laws [Zimring (1975)]. This is a representative and by no means exhaustive list of situations in which intervention models have been used.

As with the univariate model building process, we will see that the transfer function model building process involves three stages that we similarly label *identification, estimation and diagnostic checking,* and *forecasting.* Although the names of these stages are the same, the methodology is not identical, but several univariate time series concepts still will be used.

10.1 TRANSFER FUNCTION MODEL

To understand a transfer function model, we must go back to the econometric literature on distributed lag models. In a general distributed lag model we represent the current level of Y_t as a function of a number of past values of X_t:

$$Y_t = v_0 X_t + v_1 X_{t-1} + v_2 X_{t-2} + \cdots + e_t \qquad (10.1)$$

or

$$Y_t = v(B) X_t + e_t, \qquad (10.2)$$

where $v(B) = v_0 + v_1 B + v_2 B^2 + \cdots$; B is the backward shift operator defined as $BX_t = X_{t-1}$ [see Section (3.5.2)]; and e_t is a random variable with mean zero, a fixed covariance structure (it may or may not be serially correlated), and independent of $X_t, X_{t-1}, X_{t-2}, \ldots$. In equations (10.1) and (10.2) Y is the dependent variable and X is the explanatory variable, the leading indicator. For a survey covering the earlier history of distributed lag models, see Griliches (1967).

The weights $v_0, v_1, v_2, \ldots$ in (10.1) are called the *impulse response weights* and a graph of these weights is called an *impulse response function*. An example of an impulse response function is given in Figure 10.1. These weights represent the impact on Y_t of unit changes in X_t.

Without additional model information the problem of estimating the parameters $v_0, v_1, v_2, \ldots$ is not properly defined, as an *infinite* number of functionally unrelated parameters must be estimated from a *finite* set of observations. However, it is natural to require that a finite change in an explanatory variable which persists indefinitely would result only in a finite change in the dependent variable. Suppose that X is held indefinitely at the value $+1$, and observe the total change in Y. This change in Y_t is represented by the

FIGURE 10.1   The Impulse Response Function.

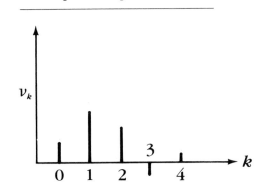

sum of the impulse response weights ν_k. Therefore we will impose the restriction that

$$\sum_{k=0}^{\infty} \nu_k = g, \qquad g \text{ finite.} \tag{10.3}$$

The value of g is called the *steady state gain* of the system as it represents the total change in Y_t for a unit change in X_t indefinitely held at the new value. Alternatively, g represents the accumulated change over time in Y_t for a unit change in X_t for one period. The former change in X_t is called a *step change,* the latter an *impulse change.* A transfer function model satisfying condition (10.3) is said to represent a *stable* system.

In addition it will be very useful to introduce two other, more subtle restrictions. First, irrespective of whatever additional restrictions, it should be possible to approximate the true distributed lag arbitrarily close. Second, adhering to the principle of *parsimony,* the number of unknown parameters should be as small as possible. It can be shown that under some general conditions [see, e.g., Jorgenson (1966)] we can approximate $\nu(B)$ by a ratio of two *finite* (rational) polynomials in B:

$$\nu(B) = \frac{\omega(B)}{\delta(B)} \tag{10.4}$$

where

$$\omega(B) = \omega_0 - \omega_1 B - \cdots - \omega_l B^l \tag{10.5}$$

$$\delta(B) = \delta_0 - \delta_1 B - \cdots - \delta_r B^r. \tag{10.6}$$

Without loss of generality we may normalize the coefficient δ_0 at unity. We also impose stationarity upon the polymials $\omega(B)$ and $\delta(B)$ implying that all roots lie outside the unit circle. In Chapter 12 we will use the relationship between $\nu(B)$ and $\omega(B)$ and $\delta(B)$ to calculate initial guess values for the parameters in the $\omega(B)$ and $\delta(B)$ polynomial, given estimates of the parameters in $\nu(B)$.

To obtain a more general form of a transfer function model, we first substitute (10.4) in (10.2):

$$Y_t = \frac{\omega(B)}{\delta(B)} X_t + e_t. \tag{10.7}$$

Above we have implicitly assumed that a change in X instantaneously affects Y. However, it is quite possible that there is a delay, or *dead time,* in the system's response to changes. To introduce this generality, let b indicate the number of periods it takes before X_t starts influencing the dependent variable. The transfer function model (10.7) may then be written as

$$Y_t = \frac{\omega(B)}{\delta(B)} X_{t-b} + e_t. \tag{10.8}$$

Multiplying both sides of this equation by $\delta(B)$ we obtain

$$\delta(B)Y_t = \omega(B)X_{t-b} + \epsilon_t \qquad (10.9a)$$

where

$$\epsilon_t = \delta(B)e_t. \qquad (10.9b)$$

Alternatively, we can rewrite (10.9) as

$$Y_t = \delta_1 Y_{t-1} + \cdots + \delta_r Y_{t-r} + \omega_0 X_{t-b} - \omega_1 X_{t-b-1} - \cdots$$
$$- \omega_l X_{t-b-l} + \epsilon_t. \qquad (10.10)$$

The error term e_t in (10.7) or (10.9b) is not necessarily white noise. However, it might be possible to represent e_t with the following, by now familiar, univariate ARIMA process, assumed to be statistically independent of the explanatory variable X_t,

$$\nabla^d e_t = \frac{\theta(B)}{\phi(B)} a_t \qquad (10.11)$$

or

$$\phi(B)\nabla^d e_t = \theta(B) a_t \qquad (10.12)$$

with

$$\phi(B) = 1 - \phi_1 B - \cdots - \phi_p B^p$$
$$\theta(B) = 1 - \theta_1 B - \cdots - \theta_q B^q, \qquad (10.13)$$

and ∇^d is the consecutive difference operator used to induce stationarity in the series e_t, and a_t is assumed to be white noise. It is also assumed that the $\phi(B)$ polynomial satisfies the stationarity conditions and that the $\theta(B)$ polynomial is invertible, that is, all roots of these polynomials should lie outside the unit circle. The univariate ARIMA model (10.11) has been the subject of the material presented in the first nine chapters of this book. We later will see that the identification of the ARIMA model also follows the same line of arguments presented in these earlier chapters.

Substituting (10.11) into (10.7) we have

$$\nabla^d Y_t = \frac{\omega(B)}{\delta(B)} \nabla^d X_t + \frac{\theta(B)}{\phi(B)} a_t. \qquad (10.14)$$

Notice that if the error process (10.11) needs differencing to induce stationarity, the same differencing carries over to the dependent and explanatory variables.

It is good practice, however, to construct a transfer function using stationary dependent and stationary explanatory variables. Therefore, there is absolutely no need to use the same difference operator for each variable. Furthermore, it is then quite possible that after these stationary transformations the error process (10.11) is a simple ARMA(p,q) process, not involving any difference operator.

As a result, the transfer function (10.14) can now more generally be written as

$$y_t = \frac{\omega(B)}{\delta(B)} x_{t-b} + \frac{\theta(B)}{\phi(B)} a_t \qquad (10.15)$$

with

$$y_t = \nabla^{d'} Y_t$$
$$x_t = \nabla^{d} X_t,$$

where d' refers to the order of consecutive differencing of the dependent variable Y_t and d refers to the order of consecutive differencing of the explanatory variable X_t, and d' and d are not necessarily of the same order. In the transfer function model construction we will use upper-case letters to denote the possibly nonstationary raw data. Lower-case letters will then refer to stationary data. This transfer function model is represented in Figure 10.2. At the top of the figure we have the transfer function structure determining the nature of the influence of the explanatory variable on the dependent variable. In the lower part we have the noise model representing a standard

FIGURE 10.2   The Transfer Function Model.

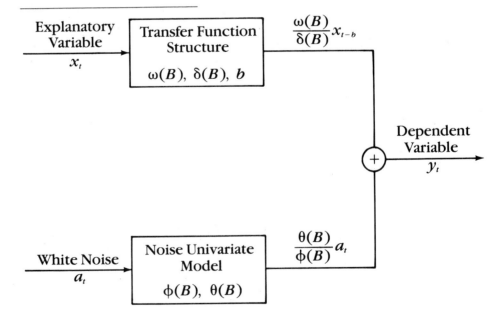

univariate ARIMA process. Finally, these two parts are put together to form the complete transfer function model.

The parameters in the polynomial $\omega(B)$ are commonly referred to as the *numerator parameters,* in $\delta(B)$ as the *denominator parameters,* in $\phi(B)$ as the (error) *autoregressive parameters,* and in $\theta(B)$ as the (error) *moving average parameters.*

For future reference, we define the following vector of parameters

$$\underline{\omega} = (\omega_0, \omega_1, \ldots, \omega_l)'$$
$$\underline{\delta} = (\delta_1, \delta_2, \ldots, \delta_r)'$$
$$\underline{\phi} = (\phi_1, \phi_2, \ldots, \phi_p)'$$
$$\underline{\theta} = (\theta_1, \theta_2, \ldots, \theta_q)'.$$

The model (10.15) can easily be extended to include several explanatory variables to obtain the following, more general, form:

$$y_t = \sum_{i=1}^{m} \frac{\omega_i(B)}{\delta_i(B)} x_{i,t-b_i} + \frac{\theta(B)}{\phi(B)} a_t \qquad (10.16)$$

with m the number of explanatory variables in the model. Again we assume that the roots of all the polynomials $\omega_i(B)$, $\delta_i(B)$, $\theta(B)$, and $\phi(B)$ lie outside the unit circle.

10.2 SPECIAL TRANSFER FUNCTION MODELS

The transfer function model is a very general model in which several commonly known econometric models are embedded. If the numerator polynomials are of order 0, if there are no denominator parameters in the model (i.e., $\delta_i(B) = 1$), and there are not autoregressive and moving average error terms (i.e., all polynomials are of 0 order), then we have the familiar *multiple regression model* (MRM):[1]

$$y_t = \omega_{10} x_{1t} + \omega_{20} x_{2t} + \cdots + \omega_{m0} x_{mt} + a_t. \qquad (10.17)$$

Notice that we assume $b_i = 0$, and that ω_{i0} denotes the zero order numerator parameter in $\omega_i(B)$. If we relax the error polynomial restriction and assume $\theta(B) = 1$, and $\phi(B) = 1 - \phi B$, we obtain the MRM *with first-order autocorrelated errors*

$$y_t = \sum_{i=1}^{m} \omega_{i0} x_{it} + e_t$$
$$e_t = \phi e_{t-1} + a_t. \qquad (10.18)$$

---

[1] The standard regression models do not necessarily require stationarity. However, spurious correlation could easily be observed with explanatory variables that contain a trend or are nonstationary in general. See Yule (1926), Granger and Newbold (1977, p. 202).

Similarly, if $\phi(B) = 1$ and $\theta(B) = 1 - \theta B$, we obtain a MRM *with first-order moving average errors.* This model is represented by

$$y_t = \sum_{i=1}^{m} \omega_{i0} x_{it} + e_t$$

$$e_t = a_t - \theta a_{t-1}. \tag{10.19}$$

Also, if only the numerator polynomial $\omega(B)$ is of nonzero order, we have the standard *distributed lag model*

$$y_t = \omega_0 x_t + \omega_1 x_{t-1} + \cdots + \omega_1 x_{t-1} + a_t. \tag{10.20}$$

In (10.20) we assumed there is only one explanatory variable. Finally, if $\delta(B) = 1 - \delta B$ and if the numerator polynomial is of 0 order, we have the MRM *with lagged dependent variable,*

$$y_t = \delta y_{t-1} + \sum_{i=1}^{m} \omega_{i0} x_{it} + a_t. \tag{10.21}$$

Currently, much research is progressing in the area of a MRM with ARIMA error terms of which (10.18) and (10.19) are special cases. A reference containing further discussion of this topic, although at a more advanced level, is Judge et al. (1980).

10.3 SEASONAL TRANSFER FUNCTION MODELS

As with the univariate time series models, we may also have seasonal parameters in a transfer function model. Similarly, we may have to induce stationarity in the data by suitable seasonal differencing. Therefore, the more general multiplicative seasonal transfer function model, assuming that there is only one explanatory variable, can be written as

$$y_t = \frac{\omega(B)\Omega(B^s)}{\delta(B)\nabla(B^s)} x_{t-b} + \frac{\theta(B)\Theta(B^s)}{\phi(B)\Phi(B^s)} a_t \tag{10.22}$$

with

$$y_t = \nabla_s^{D'} \nabla^{d'} Y_t \tag{10.23}$$

$$x_t = \nabla_s^{D} \nabla^{d} X_t. \tag{10.24}$$

The parameters d' and D' in (10.23) stand for the consecutive and seasonal difference operators, respectively, for the Y process. Similarly, d and D in (10.24) stand for the consecutive and seasonal difference operators for the X process.

The $\omega(B)$ and $\delta(B)$ polynomials have been defined in (10.5) and (10.6), and $\theta(B)$ and $\phi(B)$ in (10.13). The seasonal polynomials are defined as

$$\Omega(B^s) = \Omega_0 - \Omega_1 B^s - \Omega_2 B^{2s} - \cdots - \Omega_L B^{Ls},$$
$$\nabla(B^s) = 1 - \nabla_1 B^s - \nabla_2 B^{2s} - \cdots - \nabla_R B^{Rs},$$
$$\Theta(B^s) = 1 - \Theta_1 B^s - \Theta_2 B^{2s} - \cdots - \Theta_Q B^{Qs},$$
$$\Phi(B^s) = 1 - \Phi_1 B^s - \Phi_2 B^{2s} - \cdots - \Phi_P B^{Ps}. \tag{10.25}$$

Notice that if there are also regular numerator parameters, we normalize the transfer function with $\omega_0 \neq 1$ and $\Omega_0 = 1$. If, however, there are no regular numerator parameters, we assume $\Omega_0 \neq 1$.

10.4 OVERVIEW

We will focus the analysis on the nonseasonal transfer function model (10.15), although suitable modifications to accommodate seasonal models will be demonstrated with examples.

The *identification* process of a transfer function model involves the determination of tentative values of l and r in the polynomials (10.5) and (10.6), the values of dead time b, as well as values for p and q in the polynomials in (10.13). Just as the identification of the univariate time series model suggests initial guess values of the ARIMA parameters, the statistics used to tentatively identify these transfer function polynomial orders will be helpful in suggesting preliminary estimates of the parameters of the model.

Given the values of l, r, b, p, and q, we can then start the *estimation phase* to obtain parameter values and associated statistics. Results obtained at this stage will allow us to apply *diagnostic checks* on the model accuracy resulting in accepting or reformulating the model.

Once the transfer function model has passed the above stages we can then use the model for *forecasting* purposes. In the following chapters we will discuss in greater detail these three stages of the transfer function model building process.

CHAPTER 11

IDENTIFICATION OF TRANSFER FUNCTION MODELS

An important data analytic tool for the identification of a transfer function model is the *cross correlation function* between the dependent and an explanatory variable. The autocorrelation function, used for the identification of univariate time series models, indicates the degree of correlation within a *single* time series at different time periods, the within-series correlation. The cross correlation function measures the correlations between *two* time series at different time periods, the between-series correlation. We first will describe the basic properties of the cross correlation function and in the next section show how it can be used effectively to identify a transfer function model.

11.1 CROSS COVARIANCE AND CROSS CORRELATION FUNCTION

As seen in Chapters 2 and 3, we could describe the degree of association between a single time series at various time lags with the autocovariance γ_k and autocorrelations ρ_k. We also noticed that $\gamma_k = \gamma_{-k}$, and similarly $\rho_k = \rho_{-k}$, so that we only calculated the autocorrelation function for $k > 0$.

For a transfer function model we can also calculate the autocorrelations for the dependent as well as the explanatory variable separately, denoted by $\rho_{yy}(k)$ and $\rho_{xx}(k)$, respectively. In addition, we can measure the degree of association between an explanatory variable and the dependent variable at various time lags. These measures of association are called *cross covariances* and *cross correlations* (or *scaled covariances*), and are denoted by $\gamma_{xy}(k)$ and $\rho_{xy}(k)$, respectively. The term *cross* comes from the fact that these statistics indicate the degree of association across two variables.

Formally, the cross covariances are defined as

$$\gamma_{xy}(k) = E[(x_t - \mu_x)(y_{t+k} - \mu_y)] \qquad k = 0, \pm 1, \pm 2, \ldots, \qquad (11.1)$$

where μ_x and μ_y are the means of the stationary x and y series. Notice that, contrary to the autocovariances, the cross covariances need not be symmetric about $k = 0$. It is quite possible that x_t is a leading indicator of y_t and therefore is strongly correlated with future values of y_t; that is, $\gamma_{xy}(k) > 0$, $k > 0$, and not correlated at all with lagged values of y_t (i.e., $\gamma_{xy}(-k) = 0$, $k > 0$). So, $\gamma_{xy}(k) \neq \gamma_{xy}(-k)$. From (11.1) it can, however, be verified that $\gamma_{xy}(k) = \gamma_{yx}(-k)$. Therefore, the cross covariances measure not only the strength of a relationship but also its direction.

In equation (11.1) we used x_t and y_t and not X_t and Y_t. Again the cross covariances are only defined between two stationary time series. Of course, sample cross covariances can always be calculated based on a finite time series realization whether the series are stationary or not, although these numbers cannot be properly interpreted as cross covariances of the process. The formula used to calculate these sample covariances will be presented in Section 11.2. As was the case for autocovariances and autocorrelations, there is information contained in the calculated cross covariances allowing us to evaluate nonstationarity in a series.

The use of k in equation (11.1) is frequently a source of confusion, in particular, for cross correlations or cross covariances. Two rules should be remembered. First, a positive k in $\gamma_{xy}(k)$ denotes a lag, a negative k denotes a lead. Second, in $\gamma_{xy}(k)$, the k being written to the right of y refers to the y variable, and the value of k is *added* to the current time subscript. So, if current y is influenced by x lagged two time periods, then $\gamma_{xy}(2)$ will be nonzero. As stated above, this also implies that $\gamma_{yx}(-2) = \gamma_{xy}(2)$.

The *cross correlations* are scaled cross covariances and are defined as

$$\rho_{xy}(k) = \frac{\gamma_{xy}(k)}{\sigma_x \sigma_y} \qquad k = 0, \pm 1, \pm 2, \ldots, \qquad (11.2)$$

where σ_x and σ_y are the standard deviations of the x and y series, respectively. The graph of $\rho_{xy}(k)$, plotted for $k = 0, \pm 1, \pm 2, \ldots$, is called the *cross correlation function*, or *ccf*.

Because the cross correlation function is not symmetric, the cross correlations are calculated for both positive lags and negative lags (leads). The large lag cross correlations, $\rho_{xy}(k)$, $k > 0$, are an indication that current y_t is related to past values of the explanatory variable; the large lead cross correlations, $\rho_{xy}(k)$, $k < 0$, are an indication that y_t is a predictor of x_t. Figure 11.1 graphically shows the difference.

As will be demonstrated with the help of examples, the cross correlation function will be useful in determining

1. whether the raw data, Y_t and X_t, are stationary;
2. whether the variables, y_t and x_t, show any relationship between them; and
3. the type of relationship between y_t and x_t.

FIGURE 11.1   Lag and Lead Cross Correlations.

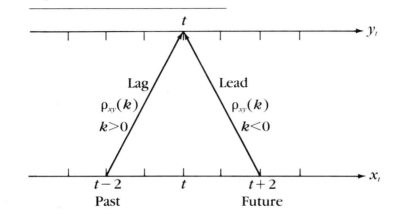

11.2 SAMPLE CROSS COVARIANCES AND CROSS CORRELATIONS

The sample cross covariances are calculated analogously to the sample auto-correlations defined in Section 4.5. An estimate of cross covariance at lag k is defined as:

$$c_{xy}(k) = \begin{cases} \dfrac{1}{n} \displaystyle\sum_{t=1}^{n-k} (x_t - \bar{x})(y_{t+k} - \bar{y}) & k = 0, 1, 2, \ldots \\ \dfrac{1}{n} \displaystyle\sum_{t=1-k}^{n} (x_t - \bar{x})(y_{t+k} - \bar{y}) & k = 0, -1, -2, \ldots, \end{cases} \tag{11.3}$$

where $\bar{x}, \bar{y}$ are the means of the stationary x series and y series, respectively, and n is the number of observations available after suitable differencing has been made to induce stationarity. Similarly, we obtain an estimate $r_{xy}(k)$ of the cross correlation coefficient $\rho_{xy}(k)$ by substituting in (11.2) the estimates $c_{xy}(k)$ for $\gamma_{xy}(k)$, $s_x = \sqrt{c_{xx}(0)}$ for σ_x, and $s_y = \sqrt{c_{yy}(0)}$ for σ_y. Therefore the sample cross correlation coefficient is defined as

$$r_{xy}(k) = \frac{c_{xy}(k)}{s_x s_y} \qquad k = 0, \pm 1, \pm 2, \ldots \tag{11.4}$$

EXAMPLE

Let us calculate the sample cross correlations at lag +1, 0, and −1 using the following series:

t	1	2	3	4	5
x_t	6	5	4	3	7
y_t	7	10	5	5	8.

The means are $\bar{x} = 5$ and $\bar{y} = 7$. The deviations are presented below:

t	1	2	3	4	5
$x_t - \bar{x}$	1	0	−1	−2	2
$y_t - \bar{y}$	0	3	−2	−2	1.

Hence,

$$\sum_{t=1}^{4} (x_t - \bar{x})(y_{t+1} - \bar{y}) = (1)(3) + (0)(-2) + (-1)(-2) + (-2)(1)$$

$$= 3,$$

and the lag 1 cross covariance is

$$c_{xy}(1) = 3/5 = 0.60.$$

The standard deviations are $s_x = \sqrt{2} = 1.41$ and $s_y = \sqrt{3.6} = 1.90$. Therefore we obtain as the lag 1 cross correlation

$$r_{xy}(1) = \frac{c_{xy}(1)}{s_x s_y} = \frac{0.60}{(1.41)(1.90)} = 0.22.$$

Similarly,

$$\sum_{t=2}^{5} (x_t - \bar{x})(y_{t-1} - \bar{y}) = -3.$$

Hence, $c_{xy}(-1) = -0.60$ and $r_{xy}(-1) = -0.22$. The zero-order cross correlation can be calculated as 0.60.

Based on these cross correlations we find that current x_t is a good predictor of current y_t and a moderate predictor of next periods y_t. At the same time, changes in y_t are associated with future changes in x_t. Therefore, x_t is not just a leading indicator.

APPROXIMATE STANDARD ERRORS

The basic results for large sample standard errors for the cross correlations rely on a formula by Bartlett (1946, 1966). It can be shown that for two independent white noise series, the approximate standard error of the cross correlations is

$$SE[r_{xy}(k)] = \frac{1}{\sqrt{n}}. \tag{11.5}$$

For more details we refer the reader to Box and Jenkins (1976, p. 376) and Haugh (1976). The latter reference contains a good discussion of the use of the cross correlations to evaluate correlations among two time series. Haugh (1976) also presents some Monte Carlo results, showing that for large $|k|$, relative to n, the approximate standard error given by

$$SE[r_{xy}(k)] = \frac{\sqrt{1 - |k/n|}}{\sqrt{n}} \tag{11.6}$$

is more accurate than (11.5) in the sense that for finite n the asymptotic distribution results give a better approximation.[1]

JOINT TEST

As was the case for the autocorrelations (see Section 4.5), rather than evaluating each cross correlation individually we may also want to see if, as a group, the first, say, 12 cross correlations are nonzero. That is, we may want to perform a joint test that all first 12 cross correlations are 0. It can be shown [see Pierce (1972) and Ljung and Box (1978)] that for cross correlations a Q-test, a 'portmanteau test,' can also be calculated and is defined as

---

[1] The program CROSSC only uses the formula (11.5) for the approximate standard error of the cross correlations.

$$Q(K) = n(n+2) \sum_{k=1}^{K} \frac{1}{n-k} r_{xy}^2(k). \tag{11.7}$$

This statistic is approximately distributed as a χ^2 (chi-square) distributed random variable with K degrees of freedom, $\chi^2(K)$. The value n represents the number of data points used in the calculation of $r_{xy}(k)$ and the value K is selected such that it can reasonably be assumed that the $r_{xy}(k)$'s for $k > K$ are negligible.

If the computed value of Q is less than the table value of the χ^2 statistic with K degrees of freedom, given a prespecified significance level the cross correlations used to calculate the test can be assumed to be not different from 0. If the computed Q statistic is larger than the χ^2 value from a χ^2 table, the cross correlations are significantly different from 0 and indicate the existence of some pattern.[2]

11.3 IDENTIFICATION USING THE CROSS CORRELATION FUNCTION

We now will show how to identify a transfer function model as specified in (10.15):

$$y_t = \frac{\omega(B)}{\delta(B)} x_{t-b} + e_t, \tag{11.8}$$

with

$$e_t = \frac{\theta(B)}{\phi(B)} a_t. \tag{11.9}$$

Specifically, we will show how to use the information in the sample cross correlations to obtain some idea of the orders l and r of the numerator and denominator polynomials in the transfer function model, the value of the delay parameter b, together with some guesses about the values of p, d, and q in the error ARIMA process. At the same time, with some careful calculations, initial guess values can be obtained for the parameter vectors ω, δ, ϕ, and θ in the transfer function model. Again in (11.8) we have assumed that the dependent and explanatory variables are stationary, possible after suitable differencing.

11.3.1 Basic Identification Rules

Initially, we will concentrate on identifying the model (11.8) assuming that there is no error (i.e., $e_t = 0$). We will proceed with the identification under

[2] Program CROSSC calls this Q statistic the CHI-SQUARE TEST and prints out the Q statistic value as well as the P-value, defined as the probability that there does not exist a significant pattern in the cross correlations. See also footnote 13, Chapter 4.

the assumption that we know the model and generate the cross correlation function based on some specific examples. This will allow us to strengthen our intuition and thus make the transition to a general rule easier. The next step then will be to introduce the error process and proceed with the complete transfer function model.

Figure 11.2 shows the cross correlations of two series,[3] y_t and x_t, each consisting of 100 observations. The x_t series is generated from a normal distribution with mean of 0 and standard deviation of two, and y_t is defined deterministically as $2x_t$. If we look at the cross correlation functions we observe that the zero lag cross correlation $r_{xy}(0) = 1$, indicating that at lag 0, y_t, and x_t are perfectly correlated. For all other time lags, the cross correlations are small. Statistically all other cross correlations are within the 95% large-sample confidence interval. Therefore, x_t influences y_t instantaneously with no lag or lead relationship.

In Figure 11.3 we present the cross correlation of a second model. The variable x_t has been generated as above, but now $y_t = x_{t-4}$. Therefore, intuitively, current x_t should be a perfect predictor of y_t four periods ahead. The results presented in Figure 11.3 confirm our intuition as all cross correlations are small, except $r_{xy}(4) = 0.92$. Therefore, for this data there is a dead time parameter $b = 4$.

Let us now introduce some more dynamics. Figure 11.4 shows the cross correlations for the model

$$y_t = 0.7y_{t-1} + x_{t-4}, \tag{11.10}$$

with x_{t-4} generated as before. We now see several large cross correlations which, starting from $r_{xy}(4)$, decline exponentially, a pattern expected from a univariate AR(1) model with a positive parameter [see also Figure 4.8(b)]. This should not come as a surprise, as (11.10) can be rewritten as

$$y_t = 0.7y_{t-1} + a_t \tag{11.11}$$

with $a_t = x_{t-4}$. Equation (11.11) is now exactly the representation of the AR(1) process discussed in earlier chapters. We could easily add additional lagged y_t terms and the cross correlation functions would still resemble the autocorrelation function of autoregressive models. In Figure 11.5 we represent the cross correlations of the model $y_t = 0.7y_{t-1} - 0.5y_{t-2} + x_{t-4}$. Again, starting from lag 4, the cross correlation function behaves like the *acf* of an AR(2) process [see Figure 4.8(f)].

Let us now evaluate the effect of an additional lagged x_t value. In Figure 11.6 we represent the cross correlation function of the following model:

$$y_t = x_{t-4} + 0.6x_{t-5}, \tag{11.12}$$

where x_t is again generated as above. Now the cross correlations are all small

[3] Appendix C, Section C.5.1 contains a detailed discussion of the statistics used in these figures.

except for lags with $k = 4$ and $k = 5$. Notice that (11.12) is an MA(1) model, and from Section 4.2.3 we know that the autocorrelation functions of an MA(1) model have a single spike. With $\theta_1 = -0.6$, this should be a positive

FIGURE 11.2   Cross Correlations Between x_t and $y_t = 2x_t$.

```
CROSS CORRELATIONS     STANDARD ERROR =  .10
ZERO LAG = 1.00
  LEADS          FUTURE           PY(t),PX(t+k)
  1-12        -.10 -.13  .18  .00  .03  .07 -.19  .10  .07 -.01  .07  .03
 13-15         .00 -.06  .00

  LAGS           PAST             PY(t),PX(t-k)
  1-12        -.10 -.13  .18  .00  .03  .07 -.19  .10  .07 -.01  .07  .03
 13-15         .00 -.06  .00

CROSS CORRELATION FUNCTION

    -1.0 -0.8 -0.6 -0.4 -0.2  0.0  0.2  0.4  0.6  0.8  1.0
     .    .    .    .    .    .    .    .    .    .    .
  -15.LEAD                  (      R      )
  -14.                      (      R.     )
  -13.                      (      R      )
  -12.                      (      .R     )
  -11.                      (      . R    )
  -10.                      (      R      )
   -9.                      (      . R    )
   -8.                      (      .  R   )
   -7.             R        (      .      )
   -6.                      (      . R    )
   -5.                      (      .R     )
   -4.                      (      R      )
   -3.                      (      .         R
   -2.                    ( R     .      )
   -1.                    (  R    .      )
    0.                      (      .      )              R
    1.                    (  R    .      )
    2.                    ( R     .      )
    3.                      (      .         R
    4.                      (    R        )
    5.                      (    .R       )
    6.                      (    . R      )
    7.             R        (      :      )
    8.                      (      .  R   )
    9.                      (      . R    )
   10.                      (    R        )
   11.                      (      . R    )
   12.                      (      .R     )
   13.                      (    R        )
   14.                      (    R.       )
   15.LAG                   (      R      )
```

FIGURE 11.3   Cross Correlations Between x_t and $y_t = x_{t-4}$.

```
CROSS CORRELATIONS     STANDARD ERROR =   .10
ZERO LAG = -.04
 LEADS          FUTURE          PY(t),PX(t+k)
 1-12         .10   .05 -.22   .13   .07 -.03   .08   .01   .02 -.05 -.02 -.06
13-15        -.09   .08 -.04

 LAGS           PAST            PY(t),PX(t-k)
 1-12         .13 -.07 -.10   .92 -.08 -.11   .16 -.01   .06   .05 -.17   .10
13-15         .08 -.03   .06
```

CROSS CORRELATION FUNCTION

```
  -1.0 -0.8 -0.6 -0.4 -0.2  0.0  0.2  0.4  0.6  0.8  1.0
    *  . . . *  . . . *  . . . *  . . . *  . . . *  . . . *  . . . *  . . . *  . . . *  . . . *
 -15.LEAD                (   R.     )
 -14.                    (    . R   )
 -13.                    (   R .    )
 -12.                    (   R.     )
 -11.                    (    R     )
 -10.                    (   R.     )
  -9.                    (    R     )
  -8.                    (    R     )
  -7.                    (    . R   )
  -6.                    (   R.     )
  -5.                    (    . R ' )
  -4.                    (    . R   )
  -3.                   R     .     )
  -2.                    (   .R     )
  -1.                    (    . R   )
   0.                    (   R.     )
   1.                    (    . R   )
   2.                    (   R .    )
   3.                    (   R .    )
   4.                    (    .     )                        R
   5.                    (   R .    )
   6.                    ( R .    )
   7.                    (    .  R)
   8.                    (   R     )
   9.                    (   .R     )
  10.                    (   .R     )
  11.                   (R   .     )
  12.                    (    . R   )
  13.                    (    . R   )
  14.                    (   R.     )
  15.LAG                 (    . R   )
```

FIGURE 11.4   Cross Correlations Between x_t and y_t, $y_t = 0.7y_{t-1} + x_{t-4}$.

```
CROSS CORRELATIONS    STANDARD ERROR =  .10
ZERO LAG =  .01
  LEADS          FUTURE          PY(t),PX(t+k)
  1-12          .08 -.01 -.08   .14  .05  .00  .03 -.04 -.06 -.10 -.09 -.11
  13-15        -.09 -.02 -.10

  LAGS           PAST            PY(t),PX(t-k)
  1-12          .11  .03 -.05   .66  .40  .20  .26  .18  .17  .16 -.02  .06
  13-15         .10  .04  .08

CROSS CORRELATION FUNCTION

   -1.0 -0.8 -0.6 -0.4 -0.2  0.0  0.2  0.4  0.6  0.8  1.0
     • . . . • . . . • . . . • . . . • . . . • . . . • . . . • . . . • . . . • . . . •
 -15.LEAD              ( R    .            )
 -14.                  (   R .             )
 -13.                  (   R .             )
 -12.                  ( R    .            )
 -11.                  (   R .             )
 -10.                  (   R .             )
  -9.                  (     R.            )
  -8.                  (     R.            )
  -7.                  (      .R           )
  -6.                  (      R            )
  -5.                  (      .R           )
  -4.                  (      .      R)
  -3.                  ( R    .            )
  -2.                  (      R            )
  -1.                  (      .  R         )
   0.                  (      R            )
   1.                  (      .   R        )
   2.                  (      .R           )
   3.                  (   R. .            )
   4.                  (      .            )              R
   5.                  (      .            )         R
   6.                  (      .      R
   7.                  (      .            ) R
   8.                  (      .   R)
   9.                  (      .   R)
  10.                  (      .   R)
  11.                  (   R. .            )
  12.                  (      .R           )
  13.                  (      .   R        )
  14.                  (      .R           )
  15.LAG               (      .  R         )
```

FIGURE 11.5   Cross Correlations Between x_t and y_t, $y_t = 0.7y_{t-1} - 0.5y_{t-2} + x_{t-4}$.

```
CROSS CORRELATIONS    STANDARD ERROR =  .10
ZERO LAG =  .07
  LEADS           FUTURE           PY(t),PX(t+k)
  1-12          .07 -.10 -.11   .12   .03   .02   .08   .03   .01 -.03 -.05 -.07
  13-15        -.03  .02 -.07

  LAGS            PAST             PY(t),PX(t-k)
  1-12          .13 -.01 -.14   .68   .48 -.09 -.17 -.08   .08   .13 -.10 -.05
  13-15         .08  .05  .05

CROSS CORRELATION FUNCTION

    -1.0 -0.8 -0.6 -0.4 -0.2  0.0  0.2  0.4  0.6  0.8  1.0
      . . . .*. . . *. . . .*. . . .*. . . *. . . .*. . *. . . .*. . . *. . . .*
  -15.LEAD              (    R .      )
  -14.                  (      R      )
  -13.                  (      R.     )
  -12.                  (      R .    )
  -11.                  (      R.     )
  -10.                  (      R.     )
   -9.                  (       R     )
   -8.                  (      .R     )
   -7.                  (       . R   )
   -6.                  (      R      )
   -5.                  (       .R    )
   -4.                  (       .   R )
   -3.                  (  R   .      )
   -2.                  (  R   .      )
   -1.                  (       . R   )
    0.                  (       . R   )
    1.                  (       .   R )
    2.                  (      R      )
    3.                  ( R    .      )
    4.                  (      .      )              R
    5.                  (      .      )         R
    6.                  (  R   .      )
    7.                  ( R    .      )
    8.                  (  R   .      )
    9.                  (       . R   )
   10.                  (       .   R )
   11.                  (  R   .      )
   12.                  (  R.         )
   13.                  (       . R   )
   14.                  (      .R     )
   15.LAG               (      .R     )
```

FIGURE 11.6   Cross Correlations Between x_t and y_t, $y_t = 0.7y_{t-4} + 0.6x_{t-5}$.

```
CROSS CORRELATIONS    STANDARD ERROR =   .10
ZERO LAG =   .02
  LEADS         FUTURE           PY(t),PX(t+k)
  1-12          .12 -.07 -.13   .16   .05   .02   .07   .01 -.01 -.05 -.04 -.10
  13-15        -.04   .05 -.10

  LAGS          PAST             PY(t),PX(t-k)
  1-12          .10   .01 -.12   .78   .42 -.14   .09   .08   .05   .07 -.13 -.01
  13-15         .13   .01   .04

CROSS CORRELATION FUNCTION

    -1.0 -0.8 -0.6 -0.4 -0.2  0.0  0.2  0.4  0.6  0.8  1.0
       *    *    *    *    *    *    *    *    *    *    *
    -15.LEAD               ( R    .      )
    -14.                   (      .R     )
    -13.                   (      R.     )
    -12.                   ( R    .      )
    -11.                   (      R.     )
    -10.                   (      R.     )
     -9.                   (      R      )
     -8.                   (      R      )
     -7.                   (      .  R   )
     -6.                   (      R      )
     -5.                   (      .R     )
     -4.                   (      .    R )
     -3.                   ( R    .      )
     -2.                   ( R    .      )
     -1.                   (      .  R   )
      0.                   (      .R     )
      1.                   (      .  R   )
      2.                   (      R      )
      3.                   ( R    .      )
      4.                   (      .      )                   R
      5.                   (      .      )              R
      6.                   (R     .      )
      7.                   (      .  R   )
      8.                   (      .  R   )
      9.                   (      .R     )
     10.                   (      .  R   )
     11.                   ( R    .      )
     12.                   (      R      )
     13.                   (      .  R   )
     14.                   (      R      )
     15.LAG                (      .R     )
```

spike [see also Figure 4.11(b)]. If we make a comparison with the memory function of an MA(1) model (see Section 3.2.4), we know that a shock at time t will disturb the observations at time t and $t + 1$, but will have no effect beyond period $t + 1$. In Figure 11.6 we see a spike at lag 4 corresponding to the initial effect and then a small spike at lag 5 corresponding to the single spike in an *acf* of an MA(1).

Based on the above examples we can conclude that past x_t terms result in single spikes in the cross correlation, while r lagged y_t terms cause the cross correlation to behave as the autocorrelations of an AR(r) process.

Therefore, values of b, r, and l can now be identified using the cross correlations as follows:

1. The dead time constant b is equal to the lag of the first significant cross correlation.
2. The denominator order r is found by examining the cross correlations for any pattern in them. This pattern, if it exists, will correspond to the *acf* of an AR(r) model. If it does not exist, $r = 0$. Thus, the order of the denominator polynomial, r, can be identified through association of the cross correlation pattern with the pattern of an autocorrelation function.
3. The numerator order l is the number of periods that the AR(r) pattern is delayed. If $r_{xy}(3)$ is significant, but the AR(r) pattern does not start until period 4, then $l = 4 - 3 = 1$. If there is no AR pattern in the *ccf*, then there are only numerator parameters with the order l one less than the number of nonzero cross correlations.

In Figures 11.3, 11.4, 11.5, and 11.6 the first significant cross correlation occurs at lag 4 so that the value of the dead time constant equals 4. In Figure 11.2 the first, and only, significant cross correlation occurs at lag 0 and therefore there is no dead time.

Both processes represented in Figure 11.2 and Figure 11.3 are white noise processes because there are no significant cross correlations except the correlation at lag 0 in Figure 11.2 and at lag 4 in Figure 11.3. Therefore, for both processes, $r = 0$ and $l = 0$.

Figure 11.4 represents an AR(1) pattern with first significant cross correlation at lag 4, corresponding to $b = 4$. Also, the AR(1) pattern seems to start at lag 4 and therefore $l = 0$. In Figure 11.5 we clearly see that the value of $b = 4$. This pattern is somewhat harder to recognize, but comparing this figure with Figure 4.8(f) we are struck by its similarity. Therefore we can easily accept that Figure 11.5 represents an AR(2) pattern, with $b = 4$ and $l = 0$.

Figure 11.6 represents a dead time $b = 4$, and two spikes corresponding to $l = 2$. The value of ω_1 is negative because the cross correlation at lag 5 is positive. This corresponds to the familiar pattern recognized in the *acf*'s of MA processes where the sign of the last significant autocorrelation is the opposite of the sign of the moving average parameter. (See also Figure 4.11.)

We would like to restate that the objective of the chapters on the transfer function model is to construct leading indicator models—that is, models where it safely can be assumed that current and past x_t's influence y_t. As a result we only analyze the information contained in the lag cross correlations and not in the lead cross correlations. If, however, the lead cross correlations are large, the user should look upon this as evidence of causal feedback. The causal feedback model is, however, more complicated to construct. We refer the reader to the references given in the opening paragraphs of Chapter 10 and in particular to the book by Granger and Newbold (1973).

What has been demonstrated with the simulated data must now be proven in general. We now will derive the theoretical cross correlations for some frequently encounted transfer function models.

The model

$$y_t = \omega_0 x_{t-b} + e_t \tag{11.13}$$

is supposed to have all 0 cross correlations, except $r_{xy}(b)$. We will first calculate the covariance $\gamma_{xy}(k)$, defined as [see (11.1)]

$$\gamma_{xy}(k) = E(x_t y_{t+k}). \tag{11.14}$$

In (11.14) we have assumed that the data x_t and y_t have mean 0. Then,

$$E(x_t y_{t+k}) = E[x_t(\omega_0 x_{t-b+k} + e_{t+k})].$$

Now, by assumption, current data is uncorrelated with the error terms e_{t+k}. So

$$\gamma_{xy}(k) = \omega_0 E(x_t x_{t-b+k}).$$

Then, assuming that x_t is a white noise series (we will discuss the case in which x_t is not white noise in Section 11.3.3),

$$\gamma_{xy}(k) = \begin{cases} \omega_0 \sigma_x^2 & \text{for } b = k \\ 0 & \text{otherwise.} \end{cases}$$

Scaling the covariance with $\sigma_x \sigma_y$ we obtain the cross correlation

$$\rho_{xy}(k) = \begin{cases} \omega_0 \sigma_x/\sigma_y & \text{for } b = k \\ 0 & \text{otherwise.} \end{cases} \tag{11.15}$$

These results correspond to the data generated in Figures 11.2 and 11.3.

Next, let us calculate the cross correlations for the following model:

$$y_t = \frac{\omega_0}{1 - \delta_1 B} x_{t-b} + e_t. \tag{11.16}$$

This is the model used in Figure 11.4. For $|\delta_1| < 1$, we can rewrite this model as an infinite series

$$y_t = \omega_0 \sum_{i=0}^{\infty} \delta_1^i x_{t-b-i} + e_t.$$

Now, the covariance $\gamma_{xy}(k)$ is obtained as

$$\gamma_{xy}(k) = E[x_t(\omega_0 \sum_{i=0}^{\infty} \delta_1^i x_{t-b+k-i} + e_t)]$$

$$= E\left(\omega_0 \sum_{i=0}^{\infty} \delta_1^i x_t x_{t-b+k-i}\right)$$

$$= \omega_0 E(x_t x_{t-b+k}) + \omega_0 \delta_1 E(x_t x_{t-b+k-1}) + \omega_0 \delta_1^2 E(x_t x_{t-b+k-2}) + \cdots .$$

For $k < b$, $x_{t-b+k-i}$ occurs before x_t for all values of i, and therefore $\gamma_{xy}(k)$ is 0. Notice we again assume that x_t is a white noise series. For $k = b$, only $x_{t-b+k-i}$ for $i = 0$ does not occur before x_t, and so $\gamma_{xy}(k) = E(\omega_0 x_t^2) = \omega_0 \sigma_x^2$ for $k = b$. For $k = b + 1$, only the second term, $i = 1$, in the infinite summation is nonzero. Therefore, $\gamma_{xy}(k) = \omega_0 \delta_1 \sigma_x^2$ for $k = b + 1$. It is easy to show that when $k = b + n$, $\gamma_{xy}(k) = \omega_0 \delta_1^n \sigma_x^2$. Dividing these convariances by $\sigma_x \sigma_y$, we obtain the cross correlations for the model (11.16) as

$$\rho_{xy}(k) = \begin{cases} 0 & k < b \\ \omega_0 \sigma_x / \sigma_y & k = b \\ \omega_0 \delta_1^i \sigma_x / \sigma_y & k = b + j, j > 0 \end{cases} \quad (11.17)$$

Again this corresponds to the results shown in Figure 11.4.

Finally, we will calculate the cross correlations for the model used in Figure 11.6,

$$y_t = \omega_0 x_{t-b} - \omega_1 x_{t-b-1} + e_t. \quad (11.18)$$

The covariance $\gamma_{xy}(k)$ is defined as

$$\gamma_{xy}(k) = E[x_t(\omega_0 x_{t-b+k} - \omega_1 x_{t-b+k-1} + e_t)].$$

For $k = b$, we obtain $\gamma_{xy}(k) = \omega_0 \sigma_x^2$; for $k = b + 1$, $\gamma_{xy}(k) = \omega_1 \sigma_x^2$. All other cross covariances are 0. Therefore the cross correlations are

$$\rho_{xy}(k) = \begin{cases} \omega_0 \sigma_x / \sigma_y & k = b \\ \omega_1 \sigma_x / \sigma_y & k = b + 1 \\ 0 & \text{otherwise.} \end{cases} \quad (11.19)$$

In Figure 11.7 we have presented four typical theoretical *ccf*. This figure should help you to further internalize the transfer function identification rules.

11.3.2 Stationarity

Up to now we have assumed that the series y_t and x_t are stationary or have been made stationary by a suitable transformation such as differencing. As

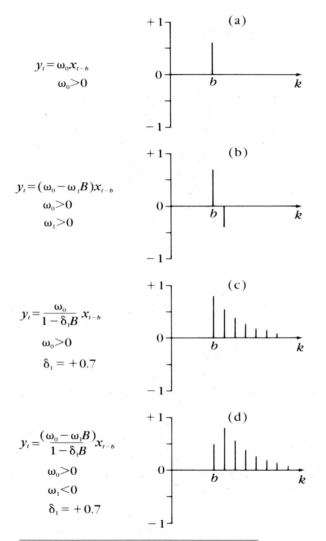

$$y_t = \omega_0 x_{t-b}$$
$$\omega_0 > 0$$

(a)

$$y_t = (\omega_0 - \omega_1 B) x_{t-b}$$
$$\omega_0 > 0$$
$$\omega_1 > 0$$

(b)

$$y_t = \frac{\omega_0}{1 - \delta_1 B} x_{t-b}$$
$$\omega_0 > 0$$
$$\delta_1 = +0.7$$

(c)

$$y_t = \frac{(\omega_0 - \omega_1 B)}{1 - \delta_1 B} x_{t-b}$$
$$\omega_0 > 0$$
$$\omega_1 < 0$$
$$\delta_1 = +0.7$$

(d)

FIGURE 11.7   Theoretical Cross Correlation Functions.

discussed in Section 4.2.2, the autocorrelations functions of a nonstationary series will not die out, even for large k. This same behavior will be observed for the cross correlation function of nonstationary series. If the cross correlations do not die out rapidly, this should be looked upon as evidence that the y_t or the x_t series is nonstationary. Determination of which variable needs differencing is made by evaluating the autocorrelation function of y_t and x_t.

In Figures 11.8(a), 11.8(b), and 11.8(c) we present the cross correlations of two independent series. In theory we should find that there are no signifi-

FIGURE 11.8(a)   Cross Correlations Between Two Independent Random Walk Series.

```
CROSS CORRELATIONS     STANDARD ERROR =   .10
ZERO LAG =   .18

LEADS          FUTURE          PY(t),PX(t+k)
1-12       .21   .23   .25   .29   .32   .36   .41   .47   .52   .57   .60   .63
13-15      .65   .67   .69

LAGS           PAST            PY(t),PX(t-k)
1-12       .15   .11   .06   .01  -.05  -.10  -.13  -.17  -.22  -.24  -.26  -.27
13-15     -.27  -.26  -.25

CROSS CORRELATION FUNCTION

    -1.0 -0.8 -0.6 -0.4 -0.2  0.0  0.2  0.4  0.6  0.8  1.0
      *....*....*....*....*....*....*....*....*....*....*
  -15.LEAD           .   (      .      )           R .
  -14.                   (      .      )           R
  -13.                   (      .      )          R
  -12.                   (      .      )          R
  -11.                   (      .      )         R
  -10.                   (      .      )        R
   -9.                   (      .      )       R
   -8.                   (      .      )      R
   -7.                   (      .      )    R
   -6.                   (      .      )   R
   -5.                   (      .      )  R
   -4.                   (      .      ) R
   -3.                   (      .      )R
   -2.                   (      .      )R
   -1.                   (      .      R      .
    0.                   (      .    R )
    1.                   (      .    R )
    2.                   (      .  R   )
    3.                   (     . R    )
    4.                   (    R      )
    5.                   (   R .     )
    6.                   (  R  .     )
    7.                   ( R   .     )
    8.                   (R    .     )
    9.                   R     .     )
   10.                  R(     .     )
   11.                  R(     .     )
   12.                 R (     .     )
   13.                 R (     .     )
   14.                 R(     .     )
   15.LAG             R(     .     )
```

FIGURE 11.8(b)   Cross Correlations Between Two Independent Random Walk Series, After Differencing the y Series.

```
CROSS CORRELATIONS     STANDARD ERROR =   .10
ZERO LAG = -.07
LEADS              FUTURE            PY(t),PX(t+k)
  1-12          -.06 -.05 -.09 -.07 -.10 -.12 -.14 -.13 -.13 -.08 -.07 -.04
 13-15          -.05 -.03 -.02


  LAGS               PAST             PY(t),PX(t-k)
  1-12          -.05 -.08 -.10 -.11 -.11 -.11 -.05 -.07 -.09 -.02 -.03 -.02
 13-15           .03  .07  .07
```

CROSS CORRELATION FUNCTION

```
     -1.0 -0.8 -0.6 -0.4 -0.2  0.0  0.2  0.4  0.6  0.8  1.0
       •....•....•....•....•....•....•....•....•....•....•
   -15.LEAD                    (       R       )
   -14.                        (       R.      )
   -13.                        (       R.      )
   -12.                        (       R.      )
   -11.                        (      R .      )
   -10.                        (      R .      )
    -9.                        (     R  .      )
    -8.                        (     R  .      )
    -7.                        (R       .      )
    -6.                        (    R   .      )
    -5.                        (      R .      )
    -4.                        (      R .      )
    -3.                        (      R .      )
    -2.                        (       R.      )
    -1.                        (      R .      )
     0.                        (      R .      )
     1.                        (       R.      )
     2.                        (      R .      )
     3.                        (      R .      )
     4.                        (     R  .      )
     5.                        (     R  .      )
     6.                        (     R  .      )
     7.                        (      R.       )
     8.                        (      R .      )
     9.                        (      R .      )
    10.                        (       R       )
    11.                        (      R.       )
    12.                        (       R       )
    13.                        (       .R      )
    14.                        (       . R     )
    15.LAG                     (       . R     )
```

FIGURE 11.8(c)   Cross Correlations Between Two Independent Random Walk Series, After Differencing Both Series.

```
CROSS CORRELATIONS    STANDARD ERROR =   .10
ZERO LAG = -.08
  LEADS          FUTURE            PY(t),PX(t+k)
  1-12          .04  .00 -.17   .06 -.01 -.07  .01  .01  .02  .12  .02  .08
  13-15         .02  .03 -.02


  LAGS           PAST              PY(t),PX(t-k)
  1-12          .11  .08  .03   .02 -.02 -.15  .06  .08 -.23  .01 -.03 -.12
  13-15        -.10  .05 -.08

CROSS CORRELATION FUNCTION

    -1.0 -0.8 -0.6 -0.4 -0.2  0.0  0.2  0.4  0.6  0.8  1.0
      *....*....*....*....*....*....*....*....*....*....*
   -15.LEAD                 (      R      )
   -14.                     (      .R     )
   -13.                     (      .R     )
   -12.                     (      .  R   )
   -11.                     (      .R     )
   -10.                     (      .  R   )
    -9.                     (      .R     )
    -8.                     (      R      )
    -7.                     (      R      )
    -6.                     (  R   .      )
    -5.                     (      R      )
    -4.                     (      .  R   )
    -3.                     (R    .       )
    -2.                     (      R      )
    -1.                     (      .R     )
     0.                     (  R   .      )
     1.                     (      .   R  )
     2.                     (      .  R   )
     3.                     (      .R     )
     4.                     (      R      )
     5.                     (      R      )
     6.                     (R    .       )
     7.                     (      .  R   )
     8.                     (      .  R   )
     9.                   R (      .      )
    10.                     (      R      )
    11.                     (      R.     )
    12.                     (   R  .      )
    13.                     (   R  .      )
    14.                     (      .R     )
    15.LAG                  (      R .    )
```

cant lag or lead cross correlations. Both series represent 100 data points independently generated as a random walk series with the noise term normally distributed with mean of 0 and standard deviation of two. In Figure 11.8(a) we present the cross correlations of the data. Notice the strange pattern in the *ccf:* As the lags or leads increase we observe stronger and stronger correlations, certainly a behavior incompatible with the dying out at large lags or leads of a *ccf* of stationary series. We now proceed by differencing the data. In Figure 11.8(b) we present cross correlations of the differenced *y* data and the raw *x* data. Now *y* is stationary, whereas *x* is still nonstationary. Although this time all the cross correlations could be viewed as insignificant, the pattern of the *ccf* is still undesirable. We still find that at large lags/ leads the cross correlations increase and, furthermore, most cross correlations are negative. Finally, in Figure 11.8(c) we present the cross correlations after differencing both series. This time we find a pattern in the cross correlations more compatible with uncorrelated series.

11.3.3 Prewhitening the Data

It should be clear that cross correlations can be a helpful tool for checking the dependencies between two time series. However, when a series, *y* or *x*, is highly autocorrelated, the cross correlation function between two time series can be difficult to interpret and even can be misleading. It is quite possible that two time series which are not related at all show high spurious correlation if each series is highly correlated with itself, a common occurrence with economic data. In Figure 11.8(a) we presented the cross correlations of two independent random walk series. Although this is not an entirely new finding [see Yule (1926), Bartlett (1935), Box and Newbold (1971), and Granger (1977)], many practitioners still just evaluate raw cross correlations. [See also Granger and Newbold (1977, p. 204, p. 232).]

In order to obtain valuable identification information from cross correlations, it is recommended to first filter, or prewhiten, the data before calculating the cross correlations. This *prewhitening* of the data amounts to first obtaining the appropriate univariate models for each series involved, and then, at the second stage, cross correlating the (residual) white noise series. Although this approach could be looked upon as a novel idea, many practitioners already do go some way along that route by removing trend and seasonal components from their data before analyzing the data, or by building a model to explain changes in the data.

The following example will further clarify this issue. Suppose that *y* and *x* are generated with the following AR(2) process:

$$y_t = \phi_{y1} y_{t-1} + \phi_{y2} y_{t-2} + e_t \qquad (11.20)$$

$$x_t = \phi_{x1} x_{t-1} + \phi_{x2} x_{t-2} + u_t. \qquad (11.21)$$

Here both e_t and u_t are assumed to be white noise processes. Next, assume that the underlying true transfer function model relating the y and x series is given by

$$y_t = \delta_1 y_{t-1} + \omega_0 x_{t-2} - \omega_1 x_{t-3} + a_t,$$

which can be rewritten as

$$y_t = \frac{(\omega_0 - \omega_1 B)B^2}{(1 - \delta_1 B)} x_t + a_t. \tag{11.22}$$

Substituting the y and x processes [(11.20) and (11.21)] in the transfer function model (11.22) we obtain

$$y_t = \delta_1(\phi_{y1} y_{t-2} + \phi_{y2} y_{t-3} + e_{t-1}) + \omega_0(\phi_{x1} x_{t-3} + \phi_{x2} x_{t-4} + u_{t-2})$$
$$- \omega_1(\phi_{x1} x_{t-4} + \phi_{x2} x_{t-5} + u_{t-3}) + a_t. \tag{11.23}$$

Therefore, although the true transfer function model is given by (11.22), empirically we would observe a much more complicated model (11.23). The joint relationship of y_t and x_t, equation (11.22), the between variation, is buried within the y_t and x_t variation, the within variation, and cannot be discovered unless all other types of variations have been eliminated. To eliminate the within variations the y_t and x_t series must be prewhitened until such variation within each of them has been eliminated as much as possible. Rather than evaluating the relationship between y_t and x_t, we should use the residuals e_t and u_t from (11.20) and (11.21) to evaluate (11.22). These residuals are defined as the prewhitened data.

In Figures 11.9(a), 11.9(b), and 11.9(c) we have analyzed the following simple model. We used the random walk x variable used in Figure 11.8 and defined $y_t = x_t$; that is, y_t is perfectly correlated with x_t. Therefore the model used is:

$$y_t = y_{t-1} + e_t$$
$$x_t = x_{t-1} + u_t$$
$$y_t = x_t. \tag{11.24}$$

In Figure 11.9(a) we present the cross correlations between y_t and x_t. Notice that the cross correlations die out very slowly, again evidence of nonstationarity. After differencing both series the cross correlations are easier to interpret [see Figure 11.9(b)]. We observe that y_t and x_t are not correlated for any lag or lead, but are in perfect instantaneous correlation. In Figure 11.9(c) we present the cross correlations after only differencing the y variable. Notice that these cross correlations are difficult to interpret and are misleading. In particular, the *ccf* does not reveal the perfect correlation between y_t and x_t.

Intuitively, the need for prewhitening also can be explained as follows. If you want to judge whether an explanatory variable can explain the behavior

FIGURE 11.9(a)   Cross Correlations Between Two Perfectly Correlated Random Walk Series.

```
CROSS CORRELATIONS    STANDARD ERROR =  .10
ZERO LAG = 1.00
  LEADS        FUTURE           PY(t),PX(t+k)
  1-12        .93  .86  .82  .75  .68  .60  .52  .46  .39  .30  .23  .15
  13-15       .07  .00 -.06

  LAGS         PAST             PY(t),PX(t-k)
  1-12        .93  .86  .82  .75  .68  .60  .52  .46  .39  .30  .23  .15
  13-15       .07  .00 -.06

CROSS CORRELATION FUNCTION

   -1.0 -0.8 -0.6 -0.4 -0.2  0.0  0.2  0.4  0.6  0.8  1.0
     *....*....*....*....*....*....*....*....*....*....*
  -15.LEAD               (    R  .      )
  -14.                   (       R      )
  -13.                   (       .  R   )
  -12.                   (       .     R)
  -11.                   (       .      )R
  -10.                   (       .      )  R
   -9.                   (       .      )    R
   -8.                   (       .      )     R
   -7.                   (       .      )       R
   -6.                   (       .      )        R
   -5.                   (       .      )          R
   -4.                   (       .      )           R
   -3.                   (       .      )            R
   -2.                   (       .      )             R
   -1.                   (       .      )              R
    0.                   (       .      )               R
    1.                   (       .      )              R
    2.                   (       .      )             R
    3.                   (       .      )            R
    4.                   (       .      )           R
    5.                   (       .      )          R
    6.                   (       .      )        R
    7.                   (       .      )       R
    8.                   (       .      )     R
    9.                   (       .      )    R
   10.                   (       .      )  R
   11.                   (       .      )R
   12.                   (       .     R)
   13.                   (       .  R   )
   14.                   (       R      )
   15.LAG                (    R  .      )
```

FIGURE 11.9(b)   Cross Correlations Between Two Perfectly Correlated Random Walk Series, After Differencing Both Series.

```
CROSS CORRELATIONS    STANDARD ERROR =   .10
ZERO LAG = 1.00
   LEADS        FUTURE          PY(t),PX(t+k)
    1-12      -.08 -.12  .16  .02  .03  .07 -.20  .11  .07  .00  .06  .03
   13-15      -.01 -.08 -.02

   LAGS         PAST            PY(t),PX(t-k)
    1-12      -.08 -.12  .16  .02  .03  .07 -.20  .11  .07  .00  .06  .03
   13-15      -.01 -.08 -.02

CROSS CORRELATION FUNCTION

      -1.0 -0.8 -0.6 -0.4 -0.2  0.0  0.2  0.4  0.6  0.8  1.0
       *....*....*....*....*....*....*....*....*....*....*
    -15.LEAD                   (     R     )
    -14.                       (   R .     )
    -13.                       (     R     )
    -12.                       (     .R    )
    -11.                       (     .R    )
    -10.                       (     R     )
     -9.                       (     . R   )
     -8.                       (     .   R )
     -7.                       R     .     )
     -6.                       (     . R   )
     -5.                       (     .R    )
     -4.                       (     R     )
     -3.                       (     .   R )
     -2.                       ( R   .     )
     -1.                       ( R   .     )
      0.                       (     .     )              R
      1.                       ( R   .     )
      2.                       ( R   .     )
      3.                       (     .   R )
      4.                       (   R       )
      5.                       (    .R     )
      6.                       (    . R    )
      7.                       R     .     )
      8.                       (     . R   )
      9.                       (     . R   )
     10.                       (   R       )
     11.                       (    .R     )
     12.                       (    .R     )
     13.                       (   R       )
     14.                       ( R   .     )
     15.LAG                    (   R       )
```

FIGURE 11.9(c)   Cross Correlations Between Two Perfectly Correlated Random Walk Series, After Differencing the y Series.

```
CROSS CORRELATIONS    STANDARD ERROR =  .10
ZERO LAG =  .20
  LEADS         FUTURE            PY(t),PX(t+k)
  1-12          .15   .10   .18   .18  .19   .22   .16   .18   .22   .20   .22   .21
 13-15          .21   .18   .16

  LAGS           PAST              PY(t),PX(t-k)
  1-12          -.15 -.13 -.07 -.14 -.14 -.15 -.17 -.10 -.13 -.16 -.15 -.17
 13-15          -.17 -.16 -.12

CROSS CORRELATION FUNCTION


   -1.0 -0.8 -0.6 -0.4 -0.2  0.0  0.2  0.4  0.6  0.8  1.0
     *....*....*....*....*....*....*....*....*....*....*
  -15.LEAD                  (       .     R)
  -14.                      (       .     R)
  -13.                      (       .     R
  -12.                      (       .     R
  -11.                      (       .     R
  -10.                      (       .     R
   -9.                      (       .     R
   -8.                      (       .     R)
   -7.                      (       .     R)
   -6.                      (       .     R
   -5.                      (       .     R
   -4.                      (       .     R
   -3.                      (       .     R)
   -2.                      (       . R   )
   -1.                      (       .     R)
    0.                      (       .     R
    1.                      (R      .     )
    2.                      ( R     .     )
    3.                      (  R    .     )
    4.                      ( R     .     )
    5.                      (R      .     )
    6.                      (R      .     )
    7.                      (R      .     )
    8.                      ( R     .     )
    9.                      ( R     .     )
   10.                      (R      .     )
   11.                      (R      .     )
   12.                      (R      .     )
   13.                      (R      .     )
   14.                      (R      .     )
   15.LAG                   ( R     .     )
```

in the y_t series, we should first eliminate all variations in each of y_t and x_t that can be explained by its own past data (i.e., create prewhitened data or residuals and then evaluate the nature of the true relationship that exists between the prewhitened series). This empirical relationship will then and only then properly reflect the true nature of transfer function models. In Section 11.3.5 we will present additional arguments for first prewhitening the data.

There is some confusion in the literature specifically about how you should calculate the prewhitened y_t series. Some authors will advocate the use of the ARIMA process of the explanatory variable in order to prewhiten the y series rather than using the y ARIMA process itself. The current concensus is that the answer depends on the nature and type of transfer function model we are constructing. In particular, if causality is an issue and the transfer function model is an equation within the context of a simultaneous equations model with the y variable referring to an endogenous variable and the explanatory variables to exogenous variables, then the y series should be prewhitened with its own process. Again we stress that causal transfer function models are beyond the scope of this book. [See, e.g., Granger and Newbold (1977).] If, on the other hand, the transfer function model is a leading indicator model, a unidirectional model not allowing for feedback, then we can prewhiten the y series with the ARIMA model of the x process.

Let us now present the theory in more general terms. Assume that possibly after suitable differencing, x_t is stationary and can be represented by a general the y series with the ARIMA model of the x process.

$$\phi_x(B)x_t = \theta_x(B)\alpha_t \tag{11.25}$$

or

$$\alpha_t = \psi_x^{-1}(B)x_t$$

with

$$\psi_x(B) = \frac{\theta_x(B)}{\phi_x(B)}. \tag{11.26}$$

We then have a series of uncorrelated white noise α_t, also called the *prewhitened* explanatory variable.

If we apply the same transformation to the y_t series, we can rewrite the transfer function model (10.15), assuming the dead time $b = 0$, as

$$\beta_t = v(B)\alpha_t + u_t, \tag{11.27}$$

where

$$\beta_t = \psi_x^{-1}(B)y_t \tag{11.28}$$

$$u_t = \psi_x^{-1}(B)\psi(B)a_t, \quad \text{with } \psi(B) = \frac{\theta(B)}{\phi(B)}. \tag{11.29}$$

The β_t is labeled the prewhitened dependent series.

On multiplying (11.27) on both sides with α_{t-k} and taking expectations, we obtain

$$\gamma_{\alpha\beta}(k) = E(\alpha_t\beta_{t+k}) = v_k\sigma_\alpha^2, \qquad k = 0, 1, 2, \ldots . \tag{11.30}$$

To derive (11.30), notice that u_t and α_t are by assumption uncorrelated, and also $E(\alpha_{t-k}\alpha_{t'}) = 0$ for all t' except $t' = t - k$ because α_t is a white noise series.

Alternatively we can solve (11.30) for v_k in terms of the cross correlations:

$$v_k = \rho_{\alpha\beta}(k) \, \sigma_\beta/\sigma_\alpha \qquad k = 0, 1, 2, \ldots . \tag{11.31}$$

From (11.31) we see that the impulse response weights, v_k, are proportional to the cross correlations of the prewhitened data α_t and β_t. Estimates of v_k are obtained by replacing the population quantities by sample values

$$\hat{v}_k = r_{\alpha\beta}(k) \, s_\beta/s_\alpha \qquad k = 0, 1, 2, \ldots . \tag{11.32}$$

11.3.4 Identification of the Noise Model

It now should be clear that given preliminary estimates of $\hat{v}(B)$, the noise series can be estimated

$$\hat{e}_t = y_t - \hat{v}(B)x_t. \tag{11.33}$$

Again in (11.33) we have assumed that the data have been made stationary by appropriate differencing or any other stationarity inducing procedure. But, as mentioned in Sections 11.3.2 and 11.3.3, the *ccf* contains valuable information that could lead us to reevaluate if the data is nonstationary. Analyzing this error series with the univariate time series methods discussed in previous chapters should allow us to identify the ARIMA process governing the noise model (11.9).

For the identification process of the transfer function model to work successfully, it is necessary that the variation in the input x_t be reasonably large relative to the variation in the noise process. Otherwise, the first stage of the identification, the identification of the orders r and l based on the cross correlations $r_{xy}(k)$, would be very difficult and tenuous at best, as the patterns of the cross correlation function would be contaminated by the noise process. If this were the case, the user would have to work with a broader class of tentative models in the estimation stage.

In Section 11.4 we will analyze an example in more detail and show how, based on the *ccf*, we can identify models that could have generated the data.

11.3.5 Several Explanatory Variables

Up to now we have discussed a transfer function model with only one explanatory variable. For the identification of a transfer function model with several explanatory variables the above steps carry over with only a few obvious modifications. In particular, let the transfer function model be written as

$$y_t = v_1(B)x_{1t} + v_2(B)x_{2t} + e_t. \qquad (11.34)$$

Then the identification process will proceed in three steps. First, we will prewhiten the x_{1t} and y_t with the x_{1t} filter in order to obtain preliminary estimates of the order of the numerator and denominator polynomials in $v_1(B)$, as well as an estimate $\hat{v}_1$ of the parameters in the $v_1(B)$ polynomial using the y_t and x_{1t} cross correlations [see equation (11.32)]. Second, we will apply the same procedure to the x_{2t} variables, leading to guess values of the order of the numerator and denominator polynomials in $v_2(B)$, as well as the estimates of the parameters in the $v_2(B)$ polynomial. Finally, replacing $v_1(B)$ and $v_2(B)$ in (11.34) with these estimates, we can obtain an estimate of the process error

$$\hat{e}_t = y_t - \hat{v}_1(B)x_{1t} - \hat{v}_2(B)x_{2t}. \qquad (11.35)$$

The error series can then be analyzed as a univariate ARIMA process in order to obtain the appropriate orders of the error autoregressive and moving average polynomials.

The above three steps would work very well if x_{1t} and x_{2t} were uncorrelated. In a regression model, if the explanatory variables are independent, we could run regression models between the dependent variable and each explanatory variable separately and still obtain unbiased estimates of the regression coefficients. However, as soon as there is correlation between the explanatory variables, separate univariate regression models would produce biased regression coefficients for the full model. This same result carries over to a transfer function model with the following two qualifications. First, we do not propose to obtain the parameter value of the polynomials, but only to obtain guess values of the orders of the polynomials. Once we have an idea about the orders of the polynomials, in the estimation phase we will calculate the parameter values based on the complete model. Second, one idea behind the prewhitening of the data is to obtain transformed data series that are less correlated, so that the values of the orders will be easier to evaluate than if we used the raw data.

Another advantage of the prewhitening phase at the identification stage is that, as shown in equation (11.32), the parameters in the $v(B)$ polynomial can be calculated very easily, as they are proportional to the cross correlations.

11.4 EXAMPLE: LYDIA PINKHAM'S VEGETABLE COMPOUND DATA

As an example we will analyze the well-known Lydia Pinkham's vegetable compound sales and advertising data. This annual data is contained in Appen-

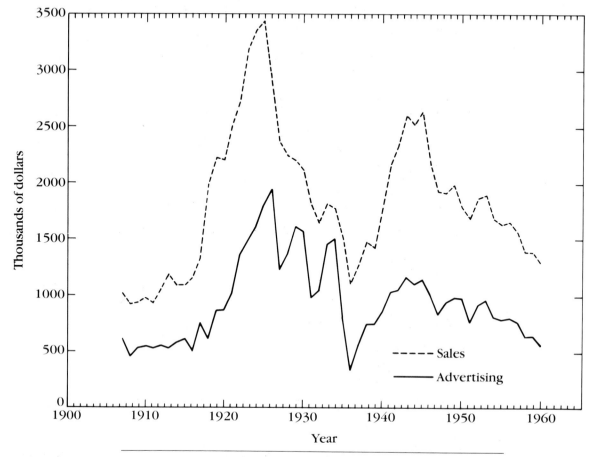

FIGURE 11.10   Lydia Pinkham's Vegetable Compound Sales—Advertising Data.

dix A and is taken from Palda (1964, p. 22). Figure 11.10 contains a plot of the sales and advertising data. The Lydia Pinkham data was first analyzed by Palda (1964) and subsequently by a number of other researchers. Some of these researchers include Clarke (1976), Caines et al. (1977), Hanssens (1977), Helmer and Johansson (1977), and Kyle (1978).

11.4.1 Univariate ARIMA Model of the Advertising Data

The first step in constructing a transfer function model is to build a univariate ARIMA model for the Lydia Pinkham advertising data. We propose only to use the annual data for the period 1907–1946, a total of 40 observations, and to save the last 14 observations for model validation. Because the construction of this univariate ARIMA model follows the same steps as discussed in previous chapters, only the final results are summarized here.

The identification of the advertising model clearly indicates that the data is nonstationary. An analysis of the *acf* and *pacf* of the first differences of the advertising data is less directive in pointing to the appropriate ARIMA model. There is quite some evidence that an AR(2) model could be appropriate, although an AR(4) model, possibly with the third-order parameter constrained to 0, cannot be ruled out *a priori*. Indeed the fourth-order partial autocorrelation is the last correlation that is just significant at the 5% level. Therefore, based on this information we should estimate both an AR(2) and an AR(4) model.

The AR(2) model results in the following estimates:[4]

$$x_t = 0.091x_{t-1} - 0.411x_{t-2}, \qquad \hat{\sigma}_a = 231.2,$$
$$(0.150) \qquad (0.149) \tag{11.36}$$

with x_t = first differences in advertising. The numbers in parentheses are estimates of the standard error of the estimates. Also, $\hat{\sigma}_a$ is an estimate of the process residual standard deviation as defined by (5.6). The residual Q statistic based on the first ten residuals has a value of 9.76, resulting in a P-value of 0.282.

The AR(4) model gives slightly better results:

$$x_t = 0.083x_{t-1} - 0.259x_{t-2} - 0.088x_{t-3} + 0.357x_{t-4}, \quad \hat{\sigma}_a = 222.3. \tag{11.37}$$
$$(0.158) \qquad (0.158) \qquad (0.158) \qquad (0.159)$$

The residual Q statistic, again based on the first ten residuals, has this time a value of only 3.00 with a P-value of 0.809. Based on these two models we would favor the AR(4) model, although some statistical improvements are still possible. Indeed the first- and third-order AR parameters in the AR(4) model are insignificant and could, therefore, be omitted from the model. More elaborate models were also evaluated but with little or no significant improvement over these two models. Also, the inclusion of a constant in both models, revealed that the constant was insignificiant. At this stage we propose to use these two models for the cross correlation analysis.

11.4.2 Cross Correlation Analysis

In Figures 11.11 and 11.12 we present the cross correlation analysis based on the two advertising models analyzed in Section 11.4.1. In the cross correlation analysis we induced stationarity by also calculating the first differences

[4] Note that although this is the model used in Helmer and Johansson (1977), the results are different because, contrary to the statement mentioned in this paper, the authors used all the data and not just the first 40 observations.

FIGURE 11.11   Cross Correlations of Lydia Pinkham Data, AR(2) Model.

```
CROSS CORRELATIONS WITH VBL ADVPNK    STANDARD ERROR =   .16
ZERO LAG =   .47
  LEADS           FUTURE          PY(t),PX(t+k)
   1-12            .40  .15  .23  .18  .21 -.06  .09  .04  .01  .02 -.13 -.22
CHI-SQUARE TEST                 P-VALUE
Q(12) =  17.3        12 D.F.   .140


  LAGS            PAST            PY(t),PX(t-k)
   1-12            .29  .10 -.09  .01 -.24 -.24 -.26 -.13 -.07 -.18 -.02 -.16
CHI-SQUARE TEST                 P-VALUE
Q(12) =  17.0        12 D.F.   .148
TWO SIDED CHI-SQUARE TEST  P-VALUE
Q(25) =  43.0        25 D.F.   .014


CROSS CORRELATION FUNCTION WITH VBL ADVPNK
   -1.0 -0.8 -0.6 -0.4 -0.2  0.0  0.2  0.4  0.6  0.8  1.0
    . . . . * . . . * . . . * . . . * . . . * . . . * . . . * . . . * . . . * . . . * . . . *
   -12.LEAD              ( R         .           )
   -11.                  (      R    .           )
   -10.                  (           R           )
    -9.                  (           R           )
    -8.                  (          .R           )
    -7.                  (          . R          )
    -6.                  (       R.              )
    -5.                  (           .    R      )
    -4.                  (           .    R      )
    -3.                  (           .     R     )
    -2.                  (           .    R      )
    -1.                  (           .           )  R
     0.                  (           .           )    R
     1.                  (           .        R  )
     2.                  (           .   R       )
     3.                  (       R   .           )
     4.                  (           R           )
     5.                  ( R        .            )
     6.                  ( R        .            )
     7.                  (R         .            )
     8.                  (       R   .           )
     9.                  (        R  .           )
    10.                  (      R    .           )
    11.                  (           R           )
    12.LAG              (       R   .           )


AUTOCORRELATIONS OF RESIDUALS USING VBL ADVPNK
  LAGS ROW SE
   1-12   .19    .03 -.15 -.10  .13  .02 -.05  .08  .00 -.04  .05 -.08  .15
CHI-SQUARE TEST                 P-VALUE
Q(12) =  3.69        10 D.F.   .960


PARTIAL AUTOCORRELATIONS       STANDARD ERROR =   .19
  LAGS
   1-12                 .03 -.15 -.09  .11 -.02 -.03  .11 -.03 -.03  .08 -.13  .18
```

FIGURE 11.12   Cross Correlations of Lydia Pinkham Data, AR(4) Model.

```
CROSS CORRELATIONS WITH VBL ADVPNK    STANDARD ERROR =   .17
ZERO LAG =   .42
  LEADS           FUTURE              PY(t),PX(t+k)
  1-12          .34   .10   .08   .09   .08 -.10   .07   .05   .03   .13 -.09 -.23
CHI-SQUARE TEST                 P-VALUE
Q(12) =  10.9          12 D.F.   .534


  LAGS            PAST                PY(t),PX(t-k)
  1-12          .29   .11 -.13 -.07 -.28 -.19 -.22 -.04   .06 -.08   .04 -.09
CHI-SQUARE TEST                 P-VALUE
Q(12) =  12.9          12 D.F.   .376
TWO SIDED CHI-SQUARE TEST  P-VALUE
Q(25) =  30.4          25 D.F.   .209


CROSS CORRELATION FUNCTION WITH VBL ADVPNK
    -1.0 -0.8 -0.6 -0.4 -0.2  0.0  0.2  0.4  0.6  0.8  1.0
     *  .  .  .  *  .  .  .  *  .  .  .  *  .  .  .  *  .  .  .  *  .  .  .  *  .  .  .  *  .  .  .  *  .  .  .  *
   -12.LEAD            (  R         .                )
   -11.                (       R    .                )
   -10.                (            .      R          )
    -9.                (            .R               )
    -8.                (            .R               )
    -7.                (            .      R          )
    -6.                (       R    .                )
    -5.                (            .      R          )
    -4.                (            .      R          )
    -3.                (            .      R          )
    -2.                (            .       R        ).
    -1.                (            .                )R
     0.                (            .                )    R
     1.                (            .            R)
     2.                (            .     R       )
     3.                (       R    .                )
     4.                (       R    .                )
     5.                (R           .                )
     6.                (    R       .                )
     7.                (    R       .                )
     8.                (           R.                )
     9.                (            .R               )
    10.                (    R       .                )
    11.                (            .R               )
    12.LAG             (       R    .                )

AUTOCORRELATIONS OF RESIDUALS USING VBL ADVPNK
  LAGS ROW SE
  1-12  .19     .02 -.19 -.12   .16   .08 -.03   .08   .04 -.07 -.04 -.19   .10
CHI-SQUARE TEST                 P-VALUE
Q(12) =  6.11          8 D.F.   .634

PARTIAL AUTOCORRELATIONS        STANDARD ERROR =   .19
  LAGS
  1-12                .02 -.19 -.12   .14   .03   .00   .15   .02 -.06   .00 -.27   .07
```

of the sales data. Indeed the large swings in the sales data suggest that the mean level of sales is not constant. The behavior depicted in Figure 11.10 could be easily taken for a random walk series.

Figure 11.11 is based on the AR(2) model[5] presented in equation (11.36), whereas the AR(4) model, equation (11.37), is used in Figure 11.12. Comparing these results we are struck by the degree of similarity in the cross correlation analysis. Therefore, the particular transfer function models that we can identify are the same.

Basically, two alternative transfer function models can be identified by comparing the sample *ccf* with the theoretical patterns.

Model T1: $(b,l,r) = (0,1,0)$. This model is supported by the fact that only the first two cross correlations are significant. Indeed, with a large-sample SE of 0.16 (or 0.17, in Figure 11.12), the third, fourth, and fifth cross correlations are very small. The dead time parameter $b = 0$ because $r_{xy}(0)$ is significant, and $r = 0$ because we claim to find no particular pattern of an AR process in the cross correlation function. The rather large values of $r_{xy}(5)$, $r_{xy}(6)$, and $r_{xy}(7)$ are somewhat disturbing. We propose, however, to ignore these correlations for now and to reassess the transfer function estimation results in this light.

Model T2: $(b,l,r) = (0,0,1)$. Again the dead time parameter $b = 0$. This time we claim to recognize an AR(1) pattern in the *ccf* which starts at the lag 0, so that $l = 0$. The AR(1) pattern is a geometric decay. With an approximate value for $r_{xy}(1) = 0.5$, we expect $r_{xy}(2) = 0.25$ and $r_{xy}(3) = 0.125$. Both in Figure 11.11 and 11.12 we must recognize that the sample cross correlations are indeed quite close to these theoretical values.

THE NOISE PROCESS

The identification process follows the univariate analysis completely. In Figures 11.11 and 11.12 we have presented the autocorrelations and partial autocorrelations of the residuals based on the stationary model formulation [see equation (11.33)]. The explicit formulation for the sales advertising is

$$\hat{e}_t = y_t - \hat{v}(B)x_t,$$

where

$$y_t = sales_t - sales_{t-1}$$
$$x_t = advertising_t - advertising_{t-1}.$$

The impulse response polynomial used is a tenth-order polynomial with the weights v_k calculated as given in (11.32). Again both prewhitened models give very similar empirical results. Indeed, both models produce residuals that can be accepted to be white noise.

---

[5] Helmer and Johansson (1977) present in their Figure 4 the lead and not, as they claim in the text, the lag cross correlations.

TRANSFER FUNCTION MODELS

Based on the above cross correlations analysis we propose to estimate and evaluate the following two transfer function models:

Let

$$y_t = sales_t - sales_{t-1}$$
$$x_t = advertising_t - advertising_{t-1};$$

then the two models are Model T1,

$$y_t = (\omega_0 - \omega_1 B)x_t + a_t; \qquad (11.38)$$

and Model T2,

$$y_t = \frac{\omega_0}{1 - \delta_1 B} x_t + a_t. \qquad (11.39)$$

11.5 SUMMARY

In this chapter we have discussed how to identify a transfer function model. We have seen that besides the autocorrelation function and partial autocorrelation function used to identify univariate time series, we now make extensive use of the cross correlation function. The *ccf* measures the degree of association between two variables at different time periods, both lags and leads. We showed that in properly measuring the degree of association it is important to prewhiten the data. The theoretical identification results were obtained after we used artifically generated data to make the results intuitively clear.

In Section 11.4 we used Lydia Pinkham's vegetable compound sales and advertising data to illustrate the identification of a transfer function model.

CHAPTER 12

ESTIMATION AND DIAGNOSTIC CHECKING OF TRANSFER FUNCTION MODELS

12.1 TRANSFER FUNCTION MODEL ESTIMATION

After identifying a particular transfer function model the next step is to estimate its parameters. Representing the transfer function model as

$$y_t = \frac{\omega(B)}{\delta(B)} x_{t-b} + e_t \tag{12.1}$$

where

$$y_t = \nabla^{d'} Y_t$$
$$x_t = \nabla^d X_t$$
$$e_t = \frac{\theta(B)}{\Phi(B)} a_t,$$

the task is to estimate the vectors of parameters $\omega = (\omega_0, \omega_1, \ldots, \omega_l)'$, $\delta = (\delta_1, \delta_2, \ldots, \delta_r)'$, $\phi = (\phi_1, \phi_2, \ldots, \phi_p)'$, and $\theta = (\theta_1, \theta_2, \ldots, \theta_q)'$. If there are several explanatory variables we will have several ω and δ vectors [see equation (10.16)]. Similarly, if the transfer function model is seasonal, there will be seasonal parameters and possibly seasonal differencing [see equation (10.22)].

As for the univariate ARIMA time series model, there are several approaches to estimating these parameters. Under the least squares method one chooses those values of the parameters which will make the sum of squared residuals as small as possible; that is, one chooses $\hat{\omega}$, $\hat{\delta}$, $\hat{\phi}$, and $\hat{\theta}$ as estimators of ω, δ, ϕ, and θ, respectively, so that

$$S(\hat{\omega}, \hat{\delta}, \hat{\phi}, \hat{\theta}) = \sum_{t=1}^n \hat{a}_t^2 \tag{12.2}$$

is a minimum. The index n represents the number of observations available, possibly adjusted for differencing. It will be useful to look at the minimization as proceeding in three stages.

First, given any initial choice of the parameter values in the $\omega(B)$ and $\delta(B)$ polynomial, we can predict the systematic part of the transfer function model as[1]

$$\hat{y}_t = \frac{\hat{\omega}(B)}{\hat{\delta}(B)} x_{t-b}. \tag{12.3}$$

For example, with $\omega(B) = \omega_0 - \omega_1 B$, and $\delta(B)$ of zero-order, (12.3) becomes

$$\hat{y}_t = \hat{\omega}_0 x_{t-b} - \hat{\omega}_1 x_{t-b-1}.$$

If $\omega(B) = \omega_0$ and $\delta(B) = 1 - \delta_1 B$, we then have for (12.3)

$$\hat{y}_t = \hat{\delta}_1 y_{t-1} + \hat{\omega}_0 x_{t-b}.$$

---

[1] The polynomial $\omega(B)$ with the parameters in the vector ω replaced by estimates is denoted by $\hat{\omega}(B)$. This notation applies to all other polynomials.

The second step is to calculate the residuals e_t using

$$\hat{e}_t = y_t - \hat{y}_t. \tag{12.4}$$

Finally, the series $\hat{e}_t$ can then be used to evaluate a_t using the univariate ARIMA model representation

$$\hat{a}_t = \frac{\hat{\phi}(B)}{\hat{\theta}(B)}\,\hat{e}_t. \tag{12.5}$$

If the dead time parameter b is unknown, in principle we can calculate the values of $\underline{\omega}$, δ, ϕ, and θ which minimize the residual sum of squares (12.2) for different values of $\bar{b}$ in the likely range and select the value of b which corresponds to the overall minimum of (12.2).

12.2 STARTING VALUES

The polynomials in (12.3) require starting values, initial conditions, for the y and x data for some periods before the beginning of the series. To see this, rewrite (12.3) for $t = 1$ as

$$\hat{\delta}(B)\hat{y}_1 = \hat{\omega}(B)x_{1-b}$$

or

$$\hat{y}_1 = \hat{\delta}_1 y_0 + \hat{\delta}_2 y_{-1} + \cdots + \hat{\delta}_r y_{1-r} + \hat{\omega}_0 x_{1-b} + \hat{\omega}_1 x_{-b} + \cdots + \hat{\omega}_l x_{1-b-l}. \tag{12.6}$$

In (12.6) we will need the starting values $y_0, y_1, \ldots, y_{1-r}$ and $x_{1-b}, x_{-b}, \ldots, x_{1-b-l}$. However, if $b = 0$, x_{1-b} would be known. This issue is identical to the starting value problem mentioned in Section 5.1.

Most of the existing computer programs will calculate y_t in (12.6) from $t = u + 1$ onwards, where u is the larger of r and $l + b$. As a result, there will be only $n - u$ residuals $\hat{e}_t$.

Next, writing (12.5) out for $t = 1$ we obtain

$$a_1 = \theta_1 a_0 + \cdots + \theta_q a_{1-q} + e_1 - \phi_1 e_0 - \cdots - \phi_p e_{1-p}.$$

If we set $a_0, a_{-1}, \ldots, a_{1-q}$ equal to their unconditional expected values of 0, we would still need p lagged e_t values, $e_0, \ldots, e_{1-p}$. Therefore, ultimately (12.2) will be evaluated from a_{u+p+1} onwards. So, we choose estimators for the parameters so that[2]

---

[2] The TS computer program TRFEST uses the Marquardt nonlinear optimizer to minimize the sum of squared residuals as defined in (12.7), with the following modification. Instead of backcasting the starting values for $y_0, y_{-1}, \ldots, y_{1-r}$, the program TRFEST uses the mean of y_t for these starting values. Therefore, the summation in (12.7) starts from $l + b + p + 1$; i.e., u is defined as $l + b$. The Marquardt optimizer is the same algorithm used in the program ESTIMA for the optimization of univariate ARIMA parameters [see Chapter 5 and Marquardt (1963)].

$$S(\hat{\underline{\omega}}, \hat{\underline{\delta}}, \hat{\underline{\phi}}, \hat{\underline{\theta}}) = \sum_{t=u+p+1}^{n} \hat{a}_t^2 \qquad (12.7)$$

is a minimum.

As with the univariate ARIMA model, the implementation of this nonlinear algorithm requires that the user also specify starting values (guess values) for the parameters to initiate the nonlinear estimation procedures. This time, these preliminary values of the parameters in the $\delta(B)$ polynomial of the transfer function model can be obtained by comparing the values of the sample cross correlations with the parameter representation of the population cross correlations and then solving for each individual parameter. The ω parameters in the $\omega(B)$ polynomial are obtained directly from the impulse response weights which are unscaled cross correlations [see equation (11.31)]. The guess values for the parameters in the noise process polynomials are obtained exactly as the guess values of the univariate ARIMA model are, by analyzing the sample residual autocorrelations.

In the next paragraphs we will present some results on how to use the cross correlations and impulse weights. We would like to indicate, however, that we have found that in most applications the number of iterations necessary to minimize (12.7) is not overly sensitive to the initial guess values. Therefore, the reader could easily skip these paragraphs and go to Section 12.3.

If the cross correlation function suggests the following model (analyzed in Section 11.3.1),

$$y_t = \omega_0 x_t + e_t$$

because only $r_{xy}(0)$ is nonzero, then a good guess value for ω_0 is $\hat{v}_1$. Indeed, combining equations (11.15) for $b = 0$ with (11.32), we find that

$$\hat{v}_0 = \omega_0.$$

For the somewhat more interesting model

$$y_t = \frac{\omega_0}{1 - \delta_1 B} x_t + e_t$$

we know from Section 11.3.1, equation (11.17), that

$$\rho_{xy}(k) = \begin{cases} 0 & k < 0 \\ \omega_0 \sigma_x / \sigma_y & k = 0 \\ \omega_0 \delta_1^j \sigma_x / \sigma_y & k = j, j > 0 \end{cases} \qquad (12.8)$$

Again combining these results with (11.32) we obtain as guess values

$$\hat{\omega}_0 = \hat{v}_0$$
$$\hat{\delta}_1 = r_{xy}(1)/r_{xy}(0) = \hat{v}_1/\hat{v}_0. \qquad (12.9)$$

Finally, for the general model the relationship between the impulse response weights and the numerator and denominator parameters can be expressed as [see Box and Jenkins (1976, p. 347)]

$$
\begin{aligned}
\nu_k &= 0 & k < b \\
\nu_k &= \delta_1\nu_{k-1} + \delta_2\nu_{k-2} + \cdots + \delta_r\nu_{k-r} + \omega_0 & k = b \\
\nu_k &= \delta_1\nu_{k-1} + \delta_2\nu_{k-2} + \cdots + \delta_r\nu_{k-r} - \omega_{k-b} & b < k \leq b + \ell \\
\nu_k &= \delta_1\nu_{k-1} + \delta_2\nu_{k-2} + \cdots + \delta_r\nu_{k-r} & k > b + \ell.
\end{aligned} \tag{12.10}
$$

In (12.10) the basic relationship to remember is that the impulse response weights ν_k follow an AR(r) process $\delta(B)\nu_k$ with $\delta(B) = 1 - \delta_1 B - \delta_2 B^2 - \cdots - \delta_r B^r$, together with a small adjustment in the ν_k for $b \leq k \leq b + \ell$ based on the numerator polynomial parameters.

For example, for the model

$$
y_t = \frac{\omega_0 - \omega_1 B - \omega_2 B^2}{1 - \delta_1 B - \delta_2 B^2} x_{t-b} + e_t,
$$

with $b = 3$, the equations (12.10) become

$$
\begin{aligned}
\nu_k &= 0 & k < 3 \\
\nu_3 &= \omega_0 & \text{(12.11a)} \\
\nu_4 &= \delta_1\nu_3 - \omega_1 & \text{(12.11b)} \\
\nu_5 &= \delta_1\nu_4 + \delta_2\nu_3 - \omega_2 & \text{(12.11c)} \\
\nu_6 &= \delta_1\nu_5 + \delta_2\nu_4 & \text{(12.11d)} \\
\nu_7 &= \delta_1\nu_6 + \delta_2\nu_5. & \text{(12.11e)}
\end{aligned}
$$

Given estimates of the impulse response weights ν_k, we could use (12.11d) and (12.11e) to solve for δ_1 and δ_2. These parameter values could then be substituted in (12.11b) and (12.11c) to solve for ω_1 and ω_2. Finally, ω_0 is obtained directly from (12.11a).

In the next section we will make use of some of these results to calculate guess values for the estimation process.

12.3 THE LYDIA PINKHAM EXAMPLE

The identification analysis in Section 11.4 of the Lydia Pinkham sales and advertising data indicated two possible transfer function models after first-order differencing of both series.

Model T1:

$$
(b,\ell,r) \times (p,d,q) = (0,1,0) \times (0,1,0)
$$
$$
y_t = (\omega_0 - \omega_1 B)x_t + a_t \tag{12.12}
$$

Model T2:

$$(b,\mathbf{1},r)\times(p,d,q) = (0,0,1)\times(0,1,0)$$

$$y_t = \frac{\omega_0}{1 - \delta_1 B} x_t + a_t \tag{12.13}$$

where

$$y_t = sales_t - sales_{t-1}$$
$$x_t = advertising_t - advertising_{t-1}.$$

Based on the first two impulse response weights[3] and the analysis in the previous section we suggest as initial values for Model T1: $\omega_0 = 0.59$, $\omega_1 = 0.36$, and for Model T2: $\omega_0 = 0.59$ and $\delta_1 = 0.65$.

In Table 12.1 we present the analysis of these two models. Most of the symbols used in the program TRFEST as represented in Table 12.1 have been explained in Section 5.3 and will not be repeated here. In addition Appendix C contains a detailed discussion of the TRFEST output. In general, the estimation results are quite acceptable, although, as we will show, model improvements are possible.

12.4 DIAGNOSTIC CHECKS

Again, once a transfer function model has been identified and its parameters estimated, it is necessary to verify that the model chosen adequately fits the data. In this section we will describe three groups of tests or diagnostic checks which can be used for this purpose. These three groups of diagnostic checks, which parallel the checks for the univariate time series model, are:

1. Residual analysis.
2. Fitting extra parameters: the underspecified model.
3. Omitting parameters: the overspecified model.

Before proceeding, we would like again to warn the reader that the above order in no way implies that the tests should be applied sequentially. It is the collection of tests that should indicate if there is sufficient evidence of model inadequacy, and not just a single test. If the transfer function model is inadequate, then some of these diagnostic checks will provide clues for the modifications necessary to improve the model specifications.

---

[3] The specific estimates of the impulse response weights depend on the advertising prewhitening model used in the calculation of the cross correlations. For the AR(2) prewhitening advertising model $\nu_0 = 0.5885$, $\nu_1 = 0.3621$, and for the AR(4) prewhitening model $\nu_0 = 0.5466$ and $\nu_1 = 0.3806$. Based on equation (12.9) we find that the AR(2) model suggests the following starting values for the Model T2 parameters: $\omega_0 = 0.59$, $\delta_1 = 0.65$, and for the AR(4) model $\omega_0 = 0.55$, $\delta_1 = 0.70$. From (12.10) we can derive that $\omega_0 = \nu_0$ and $\omega_1 = -\nu_1$. Therefore, guess values for Model T1 are $\omega_0 = 0.59$, $\omega_1 = -0.36$ for the AR(2) model, and $\omega_0 = 0.55$ and $\omega_1 = -0.38$ for the AR(4) model. The values mentioned in the text are based on the AR(2) prewhitening model.

12.4.1 Residual Analyses—Theory

If the transfer function model is properly specified, then the a_t residuals in the model should behave as white noise and be uncorrelated with the explana-

TABLE 12.1   Lydia Pinkham—Initial Model Estimates

```
        MODEL T1

PARAMETER ESTIMATES
====================

              EST        SE       EST/SE      95% CONF LIMITS
NUM( 0)      0.583     0.141      4.144       0.307     0.858
NUM( 1)     -0.215     0.140     -1.532      -0.490     0.060


EST.RES.SD =    2.1566E+02

R SQR =   .364
DF = 36
NO F-STAT, NO INTERCEPT

CORRELATION MATRIX
        NUM( 0)
 NUM( 1)    .061

AUTOCORRELATIONS OF RESIDUALS
  LAGS ROW SE
  1-10   .16    .18   .13   .04  -.03   .10   .01  -.06  -.28  -.26  -.10

CHI-SQUARE TEST              P-VALUE
Q(10) =  10.8        10 D.F.   .370

PREWHITENING TRANSFORMATION USED BEFORE CROSS CORRELATION ANALYSIS

CROSS CORRELATIONS OF RESIDUALS AND STATIONARY SERIES

STANDARD ERROR =  .16
ZERO LAG =  .07
  LEADS            FUTURE          A(t),W(t+k)
  1-10            .52   .11   .15   .15   .18   .02   .08  -.03  -.07   .21

CHI-SQUARE TEST             P-VALUE
Q(10) =  16.6        8 D.F.   .034

  LAGS             PAST            A(t),W(t-k)
  1-10            .12   .12  -.01  -.13  -.33  -.16  -.21  -.07  -.08  -.07

CHI-SQUARE TEST(+ZERO LAG) P-VALUE
Q(11) =  10.9        9 D.F.   .285
```

TABLE 12.1 (*continued*)

```
                MODEL  T2

        PARAMETER ESTIMATES
        ====================

                  EST      SE    EST/SE    95% CONF LIMITS
        NUM( 0)  0.616   0.134   4.599     0.354    0.879
        DEN( 1)  0.341   0.192   1.775    -0.036    0.719

        EST.RES.SD =    2.1171E+02

        R SQR =  .370
        DF = 37
        NO F-STAT, NO INTERCEPT

        CORRELATION MATRIX
                NUM( 0)
        DEN( 1)  -.166

        AUTOCORRELATIONS OF RESIDUALS
          LAGS ROW SE
          1-10 .16  .17  .15  .03 -.03  .10  .02 -.07 -.29 -.27 -.11

        CHI-SQUARE TEST            P-VALUE
        Q(10) =  11.9      10 D.F.   .289

        PREWHITENING TRANSFORMATION USED BEFORE CROSS CORRELATION ANALYSIS

        CROSS CORRELATIONS OF RESIDUALS AND STATIONARY SERIES

        STANDARD ERROR =  .16
        ZERO LAG =  .04
          LEADS         FUTURE          A(t),W(t+k)
          1-10           .53  .11  .16  .16  .19  .03  .08 -.02 -.08  .24

        CHI-SQUARE TEST           P-VALUE
        Q(10) =  18.3       8 D.F.   .019

          LAGS          PAST            A(t),W(t-k)
          1-10           .12  .06 -.04 -.14 -.32 -.18 -.22 -.06 -.07 -.07

        CHI-SQUARE TEST(+ZERO LAG) P-VALUE
        Q(11) =  10.5       9 D.F.   .310
```

tory variables in the model. Therefore, model inadequacy can be evaluated by examining

1. a residual plot,
2. the autocorrelation functions of the residuals, and

3. the cross correlation functions between the explanatory variables and the residuals.

It should be obvious that a residual plot always contains valuable information, some of which might not be apparent from other proposed summary measures. In particular, a peculiar sequence of residuals and outliers might be easier to spot by a careful analysis of a residual plot.

To see what information is contained in these autocorrelation and cross correlation functions, let us go back to the transfer function model as specified in (10.15),

$$y_t = \frac{\omega(B)}{\delta(B)} x_{t-b} + \frac{\theta(B)}{\phi(B)} a_t$$
$$= \nu(B)x_t + \psi(B)a_t \tag{12.14}$$

with $\psi(B)$ defined in (11.29) and $\nu(B)$ defined in (10.4) after adjustment for the dead time b.

Now, suppose that we select an incorrect model defined as

$$y_t = \nu_0(B)x_t + \psi_0(B)a_{0t}, \tag{12.15}$$

where $\nu_0(B) = \omega_0(B)/\delta_0(B)$, $\psi_0(B) = \theta_0(B)/\phi_0(B)$. The subscript 0 on the polynomial is used to distinguish these polynomials from those used in (12.14) insofar that at least one polynomial is of different order, implying different parameter values. Then we can represent the residual in the incorrectly specified model as

$$a_{0t} = \psi_0^{-1}(B)[y_t - \nu_0(B)x_t]. \tag{12.16}$$

After substitution from (12.14) for y_t, we obtain

$$a_{0t} = \psi_0^{-1}(B)[\nu(B) - \nu_0(B)]x_t + \psi_0^{-1}(B)\psi(B)a_t. \tag{12.17}$$

From (12.17) it is apparent that, if a wrong model is selected, the a_{0t} could be autocorrelated as well as possibly cross correlated with the x_t's and hence also with the prewhitened α_t's.

Two cases are of special interest:

1. Only the noise structure is incorrect; that is, the second term on the right-hand side of (12.17) is not equal to a_t.

2. Only the systematic part of the transfer function model is wrong; that is, the first term on the right-hand side of (12.17) is nonzero, and the second term equals a_t.

TRANSFER FUNCTION STRUCTURE CORRECT, NOISE STRUCTURE INCORRECT

If the systematic part of the transfer function model is correctly specified, but the noise structure is incorrect, we have $\nu(B) = \nu_0(B)$ and $\psi_0(B) \neq \psi(B)$. Therefore, from (12.17) it follows that

$$a_{0t} = \frac{\psi(B)}{\psi_0(B)} a_t. \tag{12.18}$$

As a result, the a_{0t}'s will not be cross correlated with any included explanatory variable, but will be autocorrelated, and the form of the residual autocorrelation function could indicate the necessary specification changes that should be made in the noise structure of the transfer function model.

Suppose that the error term in the misspecified transfer function model (12.15) was specified as white noise, $\psi_0(B) = 1$, but the autocorrelations of a_{0t} suggest the following autoregressive model:

$$a_{0t} = \phi_1 a_{0,t-1} + a_t$$
$$= \frac{1}{1 - \phi_1 B} a_t. \tag{12.19}$$

With $\psi(B) = \theta(B)/\phi(B)$, we have $\theta(B) = 1$ and $\phi(B) = 1 - \phi_1 B$. Then, the obvious modification to the model is to specify an ARMA(1,0) for the noise model.

TRANSFER FUNCTION STRUCTURE INCORRECT, NOISE STRUCTURE CORRECT

If only the systematic part of the transfer function model is correct, (12.17) simplifies to

$$a_{0t} = \psi^{-1}(B)[v(B) - v_0(B)]x_t + a_t. \tag{12.20}$$

Therefore, the a_{0t}'s would not only be cross correlated with the x_t's, but the a_{0t}'s could also be autocorrelated.

However, to evaluate possible model modifications it is important to calculate cross correlations between a_{0t} and the prewhitened explanatory variable. Indeed it can be shown that if x_t is autocorrelated and the residual is white noise, then the cross correlations between a_t and x_t approximately will have the same autocorrelation function as the explanatory variable x_t. Then, when the x_t's are autocorrelated, a perfectly adequate transfer function will give rise to cross correlations $r_{x\hat{a}}(k)$ which, although small in magnitude, may show pronounced *patterns*. Therefore, in order to use the cross correlation results to suggest model inadequacies, we must first prewhiten the x_t series.

To indicate the possible changes that should be introduced to improve the model, we rewrite the transfer function model in its prewhitened form [see (11.27)],

$$\beta_t = v(B)\alpha_t + u_t \tag{12.21}$$

where

$$\beta_t = \psi_x^{-1}(B)y_t$$
$$\alpha_t = \psi_x^{-1}(B)x_t$$
$$u_t = \psi_x^{-1}(B)\psi(B)a_t$$

with

$$\psi_x(B) = \frac{\theta_x(B)}{\phi_x(B)} \quad \text{and} \quad \psi(B) = \frac{\theta(B)}{\phi(B)}.$$

The error in the incorrectly specified prewhitened model is defined as

$$u_{0t} = \beta_t - v_0(B)\alpha_t. \tag{12.22}$$

Substituting for β_t from (12.21) we also have

$$u_{0t} = [v(B) - v_0(B)]\alpha_t + u_t. \tag{12.23}$$

Similar to the interpretation of the cross correlations between the prewhitened dependent and an explanatory variable [given in (11.31)], the cross correlations between u_{0t} and α_t are proportional to the differences in the coefficients of the two impulse response polynomials or, alternatively, the difference can be expressed as

$$v_k - v_{0k} = \rho_{\alpha u_0}(k)\sigma_{u_0}/\sigma_\alpha \qquad k = 0, 1, 2, \ldots. \tag{12.24}$$

Therefore the cross correlations between the prewhitened explanatory variable and the residual can be used to evaluate what changes should be made in the $v_0(B)$ polynomial and as such in the $\omega(B)$ or $\delta(B)$ polynomials so as to improve the model specification.

Of course, in order to use the misspecification analysis, we must rely on sample cross correlations $r_{\alpha u_0}(k)$. For example, if the current transfer function model is specified as

$$y_t = \omega_0 x_t + a_t \tag{12.25}$$

and an analysis of the cross correlations between the residuals and the prewhitened explanatory shows that

$$r_{\alpha u_0}(k) \neq 0 \quad \text{for } k = 0 \text{ and } k = 1,$$

this implies that the order of the numerator polynomial should be increased from order $l = 0$ to order $l = 1$. Similarly, if the $r_{\alpha u_0}(k)$'s behave as the autocorrelations of an AR(1) process, model (12.25) should be modified to include a first-order denominator polynomial resulting in

$$v(B) = \frac{\omega_0}{1 - \delta_1 B}.$$

In general, the residual cross correlations are even harder to use than the cross correlations used in Chapter 11 to identify a possible transfer function model. From a practical standpoint nonzero residual cross correlations are an indication that the transfer function polynomials $\omega(B)$ and $\delta(B)$ are incorrectly specified. If it is not clear how to modify these polynomials, the user should simply start including additional parameters and reexamine its results. The fitting of extra parameters will also be discussed in Section 12.4.4.

A final point has to be made. The user also should not too eagerly claim that the transfer function model is correctly specified as soon as the residuals are white noise and uncorrelated with the included prewhitened explanatory variables. It still is possible that the user has left out an explanatory variable that could be correlated with the residual. Therefore in building a transfer function model the cardinal rule is to make sure that all factors explaining the dependent variable are properly represented in the model.

12.4.2 Residual Analyses—Sampling Results

In practice, we do not know the transfer function models parameters exactly but must calculate autocorrelations and cross correlations based on residuals $\hat{a}_t$ computed after the model fitting. As a result, even if both the transfer function structure and the noise structure of the transfer function model are correct, the parameter estimates would not be identical to the true values and therefore the correlations based on the residuals $\hat{a}_t$ would, to a certain degree, differ from the correlations based on a_t. Therefore, we are bound to find estimated autocorrelations and cross correlations which are different from 0, even if the population values are 0.

Most of the discussion presented now parallels the univariate ARIMA residual analyses contained in Section 5.3.2. We will rely on the standard errors of these correlations to determine whether sample correlations observed to be different from 0 are indeed real nonzero correlations or are nonzero due to random fluctuations. Again it will be useful to evaluate individual autocorrelations as well as groups of autocorrelations, and similarly for the cross correlations.

We also must recommend making a residual plot before embarking on a detailed correlation analysis. As mentioned in Chapter 5, such a plot would quickly reveal a number of data and model formulation problems that a correlation analysis might not be able to detect.

INDIVIDUAL AUTOCORRELATIONS

If the transfer function model was correctly specified and the true parameter values were substituted, the residuals would be white noise and the estimated autocorrelations would be independently distributed with mean zero and variance $1/m$. The value m corresponds to the actual number of $\hat{a}_t$'s calculated and equals $n - u - p$, with u the larger of r and $l + b$.[4] The variance formula is based on Bartlett's large-sample results. In small samples, in particular for low lags, the variance of the estimated autocorrelations can be *less* than $1/m$.

GROUP OF AUTOCORRELATIONS

In order to form an overall check on the autocorrelations it is again useful to calculate the following Ljung–Box Q statistic based on the transfer function residuals $\hat{a}_t$:

[4] As explained above, in the program TRFEST u is defined as $l + b$ because the mean value of y_t is used for the past values of y_t.

$$Q(K) = m(m+2) \sum_{k=1}^{K} \frac{1}{m-k} r_{\hat{a}\hat{a}}^2(k). \tag{12.26}$$

The value m is defined above as the number of residuals used in (12.7), and the value K is again selected in such a way that it can reasonably be assumed that the $r_{\hat{a}\hat{a}}(k)$'s for $k > K$, are negligible.

If the transfer function is appropriate (i.e., if the errors are white noise), the Q statistic is approximately distributed as χ^2 with $K - p - q$ degrees of freedom. Note that the number of parameters in $\underline{\omega}$ and $\underline{\delta}$ does not influence the degrees of freedom of the residual autocorrelation Q statistic.

INDIVIDUAL CROSS CORRELATIONS

The approximate standard error (SE) of individual cross correlations is as specified in (11.5) or (11.6), with the n replaced by $m = n - u - p$, the actual number of $\hat{a}_t$ estimated that is, $SE[r_{\alpha\hat{a}}(k)] = 1/\sqrt{m}$.

GROUPS OF CROSS CORRELATIONS

As for the autocorrelations, when estimates are substituted for parameter values, the distributional properties of the cross correlations are affected in that the degrees of freedom of statistics based on these cross correlations must be adjusted. In particular the Q statistic analogue

$$Q(K) = m(m+2) \sum_{k=0}^{K} \frac{1}{m-k} r_{\alpha\hat{a}}^2(k) \tag{12.27}$$

will be approximately χ^2 distributed with $K + 1 - (r + \ell + 1) = K - r - \ell$ degrees of freedom. Notice, for reasons mentioned above, that the $r_{\alpha\hat{a}}(k)$ represent the cross correlations between the estimated residuals and the prewhitened explanatory variable itself.

Here again K is selected such that for $k > K$ both ν_k and ψ_k in (12.14) can reasonably be assumed to be negligible. Notice that $r + \ell + 1$ represents the number of parameters fitted in the transfer function model. As such, degrees of freedom in the χ^2 statistic based on the cross correlations are not influenced by the number of parameters in the noise structure of the transfer function model.

12.4.3 Overspecified Model: Omitting Parameters

A very useful check on the transfer function model adequacy is to evaluate whether the current model contains redundant parameters. These redundant parameters can be spotted by a careful use of the parameter estimate large-sample standard error (SE) and the estimate of the large-sample correlations between these parameter estimates.

We use the SE to evaluate the statistical significance of a single parameter. A parameter insignificantly different from 0 is an indication that the model may be overspecified and a simplification of the model may be possible. The simplest situation occurs when the insignificant parameter is the one of the highest order in a polynomial, in which case we should evaluate removing that parameter.

If the significant parameter is not the highest order, then we must examine the large-sample correlations between the parameter estimates to determine which one to delete from the model. If the insignificant parameter is not the highest-order parameter but is strongly correlated with the highest-order, we would evaluate a model specification without the highest-order parameter. If no such correlation exists, then we could re-estimate the transfer model with the insignificant parameter suppressed.

Finally, high correlations between parameters of different polynomials require an even more careful analysis. We refer the reader to Section 5.3.3 for a more detailed discussion of the analysis of an overspecified model.

12.4.4 Underspecified Model: Fitting Extra Parameters

To verify that the tentatively specified transfer function model contains the appropriate number of parameters, one could proceed by including additional parameters to see if this results in an improvement over the original model. However, now even more than with a univariate ARIMA model (see Section 5.3.4), one should not carelessly include additional parameters. Again the *parameter redundancy* problem will occur if we add at the same time a numerator and a denominator parameter, or at the same time an error moving average and an error autoregressive parameter. This problem was discussed in detail in Section 5.3.4.

As a general rule we recommend that you not needlessly add parameters to the model if there is no diagnostic evidence for doing so. There should be some indication contained in the different diagnostic checks discussed above that the model is underspecified before embarking on fitting extra parameters.

12.5 THE LYDIA PINKHAM EXAMPLE— DIAGNOSTIC CHECKS

A more detailed analysis of the estimation results presented in Table 12.1 indicates that, both for Models T1 and T2, there is evidence of autocorrelation in the residuals. In Section 12.4.1 we saw that the residuals can be autocorrelated as soon as the transfer function structure is incorrect, even when the noise structure is correct [see (12.20)]. An examination of the cross correlations between the prewhitened explanatory variable[5] and the residuals basi-

---

[5] The prewhitening of the explanatory variable was done using the AR(4) advertising model (11.37).

cally indicates that if there is some cross correlation, the evidence is not overwhelming. Apparently the cross correlations are somewhat larger for Model T1, although the evidence on the residual autocorrelations is better for Model T1 than for Model T2. On pragmatic grounds we therefore propose to ignore the cross correlations and to first correct for the residual autocorrelation.

An examination of the residual *acf* reveals that the residuals could follow either an AR(1) process or an MA(2) process. Again the statistical evidence is not conclusive, as none of the residuals autocorrelations are larger in absolute value than two SE. However, either we can detect a slight exponential declining pattern in the autocorrelations, evidence for an AR(1) process, or we can claim that only the first two autocorrelations are large, with then a cutoff, evidence of an MA(2) process.

In the notation of Section 12.4.1 we have

$$a_{0t} = \frac{\psi(B)}{\psi_0(B)} a_t$$

with $\psi_0(B) = 1$. With the a_{0t} following an AR(1) process, we have [see also equation (12.19)]

$$a_{0t} = \frac{1}{1 - \phi_1 B} a_t.$$

Adding this error structure to the transfer function model we obtain
Model T1.1:

$$(b, \ell, r) \times (p, d, q) = (0, 1, 0) \times (1, 1, 0)$$

$$y_t = (\omega_0 - \omega_1 B) x_t + \frac{1}{1 - \phi_1 B} a_t \qquad (12.28)$$

Model T2.1:

$$(b, \ell, r) \times (p, d, q) = (0, 0, 1) \times (1, 1, 0)$$

$$y_t = \frac{\omega_0}{1 - \delta_1 B} x_t + \frac{1}{1 - \phi_1 B} a_t. \qquad (12.29)$$

The estimation results are presented in Table 12.2. As starting guess value for the residual AR(1) term we use the value of the residual autocorrelation in Model T.1, $r_{\hat{a}\hat{a}}(1) = 0.18$.

MODELS T1.1 AND T2.1

The estimation results and diagnostic checks presented in Table 12.2 indicate that these alternative models are an improvement over the initial models. Concentrating on the residual analysis of both models we find in both the autocorrelations and cross correlations less evidence of significant correlations than before, although we might have expected even less correlation. The Q

TABLE 12.2   Lydia Pinkham—Alternative Models

```
            MODEL T1.1

PARAMETER ESTIMATES
===================

            EST       SE     EST/SE      95% CONF LIMITS
NUM( 0)    0.480    0.153    3.139      0.180    0.781
NUM( 1)   -0.189    0.140   -1.348     -0.463    0.086
 AR( 1)    0.277    0.193    1.435     -0.101    0.655

EST.RES.SD =    2.1580E+02

R SQR =   .398
DF = 34
NO F-STAT, NO INTERCEPT

CORRELATION MATRIX
        NUM( 0) NUM( 1)
 NUM( 1)  -.137
  AR( 1)  -.426    .135

AUTOCORRELATIONS OF RESIDUALS
  LAGS ROW SE
  1-10   .16  -.04   .10   .03 -.09   .11 -.02 -.02 -.23 -.19   .04

CHI-SQUARE TEST          P-VALUE
Q(10) =  6.12       9 D.F.  .728

PREWHITENING TRANSFORMATION USED BEFORE CROSS CORRELATION ANALYSIS

CROSS CORRELATIONS OF RESIDUALS AND STATIONARY SERIES

STANDARD ERROR =  .16
ZERO LAG =  .04
  LEADS          FUTURE        A(t),W(t+k)
  1-10           .49   .07   .10   .10   .17 -.02   .10 -.02 -.12   .23

CHI-SQUARE TEST          P-VALUE
Q(10) =  15.3       8 D.F.  .053

  LAGS           PAST          A(t),W(t-k)
  1-10           .12   .06 -.06 -.09 -.29 -.11 -.21 -.01 -.03 -.07

CHI-SQUARE TEST(+ZERO LAG) P-VALUE
Q(11) =  7.89       9 D.F.  .545
```

```
        MODEL T2.1

PARAMETER ESTIMATES
===================

              EST       SE      EST/SE     95% CONF LIMITS
NUM( 0)      0.520    0.156     3.333      0.214     0.825
DEN( 1)      0.359    0.249     1.440     -0.130     0.848
 AR( 1)      0.256    0.195     1.317     -0.125     0.638


EST.RES.SD =    2.1254E+02


R SQR =   .399
DF = 35
NO F-STAT, NO INTERCEPT

CORRELATION MATRIX
        NUM( 0) DEN( 1)
 DEN( 1)    .046
  AR( 1)   -.458    -.107

AUTOCORRELATIONS OF RESIDUALS
  LAGS ROW SE
  1-10   .16   -.04   .12   .02  -.09   .11  -.01  -.04  -.23  -.21   .03

CHI-SQUARE TEST          P-VALUE
Q(10) =  6.73       9 D.F.   .665

PREWHITENING TRANSFORMATION USED BEFORE CROSS CORRELATION ANALYSIS

CROSS CORRELATIONS OF RESIDUALS AND STATIONARY SERIES

STANDARD ERROR =  .16
ZERO LAG =  .01
  LEADS          FUTURE            A(t),W(t+k)
  1-10            .50   .07   .12   .12   .17   .00   .09  -.01  -.13   .25

CHI-SQUARE TEST          P-VALUE
Q(10) =  16.8       8 D.F.   .033

  LAGS           PAST              A(t),W(t-k)
  1-10            .13   .01  -.07  -.10  -.28  -.13  -.22   .00  -.03  -.07

CHI-SQUARE TEST(+ZERO LAG) P-VALUE
Q(11) =  8.01       9 D.F.   .533
```

statistic on the residual autocorrelation of Model T1.1 has a value of 6.12, with nine degrees of freedom, and a corresponding P-value of 0.728. Certainly a P-value of 0.9 or higher would have been better. Remember, however, that the Lydia Pinkham analysis is only done with 40 data points, certainly almost a minimum number of points to fruitfully construct a transfer function model.

A check on the significance of the estimated parameters in both models shows only the first numerator coefficient, ω_0, to be clearly significant. The fact that the other parameter values are positive but insignificant indicates that there are only small or no effects of advertising after one year. These results also have been found in the work of Clarke (1976). At this stage of the analysis we keep both models for further evaluation.

Figures 12.1 and 12.2 contain a residual plot of Model T1.1 and Model T2.1. There basically are no real differences in these residual plots. However, based on these plots we may want to investigate the apparent large residuals for 1918 and 1926, both about three standard deviations away from 0. Are these results suggestive of two outliers or is there an explanation that can be put forward explaining these values? In the latter case, the intervention analysis presented in Chapter 14 could be used to model these two data points.

OVERSPECIFIED AND UNDERSPECIFIED MODELS

In this section we would like to summarize the analysis of some alternative model specifications. Based on the residual autocorrelations presented in Table 12.1, we mentioned above that possibly the residuals could have been generated by an MA(2) model. The estimation of both models with an MA(2) noise structure did not significantly improve the models. We found for both models that the two MA(2) parameters were not significantly different from 0, with the second-order moving average parameter more significant than the first-order one. In Model T1, after adding the MA(2) term, ω_1 also became totally insignificant. We also found that adding a constant to the model resulted in no changes, as its estimate was insignificant.

The next step in the model validation process was to augment the specification with additional parameters. The two models examined were the adding of δ_1, a first-order denominator parameter, to Model T1.1, and the adding of δ_2, a second-order denominator parameter, to Model T2.1, leading to the following specifications:

$$y_t = \frac{\omega_0 - \omega_1 B}{1 - \delta_1 B} x_t + \frac{1}{1 - \phi_1 B} a_t \tag{12.30}$$

$$y_t = \frac{\omega_0}{1 - \delta_1 B - \delta_2 B} x_t + \frac{1}{1 - \phi_1 B} a_t. \tag{12.31}$$

In equation (12.30) we found that both the ω_1 and δ_1 parameters were highly insignificant, EST/SE $\omega_1 = -0.187$, EST/SE $\delta_1 = 0.325$, and at the same

```
NOBS = 37      MIN =    -616.9      MAX =     648.7
MEAN =   14.13     SD =    206.4    STUDENTIZED RANGE =    6.133
SKEWNESS = -5.4182E-02 KURTOSIS =    5.806


                INCREMENT =  24.
            ....................................0.........................................
1910.    19.1    .                                          .*
1911.   -45.7    .                                      .*  .
1912.   125.     .                                            .*
1913.   107.     .                                           .*
1914.  -154.     .                              .*
1915.   5.22     .                                        .*
1916.   119.     .                                           .*
1917.   45.7     .                                        .*
1918.   649.     .                                                          .*
1919.  -36.0     .                                      .*
1920.  -110.     .                               .*
1921.   257.     .                                               .*
1922.  -47.6     .                                     .*
1923.   330.     .                                                 .*
1924.  -10.5     .                                       .*
1925.  -51.9     .                                      .*
1926.  -617.     .*
1927.  -69.3     .                                     .*
1928.   13.4     .                                         .*
1929.  -171.     .                           .*
1930.  -57.8     .                                     .*
1931.   14.5     .                                         .*
1932.  -77.5     .                                    .*
1933.  -14.8     .                                      .*
1934.  -135.     .                             .*
1935.   114.     .                                           .*
1936.  -78.9     .                                    .*
1937.   160.     .                                             .*
1938.   37.0     .                                         .*
1939.  -108.     .                               .*
1940.   313.     .                                                .*
1941.   210.     .                                            .*
1942.   52.6     .                                        .*
1943.   173.     .                                            .*
1944.  -133.     .                             .*
1945.   131.     .                                           .*
1946.  -435.     .                  .*
            ....................................................................................
```

FIGURE 12.1   Residuals Lydia Pinkham Model T1.1.

time very correlated. The estimate of the correlation between the parameter estimates $\hat{\omega}_1$ and $\hat{\delta}_1$ was 0.937. This clearly shows that (12.30) is an overspecified model and that either δ_1 or ω_1 could be omitted, leading towards the earlier models T1.1 or T2.1, respectively. In estimation of (12.31) we found that the ratio of the δ_2 estimate to its standard error was only -0.199. Again

FIGURE 12.2   Residuals Lydia Pinkham Model T2.1.

```
NOBS = 38      MIN =   -631.1     MAX =     665.6
MEAN =  13.25     SD =   203.6    STUDENTIZED RANGE =   6.370
SKEWNESS = -3.3911E-02 KURTOSIS =   6.340

                    INCREMENT =    25.
        ..........................................0.........................
  1909.    4.09   .                                  *
  1910.    29.8   .                                  *
  1911.   -48.7   .                               *  .
  1912.    122.   .                                  .    *
  1913.    112.   .                                  .   *
  1914.   -155.   .                        *         .
  1915.    3.73   .                                  *
  1916.    120.   .                                  .   *
  1917.    34.6   .                                  *
  1918.    666.   .                                  .
  1919.   -49.8   .                               *  .
  1920.   -96.5   .                             *    .
  1921.    234.   .                                  .       *
  1922.   -56.2   .                               *  .
  1923.    319.   .                                  .         *
  1924.   -31.5   .                                *  .
  1925.   -67.2   .                              *    .
  1926.   -631.   . *                                .
  1927.   -67.1   .                               *  .
  1928.   -17.4   .                                 * .
  1929.   -133.   .                         *        .
  1930.   -62.5   .                               *  .
  1931.    19.1   .                                 *.
  1932.   -86.0   .                             *    .
  1933.    7.60   .                        *         .
  1934.   -133.   .                                  .    *
  1935.    112.   .                               *  .
  1936.   -73.4   .                              *    .
  1937.    191.   .                                  .      *
  1938.    71.5   .                                  .     *
  1939.   -114.   .                          *       .
  1940.    295.   .                                  .        *
  1941.    209.   .                                  .      *
  1942.    51.8   .                                  .     *
  1943.    159.   .                                  .      *
  1944.   -128.   .                        *         .
  1945.    119.   .                                  .    *
  1946.   -423.   .              *                   .
        ..........................................................................
```

this indicates that the δ_2 term should be omitted and Model T2.1 favored over equation (12.31).

Finally, it is interesting to compare the transfer function results with the results obtained from a univariate sales model. In the literature an MA(1) model with a constant term has been proposed to represent the changes in the sales [see Helmer and Johansson (1977)]. Other models have also been analyzed [see Kyle (1978)]. The estimation results for the MA(1) model are

$$\nabla s_t = 21.938 + a_t + 0.456a_{t-1}, \qquad \sigma_a^2 = 240.68.$$
$$(55.776) \qquad (0.168)$$

Although additional model improvements are not precluded, these results are quite instructive. Comparing these results with those presented in Table 12.2, it can be seen that the residual variance in the univariate model is almost 15% larger, testifying to the value of building a transfer function model to explain sales in the Lydia Pinkham example.

12.6 SUMMARY

In this chapter we have shown that the estimation of the transfer function model parameters does not introduce any basic differences with the estimation of the parameters in an ARIMA model. The most important section in this chapter, Section 12.4, contains the diagnostic checks that should be applied to validate the model accuracy. It is shown that the information contained in the residual analysis is harder to interpret, because nonzero residual autocorrelations can indicate that the transfer function structure is incorrect and/or the noise structure incorrect. We also indicated that the cross correlations between the residuals and the prewhitened explanatory variables form an important additional diagnostic check on the model. The estimation and diagnostic checks were illustrated using the Lydia Pinkham example tentatively identified in Chapter 11.

In the next chapter we will develop the forecasting methodology for transfer function models and again use the Lydia Pinkham model as an example.

CHAPTER 13

FORECASTING WITH TRANSFER FUNCTION MODELS

The forecasts that we will calculate using a transfer function model will again be minimum mean squared error forecasts. Using arguments similar to the ones presented in Section 6.3., it again can be shown that the mean of the forecast distribution is a minimum mean squared error forecast.

Depending upon how the explanatory variables are treated there are two ways to calculate the mean of the forecast distribution of a transfer function model. We could assume that the explanatory variables are also unknown and must be forecasted using their own univariate ARIMA process or, alternatively, we could forecast the output series conditional on specified values of the explanatory variables at future data points. Therefore the confidence intervals for the dependent variable will be larger in the case where the explanatory variables are also unknown and must therefore be forecasted, than in the case where the values for these variables are known.

In this chapter we first will give the general theory for forecasting with a transfer function model. Next, we will construct the confidence intervals, and in Section 13.3 we will derive the conditional forecasts. Finally, in Section 13.4 we will generate the forecasts for the Lydia Pinkham data example based on estimates obtained in Chapter 12.

13.1 EXPLANATORY VARIABLES FORECASTED

The transfer function model can be represented as [see (10.15)]

$$(1 - B)^{d'} Y_t = \frac{\omega(B)}{\delta(B)} (1 - B)^d B^b X_t + \frac{\theta(B)}{\phi(B)} a_t \tag{13.1}$$

or

$$Y_t = \frac{\omega(B)(1 - B)^d B^b}{\delta(B)(1 - B)^{d'}} X_t + \frac{\theta(B)}{\phi(B)(1 - B)^{d'}} a_t. \tag{13.2}$$

Alternatively, we can rewrite (13.1) as

$$\delta(B)\phi(B)(1 - B)^{d'} Y_t = \omega(B)\phi(B)(1 - B)^d B^b X_t + \theta(B)\delta(B) a_t \tag{13.3}$$

or as

$$\delta^*(B) Y_t = \omega^*(B) B^b X_t + \theta^*(B) a_t. \tag{13.4}$$

The polynominals in (13.4) are, respectively,

$$\delta^*(B) = \delta(B)\phi(B)(1 - B)^{d'} = 1 - \delta_1^* B - \cdots - \delta_{r^*}^* B^{r^*}$$
$$\omega^*(B) = \omega(B)\phi(B)(1 - B)^d = \omega_0^* - \omega_1^* B - \cdots - \omega_{l^*}^* B^{l^*}$$
$$\theta^*(B) = \theta(B)\delta(B) = 1 - \theta_1^* B - \cdots - \theta_1^* B - \cdots - \theta_{q^*}^* B^{q^*},$$

with

$$r^* = r + p + d'$$
$$l^* = l + p + d$$
$$q^* = q + r.$$

Notice that for forecasting purposes we write the transfer function model in level form and not in differenced form.

The mean of the forecast distribution $E(Y_{n+h})$ can now be calculated as follows. We denote with n the time period at which we make the forecast, the origin date, and with h the forecast horizon, or number of periods ahead we want to make a forecast. (This and other notations used in this chapter are identical to the notations introduced in Chapter 6.) From (13.4) we can express Y_{n+h} as, assuming $b = 0$,

$$
\begin{aligned}
Y_{n+h} = {}& \delta_1^* Y_{n+h-1} + \delta_2^* Y_{n+h-2} + \cdots + \delta_{r*}^* Y_{n+h-r*} + \omega_0^* X_{n+h} \\
& - \omega_1^* X_{n+h-1} - \cdots - \omega_{l*}^* X_{n+h-l*} + a_{n+h} \\
& - \theta_1^* a_{n+h-1} - \cdots - \theta_{q*}^* a_{n+h-q*}.
\end{aligned}
\tag{13.5}
$$

Similarly, we assume that the X_t time series can be represented by the following univariable ARIMA process [see also (11.25)]:

$$(1 - B)^d X_t = \frac{\theta_x(B)}{\phi_x(B)} a_t, \tag{13.6}$$

with a_t a white noise process, $\theta_x(B)$ the X process moving average polynominal, and $\theta_x(B)$ the X process autoregressive polynominal. Also, stationarity and invertibility are assumed for the $\phi_x(B)$ and $\theta_x(B)$ polynominals, respectively.

The expected value of Y_{n+h}, using information up to period n, $E(Y_{n+h})$, can be obtained as follows:

1. Replace the current and past errors a_{n+j}, $j \leq 0$, with the actual residuals calculated as

$$\hat{a}_t = Y_t - Y_{t-1}(1);$$

2. replace each future error a_{n+j}, $0 < j \leq h$ with its expectation which, since a_{n+j} is white noise, is just 0;

3. replace the current and past observations X_{n+j}, $j \leq 0$ with the actual observed values;

4. replace the future values of X_{n+j}, $0 < j \leq h$, with their appropriate forecasts $\hat{X}_n(j)$ obtained from (13.6);

5. replace the current and past observations Y_{n+j}, $j \leq 0$ with the actual observed values; and

6. replace the future values of Y_{n+j}, $0 < j < h$, with their appropriate forecasts

$\hat{Y}_n(j)$; therefore, we first should forecast $Y_{n+1}, Y_{n+2}, \ldots, Y_{n+h-1}$ in order to forecast Y_{n+h}.

The $\hat{X}_n(j)$ forecasts mentioned in step 4 above are also minimum mean squared error forecasts and are generated using the methods explained in Section 6.3.

We want to stress that the transfer function model (13.1) can easily be extended to include seasonal polynominals [see (10.22)] and, similarly, the univariable X process (13.6) also could include seasonal parameters. Note that it is also simple to apply the above rule to models including several explanatory variables, although the notation becomes more elaborate.

13.2 INTERVAL FORECASTS

To evaluate the variance of the forecasts, which then can be used to construct forecast confidence intervals, we combine the univariate model (13.6) with the transfer function model (13.2). Next, solving for Y_t we obtain

$$Y_t = \frac{\omega(B)\theta_x(B)B^b}{\delta(B)\phi_x(B)(1-B)^{d'}} \alpha_t + \frac{\theta(B)}{\phi(B)(1-B)^{d'}} a_t \qquad (13.7)$$

$$= \nu^*(B)\alpha_t + \psi^*(B)a_t. \qquad (13.8)$$

Comparing (13.8) with (12.14) we notice that (13.8) expresses the level of Y_t in terms of the prewhitened x_t, the α_t, and the error a_t, and not in terms of x_t and a_t directly. Therefore, the polynominals are not equal in general; that is, in general $\nu^*(B) \neq \nu(B)$ and $\psi^*(B) \neq \psi(B)$. The Y_t at time $t = n + h$ can therefore be expressed as

$$Y_{n+h} = \nu_0^* \alpha_{n+h} + \nu_1^* \alpha_{n+h-1} + \nu_2^* \alpha_{n+h-2} + \cdots$$
$$+ a_{n+h} - \psi_1^* a_{n+h-1} - \cdots. \qquad (13.9)$$

Using the rules explained in the preceding section it should be clear that we also can write the mean squared error forecast as

$$Y_n(h) = \nu_h^* \alpha_n + \nu_{h+1}^* \alpha_{n-1} + \nu_{h+2}^* \alpha_{n-2} + \cdots$$
$$- \psi_h^* a_n - \psi_{h+1}^* a_{n-1} - \cdots. \qquad (13.10)$$

Subtracting (13.10) from (13.9) the h step ahead forecast error now can be expressed as

$$Y_{n+h} - Y_n(h) = \nu_0^* \alpha_{n+h} + \nu_1^* \alpha_{n+h-1} + \cdots + \nu_{h-1}^* \alpha_{n+1}$$
$$+ a_{n+h} - \psi_1^* a_{n+h-1} - \cdots - \psi_{h-1}^* a_{n+1}. \qquad (13.11)$$

The forecast variance is now obtained as

$$\mathrm{Var}[Y_{n+h} - Y_n(h)] = E\{[Y_{n+h} - Y_n(h)]^2\} = (\nu_0^{*2} + \nu_1^{*2} + \cdots$$
$$+ \nu_{h-1}^{*2})\sigma_\alpha^2 + (1 + \psi_1^{*2} + \cdots + \psi_{h-1}^{*2})\sigma_a^2. \qquad (13.12)$$

In deriving (13.12), remember that α_t and a_t are white noise series and are uncorrelated.

Therefore, as soon as we know the first h v_i^* values, $i = 0, 1, \ldots, h - 1$, and the first $h - 1$ ψ_j^* values, $j = 1, \ldots, h - 1$, $(\psi_0^* \equiv 1)$, together with the error process variance σ_a^2 and the X innovation process variance σ_α^2, we can calculate the variance for any h step ahead forecast of Y.

These v_i^* and ψ_j^* values can be calculated from the following two polynomials by equating powers of B:

$$\delta(B)\phi_x(B)(1 - B)^{d'} v^*(B) = \omega(B)\theta_x(B)B^b \qquad (13.13a)$$

and

$$\phi(B)(1 - B)^{d'}\psi^*(B) = \theta(B). \qquad (13.13b)$$

Fortunately, the available computer programs will do this calculation automatically for any transfer function model specification.

Observe that again, as the forecast horizon is lengthened, the forecast error variances are monotonically nondecreasing:

$$\text{Var}[Y_{n+h} - Y_n(h)] - \text{Var}[Y_{n+h-1} - Y_n(h-1)] = \\ \sigma_\alpha^2 v_{h-1}^{*2} + \sigma_a^2 \psi_{h-1}^{*2} \geq 0. \quad (13.14)$$

If we assume that the error terms a_t and α_t are normally distributed, then we can characterize the whole forecasting distribution $f_{n,h}(Y)$. The forecast distribution of Y_{n+h}, $f_{n,h}(Y)$, will be that of a normally distributed random variable with mean $E(Y_{n+h}) = Y_n(h)$ and variance $\text{Var}[Y_{n+h} - Y_n(h)]$. Under these assumptions it is possible to make probability statements about future observations. Indeed, the 95% large-sample[1] confidence interval for Y_{n+h} is

$$Y_n(h) \pm 1.96 \text{ SE}[Y_{n+h} - Y_n(h)]. \qquad (13.15)$$

The SE stands for the standard error of the forecast errors defined as the square root of the variance (13.12).

13.3 CONDITIONAL FORECASTING

If future X_t's are known or, more specifically, should be considered as being known, we can still use the general framework developed above to construct conditional point forecasts and forecast confidence intervals.

Specifically, only step 4 [stated below equation (13.6)] should be slightly reinterpreted. Rather than using (13.6) to forecast X_{n+j}, $j > 0$, we would

---

[1] The qualifier large-sample is added because even for normally distributed errors terms we still must use estimates of the parameters of the transfer function model rather than their true values.

just use the *a priori* specified future values of X_{n+j}, $j > 0$. The only real modification is in the forecast variances. If we make forecasts conditional on *a priori* specified future values, then there is no X process uncertainty. Therefore the forecast variance is specified as:

$$E\{[Y_{n+h} - Y_n(h)|X_{n+j}, 0 < j \le h]^2\} = (1 + \psi_1^{*2} + \cdots + \psi_{h-1}^{*2})\sigma_a^2. \quad (13.16)$$

Also, there is no requirement that α_t or X_t is normally distributed in order that the forecast distribution be normally distributed because we now present forecasts conditional on known future X values.

13.4 LYDIA PINKHAM EXAMPLE

In Chapters 11 and 12 we analyzed the first 40 observations, spanning the period 1907 to 1946, of the Lydia Pinkham vegetable compound, reserving the last 14 observations for an evaluation of its forecasts.

The models estimated in Chapter 12 [see equations (12.28) and 12.29) and Table 12.2 for the results] are:

Model T1.1:

$$(b, \ell, r) \times (p, d, q) = (0, 1, 0) \times (1, 1, 0)$$

$$y_t = (\omega_0 - \omega_1 B)x_t + \frac{1}{1 - \phi_1 B} a_t.$$

Model T2.1:

$$(b, \ell, r) \times (p, d, q) = (0, 0, 1) \times (1, 1, 0)$$

$$Y_t = \frac{\omega_0}{1 - \delta_1 B} x_t + \frac{1}{1 - \phi_1 B} a_t.$$

In both models the data have been consecutive differences; that is, each model relates changes in sales to change in advertising:

$$y_t = sales_t - sales_{t-1}$$
$$x_t = advertising_t - advertising_{t-1}.$$

We now propose to use both models to calculate forecasts of sales conditional on *forecasted* expenditures for advertising as well as forecasts of sales conditional on *actual* expenditures for advertising. Table 13.1 contains the one step ahead forecasts and 95% confidence limits for the sales forecasts for 1947 to 1960, conditional on actual advertising expenditures for these years. All of the one step ahead forecasts in this table were generated using the parameter estimates of Model T1.1 [see Table 12.2], based on the first 40 observations (1907–1946).

In general, we feel that Model T1.1 is capable of generating quite reliable sales forecasts. On the average, the forecasts, indicate a slight, 23 units, ($000), overprediction of the sales data.

TABLE 13.1   One Step Ahead Conditional Sales Forecasts for Model T1.1

THOUSANDS OF DOLLARS

YEAR	FORECAST	ACTUAL	95% CONF. LIMIT		DEVIATION FROM ACTUAL (%)
			LOWER	UPPER	
1947	1955	1920	1532	2378	− 1.84
1948	1896	1910	1473	2319	+ 0.71
1949	1941	1984	1519	2364	+ 2.14
1950	1998	1787	1575	2421	−11.80
1951	1630	1689	1207	2053	+ 3.49
1952	1725	1866	1302	2148	+ 7.58
1953	1956	1896	1533	2379	− 3.14
1954	1825	1684	1402	2248	− 8.38
1955	1604	1633	1181	2027	+ 1.78
1956	1632	1657	1209	2055	+ 1.52
1957	1650	1569	1227	2073	− 5.17
1958	1479	1390	1056	1902	− 6.42
1959	1337	1387	914	1760	+ 3.59
1960	1355	1289	932	1778	− 5.11

The picture only changes slightly if we make sales forecasts based on one step ahead advertising forecasts. These one step ahead advertising forecasts, displayed in Table 13.2, are not very good. Although all actual values are within the 95% large-sample forecasting confidence limits, these limits

TABLE 13.2   One Step Ahead Advertising Forecasts for AR(4) Model

THOUSANDS OF DOLLARS

YEAR	FORECAST	ACTUAL	95% CONF. LIMIT		DEVIATION FROM ACTUAL (%)
			LOWER	UPPER	
1947	1035	836	599	1470	−23.75
1948	830	941	394	1266	+11.80
1949	1022	981	586	1458	− 4.20
1950	925	974	489	1361	+ 5.02
1951	891	766	455	1327	−16.33
1952	785	920	349	1220	+14.73
1953	1001	964	566	1437	− 3.89
1954	944	811	508	1379	−16.34
1955	699	789	263	1135	+11.38
1956	878	802	442	1314	− 9.46
1957	839	770	402	1274	− 8.82
1958	711	639	276	1147	−11.32
1959	627	644	192	1063	+ 2.57
1960	686	564	250	1122	−21.59

TABLE 13.3   One Step Ahead Sales Forecasts for Model T1.1 Conditional on Advertising Forecasts

		THOUSANDS OF DOLLARS			
			95% CONF. LIMIT		DEVIATION
YEAR	FORECAST	ACTUAL	LOWER	UPPER	FROM ACTUAL (%)
1947	2051	1920	1579	2523	− 6.81
1948	1843	1910	1371	2315	+ 3.51
1949	1961	1984	1489	2433	+ 1.14
1950	1974	1787	1502	2446	−10.49
1951	1690	1689	1218	2162	− 0.07
1952	1660	1866	1188	2131	+11.06
1953	1974	1896	1502	2446	− 4.09
1954	1889	1684	1417	2361	−12.17
1955	1561	1633	1089	2033	+ 4.42
1956	1668	1657	1196	2140	− 0.68
1957	1683	1569	1211	2155	− 7.25
1958	1514	1390	1042	1986	− 8.92
1959	1329	1387	857	1801	+ 4.17
1960	1413	1289	941	1885	− 9.65

are very broad, a reflection of the fact that the advertising AR(4) model presented in Chapter 11 [equation (11.37)] could be improved upon. The root mean squared advertising error (that is, the square root of the average squared error) is 103 units ($000). This corresponds to an average absolute error of 11.51%.

Nevertheless the one step ahead sales forecasts, conditional on these advertising forecasts, have not drastically deteriorated. Table 13.3 contains these forecasts with 95% large-sample confidence limits and percent deviation from actual.

TABLE 13.4   Root Mean Squared Errors of Sales Forecasts

	THOUSANDS OF DOLLARS	
	MODEL T1.1	MODEL T2.1
One step ahead		
actual advertising	91.6	191.2
forecasted advertising	119.6	210.0
Up to 14 steps ahead		
actual advertising	229.4	436.3
forecasted advertising	381.3	471.1

In Table 13.4 we have summarized the root mean squared errors for these two forecasts, together with the forecasts generated using Model T2.1 for both one step ahead forecasts and forecasts up to 14 steps ahead (i.e., 14 years into the future). Based on Model T1.1, the one step ahead root mean squared error increases by about 28 units ($000) in using forecasted advertising rates rather than actual advertising expenditures. This table also clearly shows that the root mean squared error of Model T1.1 uniformly dominates Model T2.1, both for one step ahead forecasts and forecasts up to 14 years ahead. It is also evident from these results (and, possibly, also to be expected) that the forecasts made up to 14 years ahead rather quickly break down, with root mean squared errors more than doubled. We also could have used Model T1.1 and Model T2.1 to generate one step ahead forecasts after updating the parameter estimates each time a new observation became available.

13.5 CONCLUSION

In the last four chapters we have presented the transfer function model. In Chapter 10 we expounded the theory and rationale for a transfer function model. We showed that such a model can be looked upon as a generalization of a distributed lag model or of a dynamic regression model. In Chapter 11 we discussed the cross correlation function as an important new tool, together with the autocorrelation function, for identifying a transfer function model. We again stressed that the objective of the identification stage is to narrow down the class of possible transfer function models that could have generated the data. This methodology then was applied to the Lydia Pinkham's vegetable compound data.

In Chapter 12 we presented the estimation and the diagnostic checking of a transfer function model and applied this to the proposed transfer function models for the Lydia Pinkham data. The diagnostic checks presented in this chapter are very similar to those discussed earlier in Chapter 5 for a univariate time series model. Finally, in this chapter we outlined the forecasting methodology of a transfer function model and used the Lydia Pinkham models as examples.

In the last chapter of this book (14) we will discuss a special form of transfer function model, the intervention model. We will see that both the univariate and transfer function theory must be used imaginatively in the construction of an invention model.

CHAPTER 14

INTERVENTION ANALYSIS

In this chapter we will discuss the use of both univariate **ARIMA** models and transfer function models to measure and evaluate changes, commonly called *interventions.* That is, time series analysis will be used to assess the impact of a discrete intervention on behavioral processes. The name *intervention* also is used for certain exceptional external events which have affected the variables being forecasted, such as holidays and strikes.[1] It has become common practice to define time series analysis measuring the effect of interventions as time series intervention analysis. Time series intervention analysis has been used to study the impact of new traffic laws [Campbell and Ross (1968); Glass (1968); Ross et al. (1970)]; the impact of gun control laws [Zimring (1975)]; the impact of air pollution control laws [Box and Tiao (1975); Tiao et al. (1975 a,b)] and the impact of decriminalization [Aaronson et al. (1978)]. Other applications include modeling the effect of different kinds of promotional activity on sales, representing the effect of changes in definition of economic time series and relationships between such series.

The methodology outlined in this chapter will require that the user can specify two characteristics of the intervention model. First, the user must be able to indicate the *starting point* of an event, or an intervention. Second, it is necessary to specify the general *shape* of the impact of the intervention model (i.e., one must specify *a priori* the expected nature of the impact). The most simple situation would be to use a dummy variable, a zero–one variable, with the value one at the time of the intervention. Indeed, an intervention model also is sometimes called a *structural dummy variable* model. Of course, both the starting point and the shape of the intervention model could be evaluated empirically using alternative specifications. The next section will be devoted to a discussion of several commonly used intervention models.

14.1 THE INTERVENTION MODEL

Suppose that a new law is introduced (the event) which is expected to decrease, once and for all, the number of traffic accidents by a fixed amount. Then the invention variable could be represented as

$$I_t = \begin{cases} 0 : \text{prior to the event} \\ 1 : \text{thereafter.} \end{cases} \tag{14.1}$$

Such an intervention variable would be called a *step function* and be denoted by S_t^T, with T referring to the time period at which the event starts. Figure 14.1(a) shows such a step function.

Another form of intervention variable would be

---

[1] Indeed the intervention model also has been used to model outliers. Other models to handle outliers are presented in Martin et al. (1983).

$$I_t = \begin{cases} 1 : \text{at the time of occurrence of the event} \\ 0 : \text{otherwise.} \end{cases} \qquad (14.2)$$

Such a variable could be used, for example, to denote the effect of an advertising/promotional event which only lasts for one time period. This variable

FIGURE 14.1   Responses to a Step and a Pulse Input Function.

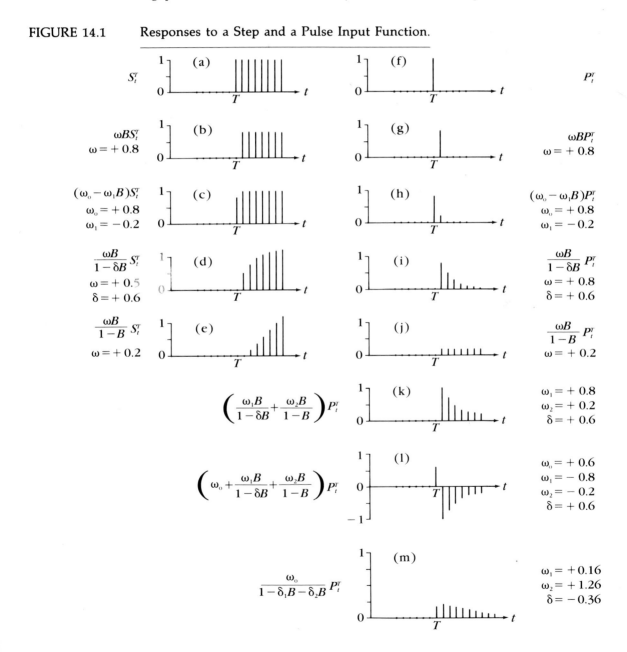

is defined as a *pulse input variable* or impulse variable and denoted with the symbol P_t^T, where T denotes the time period at which the event occurs. Graphically this is shown in Figure 14.1(f).

Note that there is an exact relationship between models (14.1) and (14.2). A step and an impulse variable are related by the following equation:

$$(1 - B)S_t^T = P_t^T. \tag{14.3}$$

Therefore an intervention model could equally well be represented with the pulse input variable P_t^T as with the step variable S_t^T. In some applied situations the empirical results will be easier to interpret using P_t^T, whereas in other situations S_t^T will be the more natural variable to use.

The *shape* of the impact can usually be classified among one of the following general forms:

1. abrupt start and the effect of the intervention of permanent duration;
2. gradual start and the effect of the intervention of permanent duration;
3. abrupt start and the intervention effect of temporary duration; or
4. gradual start and the effect of the intervention of temporary duration.

Some more elaborate interventions may have to be modeled with some combination of these four basic categories. We now will briefly discuss each of these four general intervention model forms.

ABRUPT START, PERMANENT DURATION

The simplest model would be one in which the impact of an intervention is fixed but is of unknown magnitude and starts at a known time period. Therefore, the intervention model could be represented as

$$Y_t = \omega S_t^T \tag{14.4}$$

with Y_t the output of the intervention and ω the parameter of unknown magnitude. Notice that we assume that the intervention has direct impact on the *level* of the data. We feel that this will be the most natural way to think about the impact of interventions, even in those situations where the Y series is nonstationary and requires differencing. However, in certain situations the model builder could rationalize that the intervention has an impact on the *change* of the data; that is,

$$(1 - B)Y_t = \omega S_t^T. \tag{14.5}$$

It will be apparent from further discussion in this chapter that the latter formulation can easily be accommodated.

If the effect of the intervention is felt only one period after the intervention, then (14.4) can be modified to

$$Y_t = \omega B S_t^T \tag{14.6}$$

where B is the lag operator. This model is represented in Figure 14.1(b). Of course, a more general lag can also be accommodated as

$$Y_t = \omega B^b S_t^T, \tag{14.7}$$

where b is the dead time, as defined in Chapter 10.

GRADUAL START, PERMANENT DURATION

Sometimes a step change would not be expected to produce the full impact at once but rather the response is gradually felt. Figure 14.1(d) contains an example of such a response. In this case an appropriate intervention model would be

$$Y_t = \frac{\omega B}{1 - \delta B} S_t^T. \tag{14.8}$$

Depending upon the specific value of the denominator parameter δ, (14.8) can represent several alternative situations. If $\delta = 0$, we have the constant impact model (14.6) [see Figure 14.1(b)]; if $\delta = 1$, the impact is linearly increasing without limit [see Figure 14.1(e)]. Intermediate values will accommodate intermediate situations.

ABRUPT START, TEMPORARY DURATION

In other situations it is more natural to think about the effect in terms of a pulse input variable or impulse variable. In measuring the effect of advertising on sales we might expect that sales would be up during the next month and that the effect of advertising would gradually taper off. Such behavior is represented in Figure 14.1(i), and can be represented mathematically with the following model:

$$Y_t = \frac{\omega B}{1 - \delta B} P_t^T. \tag{14.9}$$

If the impact of the abrupt onset is expected in the *same* period as the intervention, then the numerator ωB in (14.9) would be modified to just ω. If $\delta = 1$ we have an intervention model with a permanent effect, and for $\delta = 0$ we have a model with an effect lasting only one period. These situations are represented in Figure 14.1(j) and 14.1(g), respectively.

Next, if we expect a residual permanent effect after a gradually declining impact [see Figure 14.1(k)], we could model this situation as

$$Y_t = \left(\frac{\omega_1 B}{1 - \delta B} + \frac{\omega_2 B}{1 - B} \right) P_t^T. \tag{14.10}$$

Notice that (14.10) can also be looked upon as two interventions that took place at the same time, one intervention with a temporary effect and one intervention with a permanent effect.

GRADUAL START, TEMPORARY DURATION

Figure 14.1(m) shows an intervention model with the impact gradually increasing until it reaches a peak before tapering off. This impact model cannot be modeled with a low-order transfer function applied to a step or pulse model. A model that represents such a situation is

$$Y_t = \frac{\omega_0}{1 - \delta_1 B - \delta_2 B^2} P_t^T.$$
(14.11)

Indeed (14.11) is capable of modeling a two-parameter Pascal distributed lag model. For more discussion see Griliches (1967). For examples of the use of a gradual and temporary impact pattern see Izenman and Zabel (1981).

SEVERAL INTERVENTIONS

Several interventions can easily be modeled using the type of situations represented above by including several impulse and/or step variables, jointly. As mentioned above, equation (14.10) is an example of an intervention model with two pulse input variables.

14.2 IDENTIFICATION AND ESTIMATION OF AN INTERVENTION MODEL

The stochastic behavior of a series could be explained by its own past behavior together with an error model resulting in a univariate ARIMA model. Alternatively, we also may have to rely on a transfer of models to account for the impact of additional explanatory variables. If this behavior is disturbed by an intervention, both the univariate model and the transfer function model would have to include an impulse or step variable. As a result, the identification procedures for the univariate ARIMA model as outlined in Chapter 4 and for the transfer function model as outlined in Chapter 11 must be modified. The identification now will be discussed sequentially for these two cases.

14.2.1 Univariate Intervention Model

A univariate ARIMA model can be written as [see (3.81)]

$$\phi(B)w_t = \theta(B)a_t,$$

where $w_t = (1 - B)^d Z_t$. Alternatively, we can rewrite this model in level form as

$$Z_t = \frac{\theta(B)}{\phi(B)(1 - B)^d} a_t.$$
(14.12)

All symbols have been defined in Chapter 3. Notice that we use Z_t to denote the variable, and not z_t, as done in the chapters on the univariate ARIMA.

We use the upper-case letter Z to be consistent with convention for transfer function models, as these models will also be analyzed in this chapter, see Section 14.2.2. Of course a seasonal ARIMA model would also include seasonal differencing or seasonal parameters.

In order to incorporate the intervention model we would modify equation (14.12) as follows

$$Z_t = \psi(B)I_t^T + \frac{\theta(B)}{\phi(B)(1-B)^d} a_t \qquad (14.13)$$

with $\psi(B) = \omega(B)/\delta(B)$ or, alternatively, as

$$Z_t = \psi(B)I_t^T + N_t, \qquad (14.14)$$

with

$$N_t = \frac{\theta(B)}{\phi(B)(1-B)^d} a_t. \qquad (14.15)$$

The task we face is to identify the model (14.13) or (14.14). Unfortunately, just because of the intervention part, the autocorrelation function and partial autocorrelation function would not necessarily reveal the ARIMA process as all the autocorrelations would be distorted by the effect of the intervention.[2] Three methods now are available for the identification.

First, the user possibly could use the data before or after the intervention, if such a subset is sufficiently long, to identify the univariate ARIMA model using methods explained in earlier chapters. Suppose that, as a result of the analysis of the *acf* and *pacf*, we tentatively identify $p = 1$, $d = 1$, and $q = 1$; that is,

$$(1 - \phi_1 B)(1 - B)Z_t = (1 - \theta_1 B)a_t. \qquad (14.16)$$

Assume that the following intervention model is tentatively ascertained

$$Z_t = \frac{\omega_0}{1 - \delta B} S_t^T + N_t. \qquad (14.17)$$

If ω_0 and δ were both 0 (i.e., with no intervention), then $Z_t = N_t$, and from (14.16) it follows that

$$N_t = \frac{(1 - \theta_1 B)}{(1 - \phi_1 B)(1 - B)} a_t. \qquad (14.18)$$

On combining (14.18) and (14.17) we obtain the following univariate intervention model:

$$(1 - \phi_1 B)(1 - B)Z_t = \frac{\omega_0(1 - \phi_1 B)(1 - B)}{1 - \delta B} S_t^T + (1 - \theta_1 B)a_t. \quad (14.19)$$

---

[2] A similar problem is encountered for time series contaminated with outliers. See Martin et al. (1983) for robust ARIMA modeling.

Then the model can be rewritten as

$$(1 - \phi_1 B)(1 - B)Z_t = \frac{(\omega_0 - \omega_1 B)}{1 - \delta B}(1 - B)S_t^T + (1 - \theta_1 B)a_t, \quad (14.20)$$

with $\omega_1 = \omega_0 \phi_1$. Notice that (14.20) is in the form of a standard transfer function model [see (10.14)]. Note that we do not imply that the user *a priori* must be able to specify the *exact* intervention model (14.17). All that we assume is that the user should be willing to specify *some* tentative models that can then be analyzed empirically.

Second, the user may have a good understanding of the possible effect of an intervention, or an inspection of the data may suggest ways in which the intervention has affected the data pattern. Therefore the user may be able to specify the form of $\psi(B)$ in (14.14). Then we first could estimate only the intervention model as written in (14.14). Next, we can calculate the residuals N_t as

$$\hat{N}_t = Z_t - \hat{\psi}(B)I_t^T. \quad (14.21)$$

These residuals can be interpreted as cleaned data, data that we would expect to observe without the intervention. Therefore these residuals can now be used to identify the univariate ARIMA (14.15) using the autocorrelation function and partial autocorrelation function.

Next, we combine the intervention model and the residual model to obtain the overall intervention model (14.13). The parameters in this model now are jointly estimated with a transfer function estimation program. Again, a careful residual analysis might suggest model improvements that should be implemented.

Alternatively, we could go ahead and use all the data to identify the ARIMA model. This tentative model then would be estimated together with the intervention model. A careful residual analysis would be performed to check for model improvement.

Note that we should always carefully analyze the residuals to check for model adequacy, no matter how we have identified the model. As should be known by now, we never expect to specify correctly the ARIMA model by only evaluating the autocorrelation function and the partial autocorrelation function of the data.

14.2.2 Transfer Function Intervention Model

As illustrated above it is quite possible that a transfer function model may have to be augmented to include intervention variables. In general therefore, such a transfer function intervention model can be written as

$$Y_t = \psi(B)I_t^T + \frac{\omega(B)}{\delta(B)(1 - B)^{d'}}x_t + \frac{\theta(B)}{\phi(B)(1 - B)^{d'}}a_t. \quad (14.22)$$

with

$$x_t = (1 - B)^d X_t,$$

and where $\psi(B) I_t^T$ represents the intervention model, d' refers to the order of consecutive differencing of the dependent variable Y_t, and d refers to the order of consecutive differencing of the explanatory variable. Of course, seasonal parameters could be incorporated in this model. Again we assume that the intervention has an impact on the level of the data Y_t and not on the stationary series $(1 - B)^{d'} Y_t$. In (14.22) I_t^T could be a step function S_t^T or a pulse input variable P_t^T. All other parameters have the same interpretation as in the standard transfer function model (10.14).

The identification of a transfer function intervention model can again be done in more than one way. First, if there is sufficient data before or after an intervention, we could use this subset of the data to identify the transfer function model, using the cross correlation, the autocorrelation, and the partial autocorrelation functions as explained in Chapter 11. This then would allow us to specify all the necessary orders of the polynominals in (14.22) except for the intervention model itself. Next, based on an understanding of the nature of the intervention or an inspection of the data, the user must specify a tentative intervention model. With this information, the user now can estimate all the parameters of the model (14.22).

Alternatively, the user could first clean the data (i.e., eliminate the effect of the intervention). This is done by only specifying the form of the intervention model and then calculating the residuals $\hat{N}_t$ from

$$\hat{N}_t = Y_t - \hat{\psi}(B) I_t^T. \tag{14.23}$$

These residuals then can be analyzed with the methods presented in Chapter 11 to tentatively identify the orders of the polynominals in the transfer function model. Next, we jointly estimate all the parameters of the transfer function intervention model using a standard transfer function estimation program. Finally, the estimated model should be subjected to the now standard diagnostic checks on the parameter estimates and residuals to evaluate possible model over- and underparameterization.

The use of intervention models entails one important danger. Intervention variables only should be introduced to model known events that have affected the data and should not be used to arbitrarily remove large residuals. Large residuals should be looked upon as flags indicating possible model misspecification. Some of these large residuals could be the results of events external to the data, but it is advisable to iterate between the subject matter, a discussion with the company using this transfer function model, and the data before introducing *ad hoc* intervention variables to improve the model fit.

As an example of an intervention model with several intervention variables we can refer to an application discussed in Jenkins (1979, p. 67 ff.).

The analysis concerns the monthly forecasting of bad debt collection as a function of an indicator of debts outstanding at the end of each month. At the end of 1974 it was announced that as of January 1975 more severe restriction would be placed on the granting of credit. As a result there was a rush in December 1974 to take advantage of the more lenient credit facilities while they lasted. This resulted not only in a temporary increase in outstanding debt but also in an increase debt collected in December 1974. From January 1975 onwards there was a significant drop in the level of outstanding debt and in the debt collected each month. The December 1974 effect could be modeled with a pulse input variable, whereas the change in policy in January 1975 could be represented by a step variable.

FIGURE 14.2 Cincinnati Directory Assistance, Monthly Average Calls per Day.

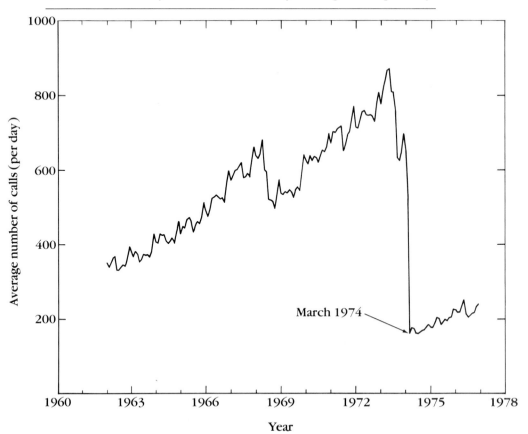

14.3 EXAMPLE: DIRECTORY ASSISTANCE IN CINCINNATI

We now will apply the above theory to a simple intervention model, more specifically a univariate intervention model. Figure 14.2 shows the average daily directory assistance calls made by all telephone subscribers in Cincinnati, Ohio from January 1962 to December 1976, a total of 180 monthly observations, as reported in McSweeny (1978) and subsequently used by McCleary and Hay (1980). Telephone company personnel long have suspected that subscribers make unnecessary use of directory assistance's services. Despite exhortations in company advertisements to use published directories for information whenever possible, directory assistance users typically request telephone numbers that have been published in standard directories for several years. The frequent use of directory assistance for published numbers, a service for which the telephone company typically does not levy a direct charge, means that the cost of providing directory assistance has to be subsidized by increasing charges for other telephone services.

In March 1974, officials at Cincinnati Bell Telephone Company initiated charges for local directory assistance calls in an attempt to reduce the frequency of those calls. Under this plan, telephone subscribers are allowed three local directory assistance calls per telephone line per month. Subscribers are then charged 20¢ for each additional call.

Cincinnati Bell has been recording each directory assistance call made within its service area since 1962. Records of the number of calls made on an average weekday, Monday through Saturday, in any given month, exist for the 146 months before and the 34 months after the directory assistance charges were introduced. It was expected that the effect of the 20¢ charge would be to reduce the number of calls. A quick look at the data plot (Figure 14.2) certainly confirms this expectation.

Because we have quite a long series before the intervention, we will use the first 146 observations to identify the model. Table 14.1 and Figure 14.3 contain autocorrelations and partial autocorrelations for different de-

TABLE 14.1   Autocorrelations and Partial Autocorrelations, Cincinnati Directory Assistance

```
SERIES WITH D = 0 DS = 0 S = 0
MEAN =     560.9     SD =     138.3    (NOBS = 146)

AUTOCORRELATIONS
  LAGS ROW SE
  1-12  .08    .97 .94 .92 .89 .87 .84 .81 .79 .76 .73 .71 .70

CHI-SQUARE TEST              P-VALUE
  Q(12) = .128E+04  12 D.F.    .000
```

TABLE 14.1 (*continued*)

```
PARTIAL AUTOCORRELATIONS        STANDARD ERROR =  .08
   LAGS
   1-12          .97 -.05   .14 -.05   .01 -.08 -.02   .03 -.09 -.01   .16   .04

SERIES WITH D = 1 DS = 0 S = 0
MEAN =    1.372      SD =      28.70     (NOBS = 145)

AUTOCORRELATIONS
  LAGS ROW SE
   1-12   .08   .03 -.19   .05   .01   .01   .02 -.07   .10   .02 -.23 -.02   .36
  13-24   .10  -.01 -.15   .03 -.05 -.06 -.01 -.07   .02   .04 -.24 -.02   .32
  25-36   .11   .00 -.09   .04 -.03 -.12   .07 -.07   .02   .00 -.15   .01   .29
  37-48   .12  -.05 -.10   .01   .02 -.06   .01 -.06   .05 -.04 -.23   .04   .28

CHI-SQUARE TEST            P-VALUE
Q(12) =  38.3       12 D.F.   .000
Q(24) =  72.2       24 D.F.   .000
Q(36) =  99.8       36 D.F.   .000
Q(48) =  134.       48 D.F.   .000

PARTIAL AUTOCORRELATIONS        STANDARD ERROR =  .08
   LAGS
   1-12          .03 -.20   .07 -.03   .03   .01 -.07   .12 -.02 -.19 -.01   .31
  13-24         -.04 -.07   .02 -.07 -.09   .00 -.03 -.08   .03 -.12   .00   .18
  25-36          .02 -.01   .01   .00 -.18   .11 -.06 -.06 -.07   .06 -.01   .13
  37-48         -.01 -.09  -.07   .11 -.04 -.06   .00   .00 -.13 -.05   .06   .05

SERIES WITH D = 1 DS = 1 S = 12
MEAN =   -2.241      SD =      28.42     (NOBS = 133)

AUTOCORRELATIONS
  LAGS ROW SE
   1-12   .09  -.04   .03   .08   .01   .09   .07   .02   .11 -.03 -.02   .07 -.29
  13-24   .10   .02 -.02  -.03 -.07   .10 -.12 -.06 -.04   .12 -.09 -.04 -.01
  25-36   .10   .03 -.03   .02   .00 -.08   .12 -.01   .01 -.05   .13 -.01   .02
  37-48   .10  -.01   .12  -.08   .02   .00   .01   .04   .04 -.01 -.12   .00 -.03

CHI-SQUARE TEST            P-VALUE
Q(12) =  18.7       12 D.F.   .096
Q(24) =  28.3       24 D.F.   .246
Q(36) =  35.8       36 D.F.   .479
Q(48) =  43.8       48 D.F.   .645

PARTIAL AUTOCORRELATIONS        STANDARD ERROR =  .09
   LAGS
   1-12         -.04   .03   .09   .01   .08   .07   .02   .10 -.04 -.04   .04 -.30
  13-24         -.02 -.03   .00 -.08   .17 -.07 -.05   .02   .15 -.13   .00 -.11
  25-36          .04 -.06   .07 -.09   .02   .09   .00 -.03   .06   .05 -.01 -.03
  37-48          .02   .02 -.04   .00 -.07   .02   .06   .05 -.03 -.08 -.03 -.04
```

FIGURE 14.3　　Sample Autocorrelation Function and Partial Autocorrelation Function, Cincinnati Directory Assistance.

```
AUTOCORRELATION FUNCTION              PARTIAL AUTOCORRELATION FUNCTION
                WITH   D = 1  DS = 1  S = 12

-0.8  -0.4   0.0   0.4   0.8          -0.8  -0.4   0.0   0.4   0.8
*  .  .  *  .  .  .  .  .  *  .  .  .     *  .  .  *  .  .  *  .  *  .  .  .  .  *
   .              ( R.   )        1  .            ( R.   )
   .              ( R    )        2  .            ( R    )
   .              ( .R   )        3  .            ( .R   )
   .              ( R    )        4  .            ( R    )
   .              ( .R   )        5  .            ( .R   )
   .              ( .R   )        6  .            ( .R   )
   .              ( R    )        7  .            ( R    )
   .              ( . R )         8  .            ( . R )
   .              ( R.   )        9  .            ( R.   )
   .              ( R    )       10  .            ( R.   )
   .              ( .R   )       11  .            ( .R   )
   .            R (   .  )       12  .          R (   .  )
   .              ( R    )       13  .            ( R    )
   .              ( R    )       14  .            ( R    )
   .              ( R.   )       15  .            ( R    )
   .              ( R.   )       16  .            ( R.   )
   .              ( . R )        17  .            ( . R
   .             (R .    )       18  .            ( R.   )
   .              ( R.   )       19  .            ( R.   )
   .              ( R.   )       20  .            ( R    )
   .              ( . R )        21  .            ( . R )
   .              ( R.   )       22  .           (R .    )
   .              ( R.   )       23  .            ( R    )
   .              ( R    )       24  .           (R .    )
   .              ( R    )       25  .            ( .R   )
   .              ( R    )       26  .            ( R.   )
   .              ( R    )       27  .            ( .R   )
   .              ( R    )       28  .            ( R.   )
   .              ( R.   )       29  .            ( R    )
   .              ( . R )        30  .            ( .R   )
   .              ( R    )       31  .            ( R    )
   .              ( R    )       32  .            ( R    )
   .              ( R.   )       33  .            ( .R   )
   .              ( . R )        34  .            ( .R   )
   .              ( R    )       35  .            ( R    )
   .              ( R    )       36  .            ( R    )
   .              ( R    )       37  .            ( R    )
   .              ( . R )        38  .            ( R    )
   .              ( R.   )       39  .            ( R.   )
   .              ( R    )       40  .            ( R    )
   .              ( R    )       41  .            ( R.   )
   .              ( R    )       42  .            ( R    )
   .              ( .R   )       43  .            ( .R   )
   .              ( .R   )       44  .            ( .R   )
   .              ( R    )       45  .            ( R    )
   .             (R .    )       46  .            ( R.   )
   .              ( R    )       47  .            ( R    )
   .              ( R.   )       48  .            ( R.   )
```

grees of differencing. As we expected, the raw (undifferenced) data clearly indicate a nonstationary process. Even the consecutively differenced series shows seasonal nonstationarity. The *acf* and *pacf* of the consecutively and seasonally differenced series (see Figure 14.3) suggest an ARIMA(0,1,0)×(0,1,1)$_{12}$ model. Writing the model in level form we have

$$Y_t = \frac{1 - \Theta B^{12}}{(1 - B)(1 - B^{12})} \, a_t. \tag{14.24}$$

As intervention model we specify the model (14.4), that is, ωS_t^T, with $T =$ March, 1974. Therefore the univariate intervention model is

TABLE 14.2   **Cincinnati Directory Assistance, Estimates**

```
DEPENDENT VBL: DIRASS
SERIES WITH D = 1 DS = 1 S = 12
MEAN =  -0.1018     SD =     53.35     (NOBS = 167)

EXPLANATORY VBL: INTERV
SERIES WITH D = 1 DS = 1 S = 12
MEAN =  -0.5089E-10 SD =    0.1094     (NOBS = 167)

PARAMETER ESTIMATES
===================
             EST        SE      EST/SE      95% CONF LIMITS
NUM( 0)  -399.258    22.689    -17.597    -443.729  -354.787   INTERV
SMA( 1)     0.891     0.037     23.942       0.818     0.964

EST.RES.SD =    2.2927E+01

R SQR =   .818
DF = 165
NO F-STAT, NO INTERCEPT

CORRELATION MATRIX
         NUM( 0)
 SMA( 1)    .018

ROOTS    REAL   IMAGINARY    SIZE
SMA      1.122    0.000      1.122

AUTOCORRELATIONS OF RESIDUALS
  LAGS ROW SE
   1-12   .08   -.03 -.03  .08 -.04  .11  .04 -.06  .07  .05 -.06 -.09  .03
  13-24   .08   -.08  .01  .02 -.10 -.02 -.04 -.03 -.04  .09 -.09 -.12  .02

CHI-SQUARE TEST            P-VALUE
Q(12) =  8.57     11 D.F.   .661
Q(24) = 18.5      23 D.F.   .732
```

$$Y_t = \omega S_t^T + \frac{1 - \Theta B^{12}}{(1 - B)(1 - B^{12})} a_t. \tag{14.25}$$

Notice again that we assume the intervention has an impact on the level of the monthly average calls per day.

In Table 14.2 we present the estimates of the model (14.25). Both parameter estimates are statistically significant and the estimate of Θ lies within the invertibility region. Diagnostic checks on the residuals indicate that the model is adequate. None of the residual autocorrelations are significant; the Q statistics also confirm this conclusion.

The interpretation of this model is obvious. After the introduction of the 20¢ directory assistance fee in March 1974, the monthly average calls per day dropped by about 40,000. (The units of the Y_t variable are 100 calls.)

14.4 CONCLUSION

In this chapter we have explained the basic intervention model, both the univariate intervention model and the transfer function intervention model. We have seen that the ideas presented in previous chapters about univariate models and transfer function models can be applied here in building a useful intervention model. The autocorrelations, partial autocorrelations, and cross correlations are still the building blocks for the model.

As an application, in Section 14.3 we used a univariate intervention model to explain the Cincinnati Directory Assistance calls, where the intervention was the introduction of a 20¢ charge. For additional applications of transfer function intervention models we refer to Box and Tiao (1975) Tiao et al. (1975 a,b), and Jenkins (1979).

APPENDIX A

DATA
BASES

TABLE A.1   Sales Data

	JAN	FEB	MAR	APR	MAY	JUN	JUL	AUG	SEP	OCT	NOV	DEC
1965	154	96	73	49	36	59	95	169	210	278	298	245
1966	200	118	90	79	78	91	167	169	289	347	375	203
1967	223	104	107	85	75	99	135	211	335	460	488	326
1968	346	261	224	141	148	145	223	272	445	560	612	467
1969	518	404	300	210	196	186	247	343	464	680	711	610
1970	613	392	273	322	189	257	324	404	677	858	895	664
1971	628	308	324	248	272							

Source: Chatfield and Prothero (1973).

TABLE A.2   Chemical Batch Yield

1–10	47	64	23	71	38	64	55	41	59	48
11–20	71	35	57	40	58	44	80	55	37	74
21–30	51	57	50	60	45	57	50	45	25	59
31–40	50	71	56	74	50	58	45	54	36	54
41–50	48	55	45	57	50	62	44	64	43	52
51–60	38	59	55	41	53	49	34	35	54	45
61–70	68	38	50	60	39	59	40	57	54	23

Source: Jenkins and Watts (1968).

TABLE A.3   Market Yield on U.S. Government Three-Month Treasury Bills (Annual Rate)

	JAN	FEB	MAR	APR	MAY	JUN	JUL	AUG	SEP	OCT	NOV	DEC
1956	2.41	2.32	2.25	2.60	2.61	2.49	2.31	2.60	2.84	2.90	2.99	3.21
1957	3.11	3.11	3.08	3.06	3.06	3.29	3.16	3.37	3.53	3.58	3.29	3.04
1958	2.44	1.54	1.30	1.13	0.91	0.83	0.91	1.69	2.44	2.63	2.67	2.77
1959	2.82	2.70	2.80	2.95	2.84	3.21	3.20	3.38	4.04	4.05	4.15	4.49
1960	4.35	3.96	3.31	3.23	3.29	2.46	2.30	2.30	2.48	2.30	2.37	2.25
1961	2.24	2.42	2.39	2.29	2.29	2.33	2.24	2.39	2.28	2.30	2.48	2.60
1962	2.72	2.73	2.72	2.73	2.68	2.73	2.92	2.82	2.78	2.74	2.83	2.87
1963	2.91	2.92	2.89	2.90	2.92	2.99	3.18	3.32	3.38	3.45	3.52	3.52
1964	3.52	3.53	3.54	3.47	3.48	3.48	3.46	3.50	3.53	3.57	3.64	3.84
1965	3.81	3.93	3.93	3.93	3.89	3.80	3.83	3.84	3.92	4.02	4.08	4.37
1966	4.58	4.65	4.58	4.61	4.63	4.50	4.78	4.95	5.36	5.33	5.31	4.96
1967	4.72	4.56	4.26	3.84	3.60	3.53	4.20	4.26	4.42	4.55	4.72	4.96
1968	4.99	4.97	5.16	5.37	5.65	5.52	5.31	5.08	5.20	5.35	5.45	5.94

Source: Federal Reserve System.

TABLE A.4   Reported Cases of Rubella (Biweekly Data)

1966	1– 9	420	550	610	910	1,115	1,210	1,190	1,080	1,175
	10–18	1,195	1,280	1,200	395	360	220	230	150	165
	19–26	150	160	110	110	140	150	190	195	
1967	1– 9	180	195	300	445	440	435	520	775	640
	10–18	775	650	760	390	200	130	100	90	20
	19–26	50	90	130	80	160	120	202	200	
1968	1– 9	125	210	300	505	500	1,500	980	1,000	900
	10–18	1,075	990	520	395	190	140	150	140	155
	19–26	194	154	162	164	160	180	185	181	

Source: Montgomery and Johnson (1976).

TABLE A.5   Nominal GNP (Quarterly Rates) in Billions of Current Dollars (Seasonally Unadjusted)

	I	II	III	IV
1947	53.4	55.9	57.5	64.5
1948	59.4	62.1	65.3	70.7
1949	61.8	62.2	64.5	68.0
1950	64.0	66.8	73.5	80.5
1951	76.6	79.9	83.2	88.7
1952	82.0	83.3	85.7	94.4
1953	87.4	91.3	90.3	95.6
1954	86.5	89.7	90.0	98.6
1955	92.6	97.4	100.4	107.6
1956	98.6	102.9	104.1	113.7
1957	104.4	109.1	110.6	117.0
1958	103.9	108.8	111.7	123.0
1959	113.1	121.4	119.3	129.9
1960	120.5	125.6	124.6	133.1
1961	120.6	128.2	129.1	142.2
1962	131.3	139.6	138.1	151.5
1963	137.8	146.1	146.5	160.2
1964	148.5	157.1	156.3	170.6
1965	158.2	169.1	168.9	188.7
1966	176.2	187.4	186.3	199.8
1967	186.5	197.2	198.4	211.7
1968	199.9	217.3	215.8	232.0
1969	217.5	232.4	234.8	246.5
1970	229.3	244.2	242.6	258.0

Source: Roberts (1974).

TABLE A.6   United States New Residential Construction in Millions of Current Dollars (Seasonally Unadjusted)

	JAN	FEB	MAR	APR	MAY	JUN	JUL	AUG	SEP	OCT	NOV	DEC
1947	556	528	545	607	701	785	874	950	1,006	1,093	1,135	1,070
1948	891	757	874	1,028	1,168	1,257	1,294	1,305	1,273	1,203	1,100	978
1949	846	731	763	844	981	1,086	1,147	1,171	1,207	1,238	1,241	1,173
1950	1,077	1,031	1,089	1,276	1,499	1,703	1,827	1,898	1,900	1,785	1,614	1,427
1951	1,289	1,188	1,229	1,288	1,324	1,399	1,428	1,409	1,400	1,397	1,330	1,200
1952	1,015	963	1,149	1,234	1,346	1,437	1,472	1,486	1,473	1,481	1,438	1,309
1953	1,131	1,057	1,206	1,363	1,431	1,570	1,577	1,550	1,514	1,481	1,420	1,294
1954	1,104	1,029	1,167	1,347	1,517	1,627	1,717	1,770	1,783	1,759	1,717	1,650
1955	1,473	1,379	1,562	1,753	1,925	2,064	2,098	2,082	2,051	1,983	1,851	1,656
1956	1,392	1,305	1,457	1,618	1,753	1,884	1,908	1,895	1,860	1,798	1,741	1,567
1957	1,324	1,206	1,350	1,486	1,604	1,718	1,767	1,796	1,787	1,761	1,694	1,513
1958	1,292	1,192	1,302	1,421	1,550	1,702	1,804	1,876	1,907	1,954	1,957	1,832
1959	1,606	1,493	1,676	1,907	2,091	2,253	2,350	2,358	2,310	2,232	2,092	1,883
1960	1,588	1,408	1,613	1,804	1,935	2,112	2,039	1,982	1,931	1,860	1,790	1,644
1961	1,378	1,221	1,459	1,720	1,860	2,059	2,053	2,053	2,055	2,041	1,974	1,807
1962	1,543	1,368	1,605	1,906	2,141	2,377	2,357	2,377	2,330	2,210	2,113	1,965
1963	1,686	1,492	1,666	1,950	2,206	2,421	2,517	2,553	2,516	2,500	2,450	2,230
1964	1,867	1,678	1,866	2,068	2,191	2,385	2,518	2,541	2,439	2,327	2,260	2,118
1965	1,834	1,639	1,782	2,000	2,203	2,429	2,550	2,561	2,473	2,377	2,284	2,136
1966	1,848	1,644	1,781	1,979	2,124	2,287	2,378	2,351	2,202	1,978	1,785	1,614
1967	1,368	1,248	1,405	1,613	1,836	2,107	2,336	2,471	2,446	2,375	2,340	2,191
1968	1,859	1,655	1,885	2,262	2,518	2,628	2,721	2,790	2,780	2,678	2,593	2,454
1969	2,133	1,940	2,195	2,540	2,810	2,962	2,974	2,880	2,763	2,648	2,482	2,288
1970	1,961	1,765	1,986	2,297	2,485	2,592	2,650	2,707	2,721	2,747	2,735	2,627

Source: U.S. Department of Commerce, *1971 Business Statistics.* See also footnote 1, Chapter 7.

TABLE A.7   United States Unemployment in Thousands of Persons

	JAN	FEB	MAR	APR	MAY	JUN	JUL	AUG	SEP	OCT	NOV	DEC
1948	2,351	2,807	2,646	2,407	2,014	2,408	2,411	2,238	2,061	1,747	2,033	2,205
1949	2,995	3,474	3,383	3,277	3,516	3,966	4,388	3,956	3,635	3,788	3,570	3,690
1950	4,648	4,791	4,313	3,665	3,263	3,551	3,316	2,613	2,490	2,055	2,356	2,409
1951	2,649	2,538	2,323	1,938	1,770	2,132	2,072	1,806	1,886	1,754	1,990	1,796
1952	2,258	2,340	2,002	1,836	1,782	2,032	2,085	1,918	1,728	1,480	1,594	1,545
1953	2,132	1,964	1,828	1,764	1,536	1,732	1,710	1,512	1,619	1,572	2,017	2,634
1954	3,561	3,983	4,037	3,846	3,658	3,681	3,679	3,470	3,433	2,929	3,115	3,009
1955	3,669	3,569	3,297	3,162	2,690	2,893	2,662	2,536	2,338	2,284	2,536	2,601
1956	3,066	3,092	3,095	2,710	2,799	3,179	2,984	2,467	2,273	2,091	2,599	2,660
1957	3,206	3,085	2,822	2,627	2,635	3,131	2,843	2,526	2,503	2,465	3,127	3,332
1958	4,464	5,116	5,155	5,064	4,832	5,223	5,098	4,607	4,058	3,750	3,785	4,068
1959	4,678	4,698	4,298	3,558	3,327	3,780	3,605	3,347	3,195	3,231	3,636	3,521
1960	4,119	3,886	4,164	3,607	3,380	4,172	3,884	3,711	3,315	3,537	3,967	4,470
1961	5,335	5,654	5,423	4,887	4,671	5,313	4,961	4,440	4,034	3,863	3,941	4,041
1962	4,621	4,481	4,323	3,863	3,592	4,219	3,829	3,842	3,455	3,234	3,726	3,760
1963	4,627	4,870	4,442	3,993	3,949	4,554	4,140	3,755	3,470	3,394	3,858	3,788
1964	4,518	4,461	4,225	3,831	3,528	4,453	3,675	3,551	3,262	3,198	3,318	3,409
1965	3,942	4,172	3,699	3,492	3,214	4,057	3,429	3,165	2,842	2,709	2,888	2,786
1966	3,244	3,109	2,990	2,730	2,793	3,592	3,050	2,821	2,503	2,465	2,579	2,655
1967	3,159	3,184	2,954	2,664	2,458	3,628	3,249	2,942	2,895	2,952	2,903	2,720
1968	3,074	3,287	2,929	2,491	2,304	3,615	3,217	2,772	2,607	2,510	2,576	2,418
1969	2,875	2,923	2,747	2,542	2,300	3,399	3,182	2,870	2,958	2,840	2,711	2,627
1970	3,406	3,794	3,733	3,552	3,384	4,669	4,510	4,220	4,292	4,259	4,607	4,636
1971	5,414	5,442	5,175	4,694	4,394	5,490	5,330	5,061	4,840	4,570	4,815	4,695
1972	5,447	5,412	5,215	4,697	4,344	5,426	5,173	4,857	4,658	4,470	4,266	4,116
1973	4,675	4,845	4,512	4,174	3,799	4,847	4,550	4,208	4,165	3,763	4,056	4,058
1974	5,008	5,140	4,755	4,301	4,144	5,380	5,260	4,885	5,202	5,044	5,685	6,106
1975	8,180	8,309	8,359	7,820	7,623	8,569	8,209	7,696	7,522	7,244	7,231	7,195
1976	8,174	8,033	7,525	6,890	6,304	7,655	7,577	7,323	7,026	6,833	7,095	7,022
1977	7,848	8,109	7,556	6,568	6,151	7,453	6,941	6,757	6,437	6,221		

Source: U.S. Department of Commerce, *Survey of Current Business.* See also footnote 1, Chapter 8.

TABLE A.8   Lydia Pinkham Vegetable Compound in Thousands of Dollars

YEAR	SALES	ADVERTISING	YEAR	SALES	ADVERTISING
1907	1016	608	1935	1518	807
1908	921	451	1936	1103	339
1909	934	529	1937	1266	562
			1938	1473	745
1910	976	543	1939	1423	749
1911	930	525			
1912	1052	549	1940	1767	862
1913	1184	525	1941	2161	1034
1914	1089	578	1942	2336	1054
			1943	2602	1164
1915	1087	609	1944	2518	1102
1916	1154	504			
1917	1330	752	1945	2637	1145
1918	1980	613	1946	2177	1012
1919	2223	862	1947	1920	836
			1948	1910	941
1920	2203	866	1949	1984	981
1921	2514	1016			
1922	2726	1360	1950	1787	974
1923	3185	1482	1951	1689	766
1924	3351	1608	1952	1866	920
			1953	1896	964
1925	3438	1800	1954	1684	811
1926	2917	1941			
1927	2359	1229	1955	1633	789
1928	2240	1373	1956	1657	802
1929	2196	1611	1957	1569	770
			1958	1390	639
1930	2111	1568	1959	1387	644
1931	1806	983			
1932	1644	1046	1960	1289	564
1933	1814	1453			
1934	1770	1504			

Source: Palda (1964), p. 23.

TABLE A.9   Average Daily Calls to Directory Assistance, Cincinnati, Ohio
in Hundreds of Calls

	JAN	FEB	MAR	APR	MAY	JUN	JUL	AUG	SEP	OCT	NOV	DEC
1962	350	339	351	364	369	331	331	340	346	341	357	398
1963	381	367	383	375	353	361	375	371	373	366	382	429
1964	406	403	429	425	427	409	402	409	419	404	429	463
1965	428	449	444	467	474	463	432	453	462	456	474	514
1966	489	475	492	525	527	533	527	522	526	513	564	599
1967	572	587	599	601	611	620	579	582	592	581	630	663
1968	638	631	645	682	601	595	521	521	516	496	538	575
1969	537	534	542	538	547	540	526	548	555	545	594	643
1970	625	616	640	625	637	634	621	641	654	649	662	699
1971	672	704	700	711	715	718	652	664	695	704	733	772
1972	716	712	732	755	761	748	748	750	744	731	782	810
1973	777	816	840	868	872	811	810	762	634	626	649	697
1974	657	549	162	177	175	162	161	165	170	172	178	186
1975	178	178	189	205	202	185	193	200	196	204	206	227
1976	225	217	219	236	253	213	205	210	216	218	235	241

Source: McCleary and Hay (1980), p. 316.

APPENDIX B

MATHEMATICAL EXPECTATIONS

B.1 MATHEMATICAL EXPECTATION

In this appendix we will only present some properties of the expectations operator needed for an understanding of its use in this book. The mathematical expectation (often called the *expected value*) of a random variable (*rv*) or uncertain quantity, say $\tilde{x}$, and denoted as $E(\tilde{x})$, is the *mean* of the probability distribution of $\tilde{x}$. We can best explain the expected value using discrete random variables. If a random variable $\tilde{x}$ can take on the discrete values $x = x_1, x_2, \ldots, x_n$, and if each value has an associated probability $f(x_i) = Prob(\tilde{x} = x_i)$, then the mathematical expectation is defined as[1]

$$E(\tilde{x}) = x_1 f(x_1) + x_2 f(x_2) + \cdots + x_n f(x_n). \tag{B.1}$$

As an example, let $\tilde{x}$ be the number observed as the roll of a fair die. Then the possible values for x are $x = 1, 2, 3, 4, 5, 6$ and its probability density function would be constant and equal to $f(x) = \dfrac{1}{6}$, $x = 1, 2, \ldots, 6$. Therefore, the expectation equals

$$E(\tilde{x}) = 1\left(\frac{1}{6}\right) + 2\left(\frac{1}{6}\right) + \cdots + 6\left(\frac{1}{6}\right) = 3\frac{1}{2}.$$

Notice that as in this example, the expectation is *not* necessarily equal to one of the possible value of the *rv*.

Transformations

1. Expectation is a *linear operation*.
   If we denote the mean as

$$E(\tilde{x}) = \mu$$

and if a and b are constants, then

$$E(a + b\tilde{x}) = a + bE(\tilde{x}) = a + b\mu. \tag{B.2}$$

If we have a collection of random variables $\{x_i, i = 1, 2, \ldots, n\}$ with expectations $\{\mu_i\}$, then the expectation of a weighted sum is the weighted sum of the expectations:

$$E\left(\sum_{i=1}^{n} a_i \tilde{x}_i\right) = \sum_{i=1}^{n} a_i E(\tilde{x}_1) = \sum_{i=1}^{n} a_i \mu_i. \tag{B.3}$$

---

[1] The expected value of a continuous variable is defined as

$$E(\tilde{x}) = \int xf(x)\,dx,$$

with $f(x)$ the probability density function of the continuous random variable $\tilde{x}$. The range of integration is defined by the range of the *rv*, x. We can view a continuous variable as a limiting case of a discrete variable, where the values that the variable can assume get more and more numerous and closer and closer to each other.

Example: Consider a set of identically distributed random variables $\{\tilde{x}_i\}$, each with mean μ. Then, the expectation of the sample mean is

$$E\left(\frac{\Sigma \tilde{x}_i}{n}\right) = \Sigma \frac{1}{n} E(\tilde{x}_i) = \frac{1}{n} n \mu = \mu.$$

In this example, $a_i = \frac{1}{n}$, $i = 1, \ldots, n$, and the summation Σ is over all values of x_i, $i = 1, 2, \ldots, n$.

2. Suppose we define another random variable $\tilde{z}$, which is equal to $\tilde{x}^2$. In the die rolling example, $\tilde{z}$ is equal to the square of the number appearing on the die. Taking the expected value of $\tilde{z}$, we obtain

$$E(\tilde{z}) = E(\tilde{x}^2) = 1\left(\frac{1}{6}\right) + 4\left(\frac{1}{6}\right) + \cdots + 36\left(\frac{1}{6}\right) = 15\frac{1}{6},$$

because the possible z values are 1, 4, 9, 16, 25, and 36, and each such value has an associated probability of $\frac{1}{6}$. We see that

$$[E(\tilde{x})]^2 = (3.5)^2 = 12.25 \neq E(\tilde{x}^2) = 15\frac{1}{6}.$$

Therefore, $[E(\tilde{x})]^2 \neq E(\tilde{x}^2)$. Putting this another way, the square of the expected value of $\tilde{x}$ is, in general, *not* equal to the expected value of $\tilde{x}^2$.

The following is an illustration of a more general result. If $\tilde{y} = g(\tilde{x})$, where $g(\tilde{x})$ is a nonlinear function of the random variable $\tilde{x}$, then, in general,[2]

$$E(\tilde{y}) = E[g(\tilde{x})] \neq g[E(\tilde{x})].$$

Another example of such a transformation is the expectation of the exponent of the random variable $\tilde{x}$, that is, $E(e^{\tilde{x}}) \neq e^{E(\tilde{x})}$.

B.2 VARIANCE AND COVARIANCE

The *variance* (Var) of an rv $\tilde{x}$ is defined as

$$\text{Var}(\tilde{x}) = E\{[\tilde{x} - E(\tilde{x})]^2\}. \tag{B.4}$$

The variance is also sometimes denoted by σ^2. Given a set of observations $\{x_i, i = 1, \ldots, n\}$, the variance is usually calculated (estimated) as

---

[2] It can be shown that if $g(\tilde{x})$ is concave over the whole range of possible x values, then

$$E[g(\tilde{x})] \leq g[E(\tilde{x})],$$

and if $g(x)$ is convex over the whole range of possible x values, then

$$E[g(\tilde{x})] \geq g[E(\tilde{x})].$$

Therefore, given that both x^2 and e^x are convex functions, $E(\tilde{x}^2) \geq [E(\tilde{x})]^2$ and $E(e^{\tilde{x}}) \geq e^{E(\tilde{x})}$.

$$\widehat{\mathrm{Var}}(x) = \frac{1}{n} \sum_{i=1}^{n} (x_i - \bar{x})^2$$

with $\bar{x}$ the sample mean of the data.

The variance of a linear function of an rv with a and b as constants equals

$$\mathrm{Var}(a + b\tilde{x}) = b^2 \mathrm{Var}(\tilde{x}). \tag{B.5}$$

Similarly, the *covariance* (Cov) between two rv's $\tilde{x}_1$ and $\tilde{x}_2$ is defined as

$$\mathrm{Cov}(\tilde{x}_1, \tilde{x}_2) = E\{[\tilde{x}_1 - E(\tilde{x}_1)] [\tilde{x}_2 - E(\tilde{x}_2)]\}$$
$$= \mathrm{Cov}(\tilde{x}_2, \tilde{x}_1). \tag{B.6}$$

Given n observations on two series $\{x_{1i}, i = 1, 2, \ldots, n\}$ and $\{x_{2j}, j = 1, 2, \ldots, n\}$, we usually calculate (estimate) the Cov as

$$\widehat{\mathrm{Cov}}(x_i, x_j) = \frac{1}{n} \sum_{i=1}^{n} (x_{1i} - \bar{x}_1)(x_{2i} - \bar{x}_2)$$

with $\bar{x}_1$ and $\bar{x}_2$ the sample means of the two series.

Let $\tilde{y} = a\tilde{x}_1 + b\tilde{x}_2$. Then the variance of $\tilde{y}$ is

$$\mathrm{Var}(\tilde{y}) = a^2 \mathrm{Var}(\tilde{x}_1) + b^2 \mathrm{Var}(\tilde{x}_2) + 2ab \mathrm{Cov}(\tilde{x}_1, \tilde{x}_2). \tag{B.7}$$

The variance of a linear combination of random variables can be obtained as follows:

$$\mathrm{Var}\left(\sum_i a_i \tilde{x}_i\right) = E\left\{\left[\sum_i a_i \tilde{x}_i - E\left(\sum_i a_i \tilde{x}_i\right)\right]^2\right\}$$

$$= E\left\{\left[\sum_i a_i \{\tilde{x}_i - E(\tilde{x}_i)\}\right]^2\right\}$$

$$= E\left\{\left[\sum_i \sum_j a_i a_j \{\tilde{x}_i - E(\tilde{x}_i)\}\{\tilde{x}_j - E(\tilde{x}_j)\}\right]\right\}$$

$$= \sum_i \sum_j a_i a_j E\{[\tilde{x}_i - E(\tilde{x}_i)] [\tilde{x}_j - E(\tilde{x}_j)]\}$$

$$= \sum_i \sum_j a_i a_j \mathrm{Cov}(\tilde{x}_i, \tilde{x}_j), \tag{B.8}$$

with $\mathrm{Cov}(\tilde{x}_i, \tilde{x}_i) = \mathrm{Var}(\tilde{x}_i)$. Similarly, it can be shown that the covariance of a linear combination is

$$\mathrm{Cov}\left(\sum a_i \tilde{x}_i, \sum b_j \tilde{x}_j\right) = \sum_i \sum_j a_i b_j \mathrm{Cov}(\tilde{x}_i, \tilde{x}_j). \tag{B.9}$$

If the random variables are uncorrelated, then (B.8) and (B.9) can be simplified as

$$\mathrm{Var}(\Sigma a_i \tilde{x}_i) = \Sigma a_i^2 \mathrm{Var}(\tilde{x}_i), \tag{B.10}$$

$$\text{Cov}(\Sigma\, a_i \tilde{x}_i,\ \Sigma\, b_j \tilde{x}_j) = \Sigma\, a_i b_i \text{Var}(\tilde{x}_i).\tag{B.11}$$

Example: Consider the variance of the sample mean of a set of independently[3] and identically distributed rv's with mean μ and variance σ^2:

$$\text{Var}\left(\frac{\Sigma\, \tilde{x}_i}{n}\right) = \Sigma\left(\frac{1}{n}\right)^2 \text{Var}(\tilde{x}_i) = \frac{1}{n^2}\, n\, \sigma^2 = \frac{\sigma^2}{n}\ .$$

B.3 CORRELATIONS

The correlation (Corr) between two rv's, $\tilde{x}_i$ and $\tilde{x}_j$, is defined as

$$\text{Corr}(\tilde{x}_i, \tilde{x}_j) = \text{Cov}(\tilde{x}_i, \tilde{x}_j)/\sqrt{\text{Var}(\tilde{x}_i)\text{Var}(\tilde{x}_j)}.\tag{B.12}$$

Therefore the Corr between two rv's is a scaled covariance between these two rv's. It indeed can be shown that any correlation must lie between $+1$ and -1. For a proof of this, see any statistics textbook [e.g., DeGroot (1975, Chapter 4)].

[3] If two rv's are independently distributed, then they are also uncorrelated, i.e., their covariance is 0. Note: The converse *only* holds for jointly normally distributed random variables.

TIME SERIES PROGRAMS: A PRIMER

C.1 PREFACE

The collection of Time Series (TS) programs is designed for interactive analyses and forecasts of time series data following the Box–Jenkins methodology. These programs are as nearly self-explanatory as possible. In many cases, all that the user needs to learn before actually starting to use the collection are (1) the small number of input conventions described in Section C.2.2 and (2) the names of the programs which perform the analyses.

During the first reading of this primer, the user should concentrate on Section C.3 ("Data Input") and the three subsections in Section C.4 dealing with program inputs. We recommend that, if possible, the user log onto a computer system while reading these sections, and perform a first run of the programs. A more thorough review of all of the material contained in this primer may be reserved for a later time.

Basically, two groups of programs are distinguished. The first group analyzes univariate time series (e.g., sales data) using ARIMA models. The second group analyzes multiple processes (e.g., sales as a function of advertising expenditures).

Three programs are used for analyzing seasonal and nonseasonal univariate time series: IDENT, ESTIMA, and FRCAST. These programs are normally used sequentially, with the user interpreting the results from each before moving on to the next program and stage of analysis. Program IDENT provides the necessary output to identify a possible time series model that could generate the data. Program ESTIMA estimates the parameters of a model. Program FRCAST generates time series forecasts conditional on the estimates obtained via ESTIMA.

There are three programs for analyzing a multiple time series model (i.e., a model including one or more explanatory variables, also called a transfer function model): CROSSC, TRFEST, and TRFFOR. These programs allow the user to carry out the identification, estimation, and forecasting of transfer function models. Program CROSSC performs pairwise analysis of a dependent variable with several explanatory variables to identify a particular transfer function model. Program TRFEST estimates the parameters of a transfer function model. Program TRFFOR computes point forecasts and confidence limits for transfer function models. The dependent variable can be predicted on the basis of given future values of the explanatory variables, or the user may ask program TRFFOR to forecast some or all of the explanatory variables using univariate ARIMA models. As with the univariate programs, these programs are normally used sequentially. These transfer function programs can also be used to generate univariate intervention models and transfer function intervention models.

C.2 INTRODUCTION[1]

C.2.1 Running the Time Series Programs

The Time Series collection consists of a number of separate programs operating under the control of a single master program.

After the user has logged onto the computer system and the computer indicates that it is under the control of the monitor by printing . (a period), the user gains access to the Time Series collection by typing the command

R TS

and then by striking the Return key.[2] After typing some introductory information the master program asks "PROGRAM?", to which the user responds by typing the name of the particular functional program desired. If the user has forgotten or does not know the exact name of the program, a list of the available programs may be obtained by typing ? (a question mark), followed by a (Carriage) Return. When the functional program chosen by the user has done its work, the master program resumes control and invites the user to specify the next functional program desired by again asking "PROGRAM?".

When the user is through using the TS collection as a whole, returning to monitor control (exiting to the monitor) is achieved by answering the question "PROGRAM?" with two dollar signs ($, mnemonic for stop).[3]

The functional programs offer the user a choice of various output options. When the user has taken all the desired options, returning to the control of the master program is achieved by typing $ in response to the question "OPTION?". No matter what functional program is being run, the user always can return immediately to the control of the master program by typing $$ in response to any question or request for input.

C.2.2 Conventions Common to All the Programs
TERMINATION OF INPUT

After typing an answer to any question posed by a TS program, the user must strike the Return key to indicate that the answer is complete.

---

[1] As the TS and AQD conventions are quite similar, users familiar with the AQD collection may defer their reading of this section. AQD is a collection of programs for interactive analysis of quantitative data developed at the Harvard University Graduate School of Business Administration, see Schlaifer (1981). For information about its availability contact the author.

[2] This is the command to be used at most of the DEC-10 computer installations. On the VAX computer the command is TS.

[3] In this or any other situation where $ is used as a signal to a TS program, it can be typed either in the usual way (by holding the Shift key down and striking the number 4 key) or, more easily, by striking the ESCape (ALTmode) key.

CORRECTING TYPING ERRORS

Typing errors cannot be corrected after Return, LineFeed, or ESCape (ALT-mode) has been struck. Before any of these keys has been struck, the last n characters typed on the current line can be erased by striking the Rubout key n times.[4] Or, everything typed on the current line can be erased by typing ConTRoL/U (^U), that is, by typing the letter U while holding the ConTRoL (CTRL) key down. On some terminals ^ may appear as an ↑ (an upper arrow).

When an incorrect answer to a question is detected after it is too late to correct it as a typing error, the user can usually recover by typing $ in response to the program's next request for input. A $ typed in this way causes a program to repeat the previous request for input, and by repeating the maneuver the user can back up as far as needed.

EXPLANATION OF A REQUEST FOR INPUT

If the user is unable to understand a question or a request for input, typing a question mark followed by Return will produce a full explanation of what is wanted. In rare cases where a wrong answer can do no harm, the program may simply repeat the original question, or instruct the user to answer YES or NO, or go ahead and act as if the user has given one of these two possible answers.

INPUT FROM FILES NOT BELONGING TO THE USER

Any file belonging to another person may be used in a TS program provided that the file is not read-protected. To designate such a file, the user must append to the file name the directory in which the file is to be found, enclosed in square brackets.[5] To designate a file in the system data library (if one exists) the user must append "/L" to the name—for example, FERIC/L.

VERBAL INPUTS OTHER THAN FILE NAMES

When a question or a request for input may be answered with a word which is *not* a file name, the legal answers usually begin with different letters. When this is true, the program will usually look only at the first letter of the user's answer. If, for example, Yes is a legal answer, the program will interpret either Yes, Y, or Yok as "YES".

INPUT OF SINGLE NUMBERS

When a program asks the user to supply a single number, the number may be typed in any convenient notation. In particular, an integer may or may

---

[4] On some CRT terminals there is a special key that allows the user to back up the pointer. The user can then retype the correct characters.

[5] On the DEC10 and DEC20, the directory is the owner's pun (PPN), e.g., [167,1234]. On the VAX this is the name of the directory or subdirectory, not to exceed eight characters, e.g., [ABT. DAT].

not be followed by a decimal point; a decimal fraction may or may not be typed with a 0 before the decimal point.

INPUT OF STRINGS OF NUMBERS

When a program asks the user to supply two or more numbers at one time, the numbers may be typed on a single line separated by a comma, spaces, or tabs. If the string is too long to fit on one line, it may be typed on any number of lines but, in this case, every line except the last must be terminated by striking the LineFeed key instead of Return. The last line must be terminated by Return in the usual manner.[6]

STRINGS OF EQUALLY SPACED NUMBERS

When the numbers in a string are equally spaced as they are in the string

```
2.2   2.4   2.6   2.8   3.0   3.2   3.4
```

the string can be more conveniently specified merely by typing the first number, the step (or the difference between any two successive numbers), and the last number, separated by colons. Thus the string listed above could be specified by simply typing

```
2.2   :   .2   :   3.4
```

The spaces before and after the colons are optional.

There is one restriction which must be observed in using a colon construction of the sort just described: nothing else may appear on the same line. The user can, however, specify a string such as

```
3     8 9 10 11 12 13     27
```

by typing on successive lines

```
3 <LF>
8 :      1    :    13    <LF>
27     <CR>
```

where <LF> denotes LineFeed and <CR> denotes Carriage Return.

SKIPPING A YES OR A NO

In many situations the answer Y(ES) or N(O) to one question leads immediately to a request for details in the form of a number or string of numbers. In almost every such situation, the user can save time by specifying the number(s) immediately, without first answering Y(ES) or N(O).

SUPPRESSION OF UNWANTED PRINTOUT

When the user has seen enough of a multiline printout, the output can be stopped by typing a CTRL/O; that is, by typing the letter O while holding

---

[6] In preparing CTL files for running TS in batch mode, every line except the last must be terminated by an ampersand(&) followed by Return.

the ConTRoL key down. If the user types a CTRL/O during a one-line message, it will suppress the printing of not only the remainder of that message but also the next request for input. This is why a program will sometimes seem to be "hung up" after a CTRL/O has been typed: the program is waiting for input and if the user does not know what the required response is, the Return key should be struck, to which the program will respond appropriately. The program will not be hung up during the printing of a multiline printout, even after typing CTRL/O during the printing of the last line.

INTERRUPTION OF PROGRAM EXECUTION

If the user wants to interrupt a program on the DEC10 or DEC20 while it is computing (not waiting for input), typing CTRL/C twice will immediately exit to the control of the monitor. The user then may continue execution of the program by typing CONTINUE.

OUTPUT TO THE LINE PRINTER OR A DISK FILE

The TS program allows printout to be directed, at the user's option, to either the user's Terminal (TeleTYpe), the Line PrinTer, or a file on the DiSK. The default option in the TS programs is that the *printed* output will appear on the TeleTYpe (TTY). The user will, however, be prompted by the question PLOTTED OUTPUT: TTY, LPT, OR DSK?, to which the user should respond by typing one of the following commands:

> T(ty): output to the Terminal (TeleTYpe)
>
> L(pt): output to the Line PrinTer
>
> D(sk): output to a file on your DiSK area.

To override the default *printed* output option, the user need only type a CTRL/E[7] at the *beginning* of a response to any question or request for input, *except* for a request for title or file name. The remainder of the response should be typed in the normal manner immediately after the CTRL/E.

If a file for output to LPT or DSK has already been opened, the program will respond with "OUTPUT TO TTY" or "OUTPUT TO LPT OR DSK" as the case may be. If no file has yet been opened, the program next asks "NAME FOR OUTPUT?", to which the user should respond with a name to be used as the header for the line printer output or as the name of the disk file. The file extension .DTP will automatically be appended to the name supplied by the user.

It is also possible still to switch the *printed* output answer from TTY to LPT or DSK and vice versa at any time. To make the switch, the user should type another CTRL/E at the *beginning* of a new response. Note, however, that output of a single program cannot be directed partially to the line printer and partially to the disk. If, for example, the user wants the *printed* output

---

[7] CTRL/E stands for typing the letter E while holding the ConTRoL key down.

to go to the LPT and answers "PLOTTED OUTPUT: TTY, LPT, OR DSK?" with "D(SK)", D(SK) will supersede L(PT) and a disk file will be created containing both the plotted and the printed options.

C.3 DATA INPUT

The Time Series programs accept as input files ASCII source files or AQD workfiles. The user could create a source file using a system editor such as EDITS, SOS, or TECO, or using tape or punched cards. Source files have to be given a file name composed of up to six letters and digits on the DEC10 or DEC20 and the file extension must be ".BJ". File names on the VAX may consist of up to eight characters. A source file is read in free format using up to 132 characters per line. It should therefore contain data on only one variable and no line numbers. For multiple time series analysis the data for each variable must be in a separate file. During the analysis it is possible to omit observations both at the start and at the end of the series. Files that cannot be read free format or that contain data on several variables should be preprocessed.

At computer installations where the AQD collection of programs is available, the user can create ADQ workfiles, as an option is built into the Time Series programs to read such workfiles. Note that there is no need to create AQD workfiles containing only one variable. The Time Series programs can select the appropriate variable from an AQD workfile if the file contains more than one variable.

C.4 UNIVARIATE TIME SERIES PROGRAMS

To illustrate how to use the univariate programs we will analyze monthly data seasonally unadjusted on United States residential (nonfarm) construction covering the period from January 1947 to December 1968 (264 observations).[8] For a more detailed analysis of this series, we refer the reader to Chapter 7.

C.4.1 Model Identification

Program IDENT allows the user to identify a possible time series model that could have generated the data. Suppose, after looking at the time series plot of the data, we decide that seasonal differencing and one or more consecu-

---

[8] *Source:* U.S. Department of Commerce, Office of Business Economics, *1971 Business Statistics,* The Biennial Supplement to the *Survey of Current Business,* pp. 49–50, 226. In Appendix A, Table A-6 we have listed the data.

```
. R  TS


PROGRAM?  IDENT

VERSION:    1-Jan-83
TITLE?  USRES
AQD FILE? N
FILE NAME? CONST
SMPL? 1 264
LOG? N
STANDARD RUN? N
MCDIF? 2
SEAS? Y
SPAN? 12
SDIF? 1


OUTPUTS:    ?
ALL DESIRED OUTPUTS MUST BE SPECIFIED IN A SINGLE LIST.
AVAILABLE OUTPUTS:
    1    PRINT AUTOCORRELATIONS AND PARTIAL AUTOCORRELATIONS
    2      "    INVERSE AUTOCORRELATIONS
    3    PLOT DATA AND DIFFERENCES
    4      "    AUTOCORRELATIONS AND PARTIAL AUTOCORRELATIONS

OUTPUTS:    1
NCORR? 24
```

FIGURE C.1	Program IDENT Input.

tive differencings are needed to induce stationarity in the data.[9] In other words—suppose we are interested in evaluating a stationary model which can be identified (characterized) by seasonal and consecutive differencing, possibly up to the order two; that is, by making transformations of the form

$$\nabla^i \nabla_{12} z_t \qquad \text{for } i = 0, 1, 2.$$

We define $\nabla_{12} z_t = z_t - z_{t-12}$, $\nabla z_t = z_t - z_{t-1}$, $\nabla^2 z_t = \nabla(\nabla z_t)$.

C.4.1.1 PROGRAM IDENT INPUT

Figure C.1 presents a protocol for analyzing the construction data using program IDENT. A discussion of this protocol follows.

As explained in Section C.2.1, upon logging onto the system the first command typed by the user on the DEC-10 installation and certain other computer installations is

```
R  TS
```

[9] In this primer we only explain the working of the programs and therefore we suppose that the user is familiar with the ARIMA time series analysis methodology.

On the VAX installation this command is simply

TS

The computer will respond with a series of messages and then will ask the user to identify the desired program by answering the question "PROGRAM?". These messages can be interrupted by typing CTRL/O.

As seen in Figure C.1, typing IDENT initiates the request for a title for this analysis. The title ("USRES" in the example) will appear at the top of each page of output directed to the line printer or to a disk file. The title is limited to 40 characters.

The program will next ask "AQD FILE?" which refers to a particular data analysis computer system available at the Harvard University Graduate School of Business Administration and several other DEC installations. If the user's data is available as an AQD workfile, Y(ES) should be typed. As discussed in Section C.3, most Time Series data input files will be ASCII source files containing data on a single variable. If this is the case, as in the example, the response should be N(O) and the file name should be specified in response to the next question ("FILE NAME?"). The extension .BJ must not be typed.

The next set of questions, referring to the specific analysis of the CONST time series to be performed with program IDENT, is discussed in Figure C.2.

The set of responses described in Figure C.2 will instruct the program to do the analyses desired on the series $\nabla_{12}z_t$, $\nabla\nabla_{12}z_t$, and $\nabla^2\nabla_{12}z_t$.

With the model specified, we inform the program of what output we want, specifying in a single list the indices in an answer to

OUTPUTS:

As before, typing a ? in response to this question will result (as in the example) in a list of the available options. In the example, the desired option is "1".

As discussed in Section C.2.2, as a default the PRINTed output will be printed on the TeleTYpe. If this default option must be modified, the user can make use of the CTRL/E option at the moment of answering the question "OUTPUTS:"

If plotted outputs are requested (say, output 4), the program will ask for the disposition of the PLOTted output "PLOTTED OUTPUT: TTY, LPT, OR DSK?". This allows the user to direct part of the output to the TeleTYpe and yet another part (the plots, for example) to the Line PrinTer or to a DiSK file.

Finally, the program will also prompt with "NCORR?". The user then should specify the Number of autoCORRelations and partial autoCORRelations to be printed by the program.

If the user asks for a plot of the data, program IDENT will request

FIGURE C.2 Program IDENT Input: Discussion.

PROGRAM REQUEST	USER RESPONSE	EXPLANATION OF REQUEST/ USER RESPONSE
SMPL?		The set of observations desired in the analysis.
	1 264	We decide to use the first 264 observations. If the first year's data were not needed, the user would have typed 13 264
LOG?	Y(ES)	A natural (to the base e) LOGarithm of the data will be taken.
	N(O)	For the construction data, no log transformation is necessary and we type N(O) to this question.
STANDARD RUN?	N(O)	Under the standard run the program IDENT will print out the first 24 autocorrelations and partial autocorrelations of the data and of the first differences of the data.
		Since we are interested in seasonal differencing as well as up to second-order consecutive differences, we respond by typing N(O).
		Because we do not select the standard run, IDENT will prompt for the desired analysis with the following four questions:
MCDIF?	2	Maximum order of the Consecutive DIFferencing is specified as 2.
SEAS?	Y(ES)	Asks if the data contains a SEASonal component, to which we respond with Y(ES).
SPAN?	12	The SPAN of the seasonality is the period of seasonal differencing. In our example, the span equals 12; that is, observations 12 periods apart exhibit similar characteristics.
SDIF?	1	The number of Seasonal DIFferencing required. In the example, we ask for one seasonal differencing.

that the CALENDAR DATE OF THE FIRST OBSERVATION be specified. This is done by typing the year and, eventually (for seasonal data), the quarter or month of the first observation, with a space after the year.

C.4.1.2 PROGRAM IDENT OUTPUT

Figure C.3 reproduces sections of the output created by the specifications given in Figure C.1. The program calculates 24 autocorrelations and partial autocorrelations together with the necessary statistics to evaluate the following three models: $\nabla_{12}z_t$, $\nabla\nabla_{12}z_t$, and $\nabla^2\nabla_{12}z_t$.

For each model, output option 1 contains:

—the Mean (MEAN), Standard Deviation (SD), and the Net number of OBServations after differencing (NOBS). These values are descriptive statistics and as such no degree of freedom adjustments are made.

—the AUTOCORRELATIONS for the specified LAGS together with the ROW Standard Errors (ROW SE). The autocorrelations are a measure of the dependence between observations separated by a particular time interval called a lag. The value of the autocorrelations lies between $+1$ and -1. The closer the autocorrelation is to $+1$ or -1, the more highly correlated are the observations separated by a particular lag. Similarly, the closer the autocorrelation is to 0, the less correlated are the observations separated by a particular time interval. Program IDENT also prints the ROW Standard Errors (ROW SE) under the null hypothesis that the true process is an MA(0), MA(12), MA(24), etc. (one less than the first lag printed on each row). The plots of autocorrelations (option 4) give an approximate 95% large-sample confidence interval for each autocorrelation under the null hypothesis that the true process is an MA model with order one less than the lag printed on each row.

—the CHI-SQUARE TEST as a "portmanteau" test of randomness. Rather than examine each autocorrelation individually, often we want to know if, say, the first 12 autocorrelations together indicate that the data exhibits nonrandomness. If so, the values of Q will be inflated. This $Q(K)$ statistic is approximately distributed as Chi-square with $(K - p - q - P - Q)$ degrees of freedom. K refers to the number of autocorrelations used to calculate the $Q(K)$ statistic. The symbols p, q, P, and Q are the number of parameters in the ARIMA model, respectively the autoregressive, the moving average, the seasonal autoregressive and the seasonal moving average parameters. In the program IDENT the number of parameters is, of course, 0. The program IDENT also prints the "P-VALUE" under the null hypothesis of random data; that is, the "tail area" probability, or significance level.[10]

—the PARTIAL AUTOCORRELATIONS for a specific lag are again a measure of the dependence that exists between observations separated by a particular time interval, but under the assumption that the data on shorter

---

[10] For a discussion of the P-value, see DeGroot (1973) and Gibbons and Pratt (1975).

FIGURE C.3   Program IDENT Output.

```
SERIES WITH D = 0 DS = 1 S = 12 <== ( D: consecutive differencing
                                       DS: seasonal differencing
                                       S: span)
MEAN =     75.29     SD =      252.2    (NOBS = 252)

AUTOCORRELATIONS
  LAGS ROW SE
  1-12   .06    .96  .88  .78  .67  .54  .40  .26  .11 -.03 -.16 -.27 -.36
  13-24  .17   -.43 -.46 -.48 -.48 -.46 -.42 -.37 -.32 -.26 -.20 -.14 -.07

CHI-SQUARE TEST              P-VALUE
Q(12) =  912.        12 D.F.   .000
Q(24) =  .135E+04    24 D.F.   .000

PARTIAL AUTOCORRELATIONS     STANDARD ERROR =  .06
  LAGS
  1-12          .96 -.57 -.05 -.13 -.17 -.22  .00 -.06 -.12 -.01  .01  .07
  13-24         .17 -.09 -.02  .00  .00 -.04 -.08 -.04 -.04  .02  .05  .00


SERIES WITH D = 1 DS = 1 S = 12
MEAN = -0.2869      SD =     66.34   (NOBS = 251)

AUTOCORRELATIONS
  LAGS ROW SE
  1-12   .06    .62  .28  .17  .18  .20  .06 -.05 -.03 -.07  ^0 <=== CTRL/0

CHI-SQUARE TEST              P-VALUE
Q(12) =  245.        12 ^0

PARTIAL AUTOCORR^0


SERIES WITH D = 2 DS = 1 S = 12
MEAN =   0.4640     SD^0

AUTOCORRELATIONS^0


CHI-SQUARE^0

PARTIAL AUT^0

OPTION? ?
AVAILABLE OPTIONS:
    M CHANGE MODEL, SAME VARIABLE
    V    "     VARIABLE
    $ STOP

OPTION? $
```

lags are also present.[11] As a measure of uncertainty, the large-sample STAN-DARD ERROR is printed and is the same for all lags.

Output option 2 provides the inverse autocorrelations [$ir(k)$'s at lag k] for each transformation as an additional tool for identifying a parasimonious ARIMA model. In evaluating the inverse autocorrelations, remember that the $ir(k)$'s of an ARIMA(p,d,q) model behave as the autocorrelations of an ARIMA(q,d,p) model; that is, the words "autoregressive" and "moving average" are interchanged. If, in evaluating the inverse autocorrelations as *if* these were autocorrelations, we determine that the model is an AR(1), we then should conclude that the appropriate model for the data is an MA(1) model. IDENT prints the $ir(k)$'s for each value of lag p in a user-specified range, and for all $k \leq p$. In practice, it is worthwhile to calculate $ir(k)$'s for a few different values of p. If the $ir(k)$'s resulting from a particular p are reliable, the values should not change drastically when p is increased by a moderate amount.

Output option 3 provides a time series plot of the data as well as of all the differenced data. If the outputs of these plots are directed to the line printer or to a disk file, the user has the option of selecting a wide or narrow plot. A narrow plot only will make use of the first 72 columns of a page.

Finally, under option 4, we obtain a plot of the autocorrelations and partial autocorrelations. As discussed above, an approximate 95% large-sample confidence interval is also given for each autocorrelation under the null hypothesis that the true order of the autoregressive process is one less than the order of the autocorrelation being evaluated.

At the end of the output, the program will ask

OPTION?

allowing the user to reanalyze the same data but using a different model (Option M), for example, to change the consecutive differencing, to reuse the program IDENT but with a new variable or a different subset of observations of the same variable (Option V), or just to stop the program (Option $).

C.4.2 Model Estimation

After identifying a potential model the user will call for program ESTIMA to estimate the model parameter values. The parameter estimation can be done using one of the following two methods: the Least SQuares estimation

---

[11] Let w_t be the stationary data at time t; then, the autocorrelation at lag k specifies the simple correlation between w_t and w_{t-k}, whereas the partial autocorrelation at lag k gives the correlation between w_t and w_{t-k}, given $w_{t-1}, \ldots, w_{t-k+1}$. Therefore the partial autocorrelation is really a partial correlation.

method (LSQ), or the Maximum Likelihood Estimation method (MLE). The MLE method implemented in ESTIMA makes use of the approach discussed in Ansley (1979). If the user selects the LSQ method, program ESTIMA will first calculate starting values for the lagged data and error terms based on the assumed process that is generating the data; that is, the program will *back forecast* or backcast these values. This "unconditional" approach will yield estimates which are very nearly maximum likelihood. The program ESTIMA will backcast 104 data points.

Suppose that the autocorrelations and partial autocorrelations calculated in program IDENT using the differenced series $\nabla\nabla_{12}z_t$ suggest a model formulation which includes a first- and a second-order autoregressive component, AR(1) and AR(2), and a first-order seasonal moving average component, SMA(1). We select this model only to demonstrate some of the capabilities of the program ESTIMA, and the reader should not assume that this formulation is the best representation of the construction series.

C.4.2.1 PROGRAM ESTIMA INPUT

The input section of program ESTIMA is given in Figure C.4. Except for the question "METHOD (LSQ OR MLE)?", the first group of questions about the data file is similar to those asked by program IDENT. However, in the model specification section, rather than asking "MCDIF?", program ESTIMA prompts for the *exact order* of Consecutive DIFferencing and *not* the maximum order. Therefore, in answering the question "CDIF?", we type a 1. In the next group of questions (discussed in Figure C.5) ESTIMA prompts the user for the parameter specifications as well as guess values to start off the nonlinear least squares optimization routine.[12]

C.4.2.2 PROGRAM ESTIMA OUTPUT

Figure C.6 presents the output option 1 of program ESTIMA. After printing information based on the input section [such as initial values, model differencing, mean (MEAN), Standard Deviation (SD), and Net number of OBServations after differencing (NOBS)], the program ESTIMA gives the INITial value of the Sum of Squared Residuals (INIT SSR) as calculated based on the guess values.

A successful termination of the nonlinear optimization procedure will be indicated by

```
REL. CHANGE IN SSR <=    0.1000E-05
```

or

```
REL. CHANGE IN EACH PARAMETER <=    0.1000E-03
```

---

[12] The program uses the Marquardt iterative nonlinear optimizer. See Marquardt (1963).

followed by the minimum value of the SSR as obtained by the final results (FINAL SSR) and the number of iterations. Intermediate iteration results for the parameter estimates and SSR are obtained by requesting output option 2.

FIGURE C.4   Program ESTIMA Input.

```
PROGRAM? ESTIMA

VERSION:   1-Jan-83
TITLE? USRES

METHOD (LSQ OR MLE)? L
AQD FILE? N
FILE NAME? CONST
SMPL? 1 264

MODEL SPECIFICATION :
LOG? N
CDIF? 1
SEAS? Y
SPAN? 12
SDIF? 1

CONSTANT? N
ORDERS AR PARAM? 1 2
GUESS VALUE  1? .4
GUESS VALUE  2? .3
ORDERS SEAS AR PARAM? N
ORDERS MA PARAM? N
ORDERS SEAS MA PARAM? 1
GUESS VALUE  1? .2

OUTPUTS:   ?
ALL DESIRED OUTPUTS MUST BE SPECIFIED IN A SINGLE LIST.
AVAILABLE OPTIONS:
   1   PRINT ESTIMATES
   2     "    ITERATION RESULTS
   3     "    BACK FORECASTED RESIDUALS (ONLY FOR LSQ)
   4     "    RESIDUALS OF THE SERIES
   5     "    RESIDUAL AUTOCORRELATIONS
   6     "    RESIDUAL CROSS CORRELATIONS
   7     "    FITTED VALUES
   8   PLOT RESIDUALS
   9     "    RESIDUAL AUTOCORRELATIONS
  10     "    RESIDUAL CROSS CORRELATIONS
  11     "    PERIODOGRAM OF RESIDUALS
  12  FILE RESIDUALS

OUTPUTS:   1
```

FIGURE C.5   Program ESTIMA Input: Discussion.

PROGRAM REQUEST	USER RESPONSE	EXPLANATION OF REQUEST/ USER RESPONSE
CONSTANT?	N(0)	We do not want an intercept in the model.
ORDERS AR PARAM?	1 2	First- and second-order autoregressive parameters are to be included in the model. If the single number 2 were typed, the programs would estimate a second-order autoregressive model, but with the first-order parameter suppressed (that is, put equal to 0). Note that there is a space between 1 and 2.
GUESS VALUE 1? GUESS VALUE 2?	.4 .3	We specify that the guess value to start off the optimization routine is 0.4 for the first-order and 0.3 for the second-order parameter.
		These guess values are checked to see if they imply a stationary or invertible polynomial. If not, the program will prompt the user for alternative guess values.
ORDERS SEAS AR PARAM? ORDERS MA PARAM?	N(0) N(0)	As the model does not involve seasonal autoregressive or moving average parameters, these questions are answered twice with N(O).
ORDERS SEAS MA PARAM? GUESS VALUE 1?	1 .2	We specify that the first-order seasonal moving average parameter is to be included and initialized with guess value $= 0.2$.
OUTPUTS:		In answer to this question, the user selects specific output options.

FIGURE C.6   Program ESTIMA Output.

```
NOBS = 264

INITIAL VALUES
 AR( 1)   0.4000E+00
 AR( 2)   0.3000E+00
SMA( 1)   0.2000E+00

MODEL WITH D = 1 DS = 1 S = 12
MEAN = -0.2869    SD =   66.34    (NOBS = 251)

INIT SSR =   0.7079E+06
REL. CHANGE IN SSR <=   0.1000E-05
FINAL SSR =  0.4631E+06
5 ITERATIONS


LEAST SQUARES METHOD
PARAMETER ESTIMATES
====================
             EST      SE      EST/SE     95% CONF LIMITS
 AR( 1)     0.710    0.061    11.641     0.590    0.829
 AR( 2)    -0.238    0.061    -3.931    -0.357   -0.119
SMA( 1)     0.682    0.047    14.527     0.590    0.774


EST.RES.SD =    4.2361E+01
EST.RES.SD(WITH BACK FORECAST) =    4.3211E+01


R SQR =   .597
DF = 248
NO F-STAT, NO INTERCEPT

CORRELATION MATRIX
          AR( 1)  AR( 2)
 AR( 2)   -.561
SMA( 1)   -.077    -.029

ROOTS   REAL   IMAGINARY   SIZE
 AR     1.490    1.406      2.049
 AR     1.490   -1.406      2.049
SMA     1.466    0.000      1.466

OPTION? ?
AVAILABLE OPTIONS:
    F FILE OUTPUT FOR PROGRAM FRCAST
    M CHANGE MODEL OR ESTIMATION METHOD, SAME  VARIABLE
    V    "    VARIABLE
    $ STOP

OPTION? F
OUTPUT FILED
```

For any coefficient estimate, the program ESTIMA prints the following statistics:

EST: the point ESTimate of the true coefficient of the ARIMA model;

SE: the estimate of the Standard Error of the estimates (the square root of an unbiased estimate of the sampling variance);[13]

EST/SE: the ratio of the point ESTimate to its Standard Error;

95% CONF LIMITS: the limits of the approximate 95% large-sample confidence interval.

After printing the statistics described above for each parameter, ESTIMA gives several measures of goodness of fit.

1. EST.RES.SD: an ESTimate of the Standard Deviation of the RESiduals in the ARIMA model. These residuals are also called the random shock or white noise series. This estimate is the square root of an unbiased estimate of the corresponding variance.
2. EST.RES.SD (WITH BACK FORECAST): an ESTimate of the true or process RESidual Standard Deviation. This estimate also uses the back forecasted residuals (see option 3) and is the square root of an unbiased estimate of the process residual variance.
3. R SQR: an estimate of the fraction of the variance of the sample values of the transformed data $[\nabla\nabla_{12}z_t = w_t]$ accounted for by the fitted ARIMA model. The R SQR is also called the coefficient of determination.

For a model with an intercept the R SQR statistic is calculated as

$$R\ SQR = 1 - SSR/SQ(\dot{w})$$

where SSR is the Sum of Squared Residuals excluding the back forecasted residuals and $SQ(\dot{w})$ is defined as the sum of squared deviations of the stationary data w_t from its mean. If the model does not contain an intercept, R SQR is still calculated as specified above, but now $SQ(\dot{w})$ is defined as the sum of the stationary data squared with no subtraction of the mean.

Program ESTIMA prints out the number of Degrees of Freedom (DF); that is, the difference between the number of observations after differencing and the number of parameters. As an additional overall goodness-of-fit measure of the complete model, the value of the F statistic is given as well as its "P-VALUE" equal to the tail area probability of the hypothesis that all coefficients are 0, except the intercept. Finally, ESTIMA provides an estimate of the sampling CORRELATION MATRIX of the coefficient estimates.

As another check on the model adequacy, the program ESTIMA prints out the ROOTS of the estimated polynomials as well as the SIZE of these roots. The size is defined as the square root of the sum of the squares of

---

[13] When we use the term *unbiased* we imply that we have used a correction for degrees of freedom.

the real and the imaginary part of each root. For stationarity and invertibility to hold, the size of each root should be larger than one (should be outside the unit circle). Roots of the AR polynomials with size close or equal to one are a possible indication that additional differencing may be required, whereas roots close or equal to one of the MA polynomials are a possible indication of overdifferencing.

Other ESTIMA output options contain the following information:

Option 2: Iteration results. Under this option, program ESTIMA prints out the values of the estimates of the coefficients as well as the SSR at each iteration.

Option 3: Back forecasted residuals (LSQ method only). Consistent parameter estimates, unconditional on the starting values, are obtained by forecasting the data backwards; that is, forecasting the data for a number of periods before the actual start of the series. Given the back forecasted data, we then can also evaluate the residuals for time periods before the actual start of the sample period, the back forecasted residuals. Under this option, the back forecasted residuals are printed out.

Option 4: Residuals. The estimates of the white noise series or random shocks are printed out under this option.

Option 5: Residual autocorrelations. Under this option the autocorrelations of the residuals and first differenced residuals are given, together with measures of uncertainty. If the model specification is adequate, the residuals should approximately behave as a white noise series. Therefore, one might expect that the residuals could indicate the existence and the nature of model adequacy. For example, a large residual autocorrelation at lag 1 might indicate that a first-order moving average term ought to be added to the model. In order to evaluate these autocorrelations up to a specific order, the program ESTIMA also gives a CHI-SQUARE TEST as a "portmanteau" test of the hypothesis of model adequacy.

Option 6: Residual cross correlations. In theory, white noise disturbances are correlated with the current values and future values of the differenced series w_t (LEADS) but not necessarily with past values (LAGS). The series w_t is the stationary data actually used in the estimation program and could represent differences of the raw data, as is the case in our example. Therefore, as an additional model check, sample cross correlations between residuals, a_t, and the future values of w_t [a_t with w_{t+k}] and past values of w_t [a_t with w_{t-k}] are given, together with measures of uncertainty.

Option 7: Fitted values. The program ESTIMA gives the fitted values of the undifferenced series. These fitted values are equal to the one step ahead forecasts. More forecast information is obtained by running program FRCAST.

Options 8, 9, and 10 provide plots of the residuals, residual autocorrelations, and residual cross correlations.

Option 11: Program ESTIMA plots the normalized cumulative periodogram[14] of the residuals together with 90% large-sample Kolmogorov–Smirnov confidence limits. For random residuals, the plot of the normalized cumulative periodogram against the frequency would be scattered about a straight line joining the points (0, 0) and (0.5, 1). Model inadequacies would produce nonrandom residuals whose cumulative periodogram would show systematic deviations from this line. In particular, cycles or periodicities in the residuals tend to produce a series of neighboring values which form bulges about the expected line. Insufficient differencing results in larger values than expected in the low frequencies of the periodogram. To be able to judge large departures from the straight line for a white noise series, program ESTIMA gives a 90% large-sample Kolmogorov–Smirnov confidence interval. Each time a value of the periodogram is outside this interval, a "+" is printed rather than an "*".

At the end of the output, the program will again ask

OPTION?

Besides option M (change the model or estimation method) and option V (change the variable) discussed in Section C.4.1.2, program ESTIMA has an additional and very useful option of filing the basic results of ESTIMA for forecasting purposes. If the user specifies option F, the FRCAST program automatically will be started.

C.4.3 Model Forecasting

After defining a suitable model, program FRCAST is used to generate forecasts. In the example given below the forecasts will be calculated using the filed results of the model formulated and estimated in Figures C.4 and C.6.

C.4.3.1 PROGRAM FRCAST INPUT

Figure C.7 reproduces the input section of the FRCAST program. Given that we have filed all relevant information of the model estimated above, program FRCAST is automatically started using this information. However, in order to compare forecasted and actual values, program FRCAST can read in additional observations besides the ones used in program ESTIMA. However, the calculation of the forecast large-sample standard errors depends

---

[14] The normalized cumulative periodogram is calculated using the Fast Fourier Transform. For references see Singleton (1967) and Brigham (1974).

```
SMPL? 1 276

FORECAST INFORMATION :
CALENDAR DATA OF FIRST OBS.? 1947 1
HORIZON? 12
FIRST ORIGIN? 264
LAST ORIGIN? 264

OUTPUTS:  ?
ALL DESIRED OUTPUTS MUST BE SPECIFIED IN A SINGLE LIST.
AVAILABLE OPTIONS:
  1  PRINT FORECASTS WITH 95% CONFIDENCE INTERVAL
  2    "    ERROR ANALYSIS (THEIL DECOMPOSITION)
  3    "    FORECAST STANDARD ERRORS AND ERROR LEARNING COEFF.
  4    "    ANTILOG FORECASTS (ONLY IF LOGTRANSFORMATION ASKED)
  5  PLOT FORECASTS

OUTPUTS:  1
```

FIGURE C.7   Program FRCAST Input.

on the number of observations actually used to estimate the parameters. In the example we will compare the forecast for the next 12 months with the observed values and therefore, in an answer to

SMPL?

we specify

1 276

If no information had previously been filed, program FRCAST would start off the input section by asking questions about the data file and the model specification similar to those asked in program ESTIMA, see Figure C.4. However, this time it would not only ask SMPL? to indicate which data to read in, but it would also ask

NOBS USED IN ESTIMA?

to allow for a correct calculation of the forecast large-sample standard errors.

The questions in the FORECAST INFORMATION section are discussed in Figure C.8.

C.4.3.2 PROGRAM, FRCAST OUTPUT

Output option 1 is given in Figure C.9. After reproducing part of the input section, program FRCAST lists for each observation in the forecast horizon the point forecast, the upper and lower limits of a 95% large-sample confidence interval and, if actual observations are available, actual, error (actual minus forecast), and percentage error values.

FIGURE C.8   Program FRCAST Input: Discussion.

PROGRAM REQUEST	USER RESPONSE	DEFINITION OF REQUEST/ USER RESPONSE
CALENDAR DATE OF FIRST OBS.?	1947 1	This allows the program FRCAST to clearly identify the forecast periods in the output section.
HORIZON?	12	The forecasts are made for the next 12 months.
FIRST ORIGIN?	264	The point of origin of the forecast is observation 264 and the first forecast will be made for period 265. In making forecasts the actual observations are used only up to the first origin; therefore, in forecasting period 266, the forecasted value for period 265 is used irrespective of whether actual data is available.
LAST ORIGIN?	264	The program FRCAST can advance the point of origin each time by one observation and then make forecasts for the next number of periods specified in the HORIZON. In this way, all actual observations up to and including observation 264 are used. Note that after incorporating one additional actual observation in the forecast, the parameters are not re-estimated. This adaptive forecasting continues until all the data up to the LAST ORIGIN are included. In the example, only forecasts for one origin data are desired, and therefore the same observation number 264 is specified.
OUTPUTS:		Program FRCAST prompts the user for the output options and their disposition.

```
HORIZON 12 PERIOD(S)
STARTING AT 264

POINT ESTIMATES

   AR( 1)    0.7099E+00
   AR( 2)   -0.2382E+00
  SMA( 1)    0.6823E+00

MODEL WITH D = 1 DS = 1 S = 12
```

OBS	95% LOWER CONF LIMIT	FORECAST	95% UPPER CONF LIMIT	ACTUAL	ERROR (A–F)	%ERROR (A–F)/A
ORIGIN = 1968.12						
1969.1	2090.	2174.	2257.	2133.	−40.52	−1.900
1969.2	1836.	2000.	2165.	1940.	−60.45	−3.116
1969.3	1952.	2184.	2416.	2195.	10.84	.4939
1969.4	2165.	2450.	2735.	2540.	89.83	3.537
1969.5	2331.	2659.	2987.	2810.	151.2	5.381
1969.6	2478.	2842.	3206.	2962.	120.4	4.064
1969.7	2564.	2961.	3357.	2974.	13.38	.4500
1969.8	2587.	3013.	3439.	2880.	−132.8	−4.610
1969.9	2508.	2962.	3416.	2763.	−198.8	−7.195
1969.10	2376.	2856.	3337.	2648.	−208.5	−7.874
1969.11	2263.	2769.	3275.	2482.	−286.9	−11.56
1969.12	2087.	2616.	3146.	2288.	−328.2	−14.34

```
OPTION? ?
AVAILABLE OPTIONS:
    F CHANGE FORECAST OPTIONS, SAME VARIABLE/MODEL
    V     "    VARIABLE/MODEL
    $ STOP

OPTION? $
```

FIGURE C.9   Program FRCAST Ouput.

Output option 2 performs an error analysis to measure the accuracy of the forecasts. Specifically, several forms of mean squared errors are calculated. In addition, two forms of Theil's Inequality Coefficient are presented together with two different decompositions as developed by Theil. For more information about this option, see Maddala (1977), p. 343 ff. Note, however, that the program calculates all statistics based on the actual and forecasted values and not on relative changes of these two quantities.

Under option 3, the user obtains the error learning coefficients and forecast standard errors. The error learning coefficients (PSI WEIGHT) indicate by what amount the forecasts are adjusted based on a new observation and the resulting current one step ahead error. Defining $z_n(h)$ as the forecast made at period n for period $n + h$, that is, h steps ahead, then the change in the forecast for period $n + h$ given the observation z_{n+1} is a fraction of the current forecast error

$$z_{n+1}(h - 1) - z_n(h) = \text{PSI}(h - 1)\,[z_{n+1} - z_n(1)]$$

where $\text{PSI}(h - 1)$ is the $h - 1$ steps ahead PSI WEIGHT. For an AR(1) model, these error learning coefficients decrease as h increases, reflecting that the current error provides less information the further we look into the future.

The PRED. SE are the large-sample Standard Errors for PREDictions made h steps ahead. These are used in constructing the 95% large-sample confidence limits under option 1.

If a logarithmic transformation is specified, the user also can obtain antilog forecasts by requesting option 4. These antilog forecasts are not just the antilog of the forecasts, but involve a correction based on the error variance.

Option 5 provides a plot of the forecasts for the specified origin dates, starting at a user-specified number of periods before these origin dates. Over the forecast horizon, actual (if available) and forecasted values are plotted, as well as 95% large-sample confidence limits.

Finally, the program FRCAST prompts the user with OPTION?, allowing the user to retain all information read in from ESTIMA and only to the change the forecast specifications (option F) or to specify option V to indicate that a new model or variable should next be used.

C.5 TRANSFER FUNCTION PROGRAMS

To illustrate the use of the transfer function programs, we will analyze the well-known advertising-sales data of the Lydia Pinkham vegetable compound [see Palda (1964), p. 22].

The annual data is composed of sales and advertising data between 1907 and 1960. Only the first 40 annual observations between 1907 and 1946 are used in the program TRFEST. The last 15 observations are saved for a test of the forecasting ability of the model. This data has been the source of many analyses. For a survey, see Helmer and Johansson (1977).

C.5.1 Model Identification

Program CROSSC allows the user to calculate the cross correlations between a dependent variable and each of a set of explanatory variables. In addition, the program can calculate several other identification tools and statistics useful to identify a particular transfer function model. These are autocorrelations of the prewhitened dependent variable, impulse response weights, residuals, and residual autocorrelations and partial autocorrelations. Both printing and plotting capabilities are part of the output options of the CROSSC program. Finally, the prewhitened data can also be filed under a user specified file name. The file extension is by default .BJ, so that these files can be used by any other program in the TS collection.

C.5.1.1 PROGRAM CROSSC INPUT

Suppose that the user has decided to prewhiten the sales and advertising data using the advertising prewhitening filter composed of first consecutive differences and an AR(2) model with the first-order autoregressive parameter = 0.074 and the second-order autoregressive parameter = −0.407. Then, Figure C.10 would contain the protocol of the input for the CROSSC program.

The set of responses as described in Figure C.11 will instruct the program CROSSC to prewhiten the sales and advertising data with an AR(2) process after first differencing the data, and to calculate any of the outputs specified in a single list in an answer to the prompt

OUTPUTS:

As before, typing a ? in response to this prompt will result (as in the Figure C.10) in a list of the available options. We specify the options 1, 3, and 4. In response to these output specifications the program asks for the NumbeR of autocorrelations and CROSS CORrelations (NR CROSS COR.?) and the Number of impulse response WEIGHTS (NWEIGHTS?). This number of impulse response weights is also used for calculating the residuals, e_t:

$$e_t = y_t - v_0 x_t - v_1 x_{t-1} - \ldots - v_g x_{t-g}$$

with v_i the impulse response weights and g = NWEIGHTS.

As discussed in Section C.2.2, as a default the PRINTed output will be printed on the TeleTYpe. If this default option must be modified the user should make use of the CTRL/E option at the moment of answering the prompt OUTPUTS:.

If PLOTted outputs are requested, the user will again be given the option to specify where plots have to go by answering the question "PLOTTED OUTPUT: TTY, LPT, OR DSK?".

Finally, at the moment of filing prewhitened data, the user will be prompted for the file name for the prewhitened dependent and each explanatory variable.

C.5.1.2 PROGRAM CROSSC OUTPUT

Figure C.12 reproduces the output created by the specifications given in Figure C.10. Program CROSSC first prints information based on the input section, such as model differencing, mean (MEAN), Standard Deviation (SD), Net number of OBServations after differencing (NOBS) for each variable, as well as initial values of the parameters in the prewhitening model. These descrip-

FIGURE C.10   Program CROSSC Input.

```
PROGRAM? CROSSC

VERSION:    1-Jan-83
TITLE? LYDIA PINKHAM
AQD FILE? N
NVBLS? 2
SMPL? 1 40


DEPENDENT VBL
FILE NAME? SALPNK
LOG? N
CDIF? 1
SEAS? N
PREWHITEN DEP. VBL. WITH OWN PROCESS? N

EXPLANATORY VBL: 1
FILE NAME? ADVPNK
LOG? N
CDIF? 1
SEAS? N

CONSTANT? N
ORDERS AR PARAM? 1 2
ESTIMATE  1? .074
ESTIMATE  2? -.407
ORDERS MA PARAM? N

OUTPUTS:    ?
ALL DESIRED OUTPUTS MUST BE SPECIFIED IN A SINGLE LIST.
AVAILABLE OPTIONS:
    1   PRINT PREWHITENED AUTOCORR. AND CROSS CORR.
    2     "    PREWHITENED DATA
    3     "    IMPULSE RESPONSE WEIGHTS
    4     "    RESIDUAL AUTOCORR. AND PARTIAL AUTOCORR.
    5     "    RESIDUALS
    6   PLOT PREWHITENED AUTOCORR. AND CROSS CORR.
    7     "    PREWHITENED DATA
    8     "    RESIDUAL AUTOCORR. AND PARTIAL AUTOCORR.
    9     "    RESIDUALS
   10   FILE PREWHITENED DATA
   11     "    RESIDUALS

OUTPUTS:  1 3 4
NR CROSS COR.? 12
NWEIGHTS? 10
```

FIGURE C.11   Program CROSSC Input: Discussion.

PROGRAM REQUEST	USER RESPONSE	EXPLANATION OF REQUEST/ USER RESPONSE
AQD FILE?	N(O)	The user does not have the data in an AQD file, but has the data in separate .BJ files.
NVBL?	2	The total number of variables to be analyzed, specifically, the dependent variable plus the number of explanatory variables.
		If the user answers Y(ES) to the question AQD FILE?, the program next will prompt for the AQD workfile name, followed by a request to specify the indices of the variables to be analyzed. The user then must first specify the index of the dependent variable, followed by the indices of the explanatory variables.
SMPL?	1 40	We decided to use only the first 40 data points.
		Note that only one SMPL can be used for all the variables. If, therefore, some variable must be transformed with a different difference operator, the answer to SMPL? should be guided by the variable with the greatest difference operator and all other variables should have eventually 0's inserted at the beginning of the file to compensate for a different initial length. If, say, advertising has $D = 1$, and sales should not be differenced, then the first observation will not be used even if the user specifies SMPL? 1 40.

FIGURE C.11 (*continued*)

PROGRAM REQUEST	USER RESPONSE	EXPLANATION OF REQUEST/ USER RESPONSE
DEPENDENT VBL? FILE NAME?	SALPNK	These questions are similar to those prompted by the program ESTIMA.
LOG? CDIF?	N(0) 1	
PREWHITEN DEP. VBL.	WITH OWN PROCESS? N(0)	The user can request that the dependent variable should be prewhitened with its own process rather than with the process of an explanatory variable. This is particularly important for causality analysis.
		If the user were to answer Y(ES), the program would then prompt for the parameter information as done below for transfer function parameters of the explanatory variable.
EXPLANATORY VBL: FILE NAME? LOG? CDIF? SEAS?	1 ADVPNK N(0) 1 N(0)	The user specifies the advertising data-file, followed by the stationary transformations.
CONSTANT? ORDERS AR PARAM? ESTIMATE 1? ESTIMATE 2? ORDERS MA PARAM?	N(0) 1 2 .074 −.407 N(0)	Next, the particular prewhitening model is specified as an AR(2) process. If the user had answered Y(ES) to SEAS?, the program also would have prompted for seasonal AR and MA parameters.
		The program CROSSC has the same flexibilities in specifying the parameters as the univariate programs; that is, parameters also can be suppressed.

FIGURE C.12   Program CROSSC Output.

```
SMPL:      1   40

DEPENDENT VBL: SALPNK
SERIES WITH D = 1 DS = 0 S = 0
MEAN =     29.77    SD =     258.0   (NOBS = 39)

EXPLANATORY VBL: ADVPNK
SERIES WITH D = 1 DS = 0 S = 0
MEAN =     10.36    SD =     247.1   (NOBS = 39)

EXPLANATORY VBL: ADVPNK

INITIAL VALUES
 AR(  1)   0.7400E-01
 AR(  2)  -0.4070E+00

AUTOCORRELATIONS OF PREWHITENED DEPENDENT VBL USING VBL ADVPNK
  LAGS ROW SE
  1-12  .16   .47   .42   .18   .04   .05  -.16  -.14  -.21  -.27  -.13  -.29  -.23

CHI-SQUARE TEST          P-VALUE
Q(12) =  34.0      12 D.F.   .001

CROSS CORRELATIONS WITH VBL ADVPNK   STANDARD ERROR =  .16
ZERO LAG =  .48
  LEADS          FUTURE          PY(t),PX(t+k)
  1-12           .40   .15   .23   .18   .21  -.05   .08   .04   .01   .02  -.13  -.23

CHI-SQUARE TEST          P-VALUE
Q(12) =  17.6      12 D.F.   .127

  LAGS           PAST            PY(t),PX(t-k)
  1-12           .30   .10  -.09   .01  -.24  -.24  -.26  -.13  -.07  -.17  -.02  -.17

CHI-SQUARE TEST          P-VALUE
Q(12) =  17.4      12 D.F.   .135
TWO SIDED CHI-SQUARE TEST  P-VALUE
Q(25) =  44.1      25 D.F.   .011

IMPULSE RESPONSE WEIGHTS OF VBL ADVPNK
  LEADS          FUTURE          V(-k)
  1 -6   0.5054      0.1879      0.2840      0.2249      0.2671    -6.3947E-02
  7-10   0.1056      4.9136E-02 1.4940E-02 2.1765E-02

ZERO LAG =   0.6009
  LAGS           PAST            V(+k)
  1 -6   0.3707      0.1264     -0.1121     7.7438E-03-0.3035      -0.3041
  7-10  -0.3292     -0.1631     -9.3266E-02-0.2161
```

FIGURE C.12 (*continued*)

```
29 RESIDUALS USING CURRENT AND 10 PAST IMPULSE WEIGHTS OF VBL ADVPNK

AUTOCORRELATIONS OF RESIDUALS USING VBL ADVPNK
   LAGS ROW SE
   1-12   .19    .03 -.16 -.10   .13   .02 -.05   .08   .01 -.04   .05 -.08   .15

CHI-SQUARE TEST              P-VALUE
Q(12) =  3.88        10 D.F.   .953

PARTIAL AUTOCORRELATIONS      STANDARD ERROR =   .19
   LAGS
   1-12             .03 -.16 -.09   .12 -.02 -.03   .11 -.02 -.02   .08 -.13   .18

OPTION? ?
AVAILABLE OPTIONS:
   M CHANGE MODEL, SAME VARIABLE
   V    "       VARIABLE
   $ STOP

OPTION? $
```

tive statistics are the same as those given in other TS programs and are described in Section C.4.1.2.

The AUTOCORRELATIONS OF PREWHITENED DEPENDENT VBL are autocorrelations calculated based on the values of the prewhitened dependent variables as transformed by the explanatory variable 1 (advertising, ADVPNK) prewhitening model. If the user had requested that the dependent variable be prewhitened with its own process, the program CROSSC then would calculate the autocorrelations of such a transformed dependent variable and label the autocorrelations as such. The associated statistics (ROW SE, CHI-SQUARE TEST, P-VALUE) have also been explained in Section C.4.1.2.

The CROSS CORRELATIONS WITH VBL 1 (Advertising) gives the correlations between the prewhitened dependent variable (PY) and the prewhitened explanatory variable (PX) at different lag values. The lag values represent LEADS (FUTURE) as well as LAGS (PAST). Specifically the LEAD cross correlations give the correlations between PY_t and PX_{t+k} for different positive values of k and as such indicate, very roughly, the degree of prediction of future PX values that can be made by PY. Similarly, the LAG cross correlations measure the correlations between PY_t and PX_{t-k} for different positive values of k and as such measure the degree with which values of PY can be predicted by lagged PX. In addition the ZERO LAG cross correlation which measures the correlation between PY_t and PX_t is printed. The associated test statistics have the same interpretations as with the autocorrelations.

The IMPULSE RESPONSE WEIGHTS of the explanatory variable are proportional to the cross correlations, with the constant of proportionality equal to the ratio of the standard deviation of the prewhitened explanatory variable and the standard deviation of the prewhitened dependent variable. These impulse response weights represent the weights in a two-sided finite distributed lag model of PX on PY.

As specified above, the current and past impulse weights are then used in calculating the residuals. Output option 4 provides the autocorrelations and partial autocorrelations with associated test statistics based on these residuals.

The other options provide additional information. The prewhitened data, both the prewhitened dependent and the prewhitened explanatory variables, can be printed (Output option 2), plotted (Output option 7), and filed for future analysis (Output option 10). Similarly, the residuals calculated as described above can be printed (Output option 5) or plotted (Output option 9).

Under output option 6 we obtain a plot of the autocorrelation function and cross correlation function of output option 1. Note that there is no need to ask for PRINTed output options when the user only wants to see PLOTted options. Finally, under option 8, the residual autocorrelation function and partial autocorrelation function can be plotted. Each of the correlation function plots contains an approximate 95% large-sample confidence interval under the null hypothesis that the true order of the correlation process is one less than the order of the correlation being evaluated.

At the end of the output, the program will ask

OPTION?

allowing the user to reanalyze the same data but using a different model (option M), for example, to change the consecutive differencing, to restart the program CROSSC but with a new variable or a different subset of observations of the same variable (option V), or just to stop the program (option $).

C.5.2 Model Estimation

After identifying a potential transfer function model using the CROSSC program the user now must estimate and evaluate this model using the program TRFEST.

Contrary to the program ESTIMA, the program TRFEST does not make use of the back forecasting technique to generate initial values for the dependent variable and the residuals. The initial values for the dependent variable are set equal to the sample mean of the, possibly differenced, dependent variable, and the initial values for the residuals are set equal to their unconditional expected value of 0.

Suppose that, based on the information presented in the CROSSC program, we conclude that after first consecutive differencing both the dependent variable sales and the explanatory variable advertising, we ought to estimate

FIGURE C.13   Program TRFEST Input.

```
PROGRAM? TRFEST

VERSION:   1-Jan-83
TITLE? LYDIA PINKHAM
AQD FILE? N
NVBLS? 2
SMPL? 1 40

MODEL INFORMATION

DEPENDENT VBL
FILE NAME? SALPNK
LOG? N
CDIF? 1
SEAS? N

EXPLANATORY VBL: 1
FILE NAME? ADVPNK
LOG? N
CDIF? 1
SEAS? N
LAG? N

PREWHITENING? N

TRANSFER FUNCTION PARAMETERS:
ORDERS NUMERATOR PARAM? 0
GUESS VALUE  0? .6
ORDERS DENOMINATOR PARAM? 1
GUESS VALUE  1? .84

ERROR PARAMETERS

CONSTANT? N
ORDERS AR PARAM? 1
GUESS VALUE  1? .33
ORDERS MA PARAM? N

OUTPUTS: ?
ALL DESIRED OUTPUTS MUST BE SPECIFIED IN A SINGLE LIST.
AVAILABLE OPTIONS:
   1  PRINT ESTIMATES
   2    "    ITERATION RESULTS
   3    "    RESIDUALS
   4    "    RESIDUAL AUTOCORRELATIONS
   5    "    RESIDUAL CROSS CORRELATIONS
   6  PLOT RESIDUALS
   7    "   RESIDUAL AUTOCORRELATIONS
   8    "   RESIDUAL CROSS CORRELATIONS
   9    "   RESIDUAL PERIODOGRAM
  10    "   DISTRIBUTED LAG WEIGHTS
  11 FILE RESIDUALS

OUTPUTS: 1 4
NR AUTOCOR.? 12
```

a transfer function model with a zero-order numerator polynomial, NUM(0), a first-order denominator polynomial, DEN(1), no constant, and a first-order residual autoregressive process, AR(1). We select this model only to demonstrate some of the capabilities of the program TRFEST, and the reader should not assume that this formulation is the best representation of the Lydia Pinkham vegetable compound.

C.5.2.1 PROGRAM TRFEST INPUT

The input section of program TRFEST is given in Figure C.13. The first group of questions about data file and model transformation of the dependent variable is similar to those questions asked by the program CROSSC. In the next group of questions (explained in Figure C.14) program TRFEST prompts the user for the model information of each explanatory variable, starting with the transformations and dead time. Next, the program prompts for the parameter specification of the transfer function and error process, as well as guess values to start off the nonlinear least squares optimization routine. As with the program ESTIMA, the program TRFEST uses the Marquardt iterative nonlinear optimizer.

C.5.2.2 PROGRAM TRFEST OUTPUT

Figure C.15 presents the results for output options 1, and 4. After printing the information based on the input section, such as model differencing, mean (MEAN), Standard Deviation (SD), and Net number of OBServations after differencing (NOBS) for each variable in the model, the program TRFEST specifies the transfer function model and initial values of each parameter. Next, the program TRFEST repeats NOBS, prints the total NumbeR of PARAMeters to be estimated, and gives the INITial value of the Sum of Squared Residuals (INIT SSR) as calculated based upon the specified guess values.

A successful termination of the nonlinear estimation is again specified as in the program ESTIMA (see Section C.4.2.2). Intermediate iteration results for the parameter estimates and SSR are obtained by requesting output option 2.

For any coefficient estimate, the program TRFEST prints again EST, SE, EST/SE and 95% CONF LIMITS, together with several measures of goodness of fit. For an explanation of these statistics, see Section C.4.2.2.

The labels used to denote the transfer function model parameters are:

NUM for numerator parameters,
SNM for seasonal numerator parameters,
DEN for denominator parameters, and
SDN for seasonal denominator parameters.

In the program TRFEST, there is, however, only one EST.RES.SD available, as no back forecasted residuals are calculated. Also, because TRFEST

FIGURE C.14   Program TRFEST Input: Discussion.

PROGRAM REQUEST	USER RESPONSE	EXPLANATION OF REQUEST/ USER RESPONSE
EXPLANATORY VBL: 1		
LOG?	N(0)	The first three prompts should, by now, be familiar to the user.
CDIF?	1	
SEAS?	N(0)	
LAG?	N(0)	The prompt LAG? allows the user to specify a value for the "dead time", the number of time intervals before we expect the explanatory variable to have any impact on the dependent variable. If there are no delays, answer N(O) or 0.
PREWHITENING?	N(0)	If the user wants to have the explanatory variables prewhitened for residual cross correlation analysis, answer Y(ES). If the user does answer Y(ES), the program then will prompt for the univariate ARIMA process and its parameter values to be used for the prewhitening of the explanatory variables.
TRANSFER FUNCTION PARAMETERS: ORDERS NUMERATOR PARAM? 0		Because the transfer function model is normalized such that the zero-order coefficient has a value different from 1 in the regular, as opposed to the seasonal, numerator polynomial, there should always be at least a zero-order (regular) numerator polynomial. If, however, there is a "dead time", the user should specify this in an answer to the above-discussed prompt LAG?
GUESS VALUE 0?	.6	We specify that the guess value to start off the optimization is 0.6 for the zero-order numerator parameter. This value can be derived from the values of the impulse weights calculated in the program CROSSC.

PROGRAM REQUEST	USER RESPONSE	EXPLANATION OF REQUEST/ USER RESPONSE
		As with the program ESTIMA, these guess values are automatically evaluated to see if the polynomial is stationary or invertible. Eventually alternative guess values are requested.
ORDERS DENOMINATOR PARAM? 1 GUESS VALUE 1? .84		We specify that the first-order denominator parameter is to be included and initialized with guess value = 0.84.
ERROR PARAMETERS		If there were more explanatory variables present in the transfer function model, the program first would prompt for the transformations, transfer function model parameters and guess values of these other explanatory variables.
CONSTANT? ORDERS AR PARAM? GUESS VALUE 1? ORDER MA PARAM?	N(0) 1 .33 N(0)	We specify that there is no constant in the model and that the error process has to include a first-order autoregressive parameter with guess value = 0.33, and no moving average parameters.
OUTPUTS:		In answer to this prompt, the user selects the specific output options for this model.

FIGURE C.15   Program TRFEST Output.

```
SMPL:     1   40

DEPENDENT VBL: SALPNK
SERIES WITH D = 1 DS = 0 S = 0
MEAN =    29.77    SD =    258.0    (NOBS = 39)

EXPLANATORY VBL: ADVPNK
SERIES WITH D = 1 DS = 0 S = 0
MEAN =    10.36    SD =    247.1    (NOBS = 39)
```

FIGURE C.15 (*continued*)

```
INITIAL VALUES

EXPLANATORY VBL: ADVPNK

TRANSFER FUNCTION PARAMETERS
NUM( 0)   0.6000E+00
DEN( 1)   0.8400E+00

ERROR PARAMETERS
 AR( 1)   0.3300E+00

NOBS = 39 ; NR OF PARAM = 3

INIT SSR =   0.2325E+07
REL. CHANGE IN SSR <=   0.1000E-05
FINAL SSR =   0.1581E+07
8 ITERATIONS

PARAMETER ESTIMATES
====================
             EST     SE      EST/SE    95% CONF LIMITS
NUM( 0)     0.520   0.156    3.332     0.214    0.825    ADVPNK
DEN( 1)     0.359   0.249    1.439    -0.130    0.847
 AR( 1)     0.256   0.195    1.317    -0.125    0.638

EST.RES.SD =   2.1254E+02

R SQR =   .399
DF = 35
NO F-STAT, NO INTERCEPT

CORRELATION MATRIX
         NUM( 0) DEN( 1)
 DEN( 1)   .046
  AR( 1)  -.458   -.107

ROOTS   REAL  IMAGINARY  SIZE
DEN    2.788   0.000     2.788   ADVPNK
 AR    3.900   0.000     3.900

AUTOCORRELATIONS OF RESIDUALS
  LAGS ROW SE
  1-12  .16  -.04   .12   .02  -.09   .11  -.01  -.04  -.23  -.21   .03  -.22   .04

CHI-SQUARE TEST            P-VALUE
Q(12) = 9.62        11 D.F.   .565

OPTION? ?
AVAILABLE OPTIONS:
    M  CHANGE MODEL, SAME VARIABLES
    V     "     VARIABLES AND/OR DATAFILE
    $  STOP

OPTION? $
```

does no back forecasting, the degrees of freedom are no longer equal to the difference between NOBS and the number of parameters, but NOBS is decreased by the sum of the dead time value (LAG) and the maximum numerator polynomial order and the maximum error autoregressive polynomial order. In our example, DF = 36 is calculated as NOBS (39) − LAG (0) − NR OF PARAM (3) − maximum order AR (1). Finally, under option 1, TRFEST provides an estimate of the sampling correlation matrix of the coefficient estimates.

Next, as another check on the model adequacy, the program prints out the roots of the estimated polynomials as well as the size of these roots. For stationarity and invertibility to hold, the size of each root should be larger than one.

Under option 4, the program gives the residual autocorrelations and associated Chi-square test for model adequacy. Under option 5, the user obtains the cross correlations between the residuals and a (possibly) prewhitened, stationary explanatory variable. Options 4 and 5 both are used to evaluate model inadequacy. If the transfer function model is correct, but the error model is incorrect, we expect the residuals to be autocorrelated. If the transfer function model is incorrect, then the residuals would be cross correlated with past (LAGS) values of the stationary explanatory variables as well as being autocorrelated.

The other TRFEST output options contain the following information:

Option 2: Iteration results. Under this option, program TRFEST prints out the values of the estimates of the coefficients as well as the SSR at each iteration.

The estimates of the white noise residuals are printed under option 3, are plotted under option 6, and can be filed with option 11.

Options 6 and 7 provide plots of the residual autocorrelation function and residual cross-correlation function.

Option 9 allows the user to obtain a plot of the normalized cumulative periodogram of the residuals together with 90% large-sample Kolmogorov–Smirnov confidence limits. For further information about this output, see Section 4.2.2.

Finally, under option 10 a plot is obtained of the distributed lag weights of each explanatory variable. These distributed lag weights are obtained by solving the ratio of the numerator polynomial and the denominator polynomial in terms of a general (infinite) distributed lag polynomial.

At the end of the output, the program TRFEST will again ask for

OPTION?

The available options are M (change model, same variables), option V (change variables and/or data file), and $ for stop.

C.5.3 Model Forecasting

After defining a suitable model, program TRFFOR is used to generate forecasts. In generating forecasts for the dependent variable program TRFFOR can either forecast the future values of each and every explanatory variable or can use future values stored in the data file of these explanatory variables.

C.5.3.1 PROGRAM TRFFOR INPUT

Figure C.16 reproduces the input section of the TRFFOR program. Because most of these input questions are identical to questions asked in other programs, we will only summarize these questions and refer to these earlier programs. The first group of questions is identical to those questions asked in the program TRFEST, with the exception of the additional questions about the number of observations (NOBS USED IN TRFEST?.) This is done so that the user can compare forecasted values with actual data. We refer to the program FRCAST, Section C.4.3.1, for a more detailed discussion.

Next the program asks the user if the explanatory variables must be forecasted. If the user specifies NO, the program will read the future data from the data file specified above in the input section of the program. Here the user specifies that the explanatory variable should be forecasted using an AR(4) univariate time series model. Consequently, the program prompts the user for the ESTIMATEs of the autoregressive parameters.

The next series of questions allows the user to specify the specific transfer function model. In Figure C.16, we specified a second order numerator model, NUM(2), and an AR(1) process for the error term, with no constant term in the model.

FIGURE C.16   Program TRFFOR Input.

```
VERSIO      1-Jan-83
TITLE? LYDIA PINKHAM FORECAST
AQD FILE? N
NVBLS? 2
SMPL? 1 54

MODEL INFORMATION
NOBS USED IN TRFEST? 40

DEPENDENT VBL
FILE NAME? SALPNK
LOG? N
CDIF? 1
SEAS? N
```

```
EXPLANATORY VBL: 1
FILE NAME? ADVPNK
LOG? N
CDIF? 1
SEAS? N
LAG? N


VBL -ADVPNK- TO BE FORECASTED? Y

CONSTANT? N
ORDERS AR PARAM? 1 2 3 4
ESTIMATE  1? .08296
ESTIMATE  2? -.2587
ESTIMATE  3? -.08765
ESTIMATE  4? .3567
ORDERS MA PARAM? N

TRANSFER FUNCTION PARAMETERS:
ORDERS NUMERATOR PARAM? 0 1
GUESS VALUE  0? .48049917
GUESS VALUE  1? -.18872084
ORDERS DENOMINATOR PARAM? N

ERROR PARAMETERS

CONSTANT? N
ORDERS AR PARAM? 1
GUESS VALUE  1? .27706643
ORDERS MA PARAM? N

FORECAST INFORMATION :
CALENDAR DATE OF FIRST OBS.? 1907
HORIZON? 14
FIRST ORIGIN? 40
LAST ORIGIN? 40

OUTPUTS:  ?
ALL DESIRED OUTPUTS MUST BE SPECIFIED IN A SINGLE LIST.
AVAILABLE OPTIONS:
   1  PRINT FORECASTS WITH 95% CONFIDENCE INTERVAL
   2    "    ERROR ANALYSIS (THEIL DECOMPOSITION)
   3    "    FORECAST STANDARD ERRORS AND ERROR LEARNING COEFF.
   4    "    ANTILOG FORECASTS (ONLY IF LOGTRANSFORMATION ASKED)
   5    "    EXPLANATORY VBL FORECASTS
   6  PLOT FORECASTS

OUTPUTS:   1 5
```

Finally, the last group of questions, labeled FORECAST INFORMA-TION, is again identical to those specified in the program FRCAST; see Section C.4.3.1.

The different OUTPUT options are discussed in the next section.

C.5.3.2 PROGRAM TRFFOR OUTPUT

Figure C.17 presents the output of OUTPUT option 1 and 5 of the program TRFFOR. After reproducing part of the input section, program TRFFOR lists for each observation in the forecast horizon the point forecast, the upper and lower limits of a 95% large-sample confidence interval, and, if actual observations are available, actual, error (actual minus forecast), and percentage error values. Also as a result of requesting OUTPUT option 5, the program prints in Figure C.17 the point forecasts for the explanatory variable ADVPNK, using the AR(4) model specified in the input section.

The TRFFOR output options 2, 3, 4, and 6 correspond to those specified in the TS program FRCAST and have been discussed in Section C.4.3.2 above.

In the program TRFFOR there are no other final OPTIONs but the STOP option.

FIGURE C.17   Program TRFFOR Output.

```
HORIZON 14 PERIOD(S)
STARTING AT 40
SMPL:      1    54

DEPENDENT VBL: SALPNK
SERIES WITH D = 1 DS = 0 S = 0
MEAN =     29.77    SD =     258.0    (NOBS = 39)

EXPLANATORY VBL: ADVPNK
SERIES WITH D = 1 DS = 0 S = 0
MEAN =     10.36    SD =     247.1    (NOBS = 39)

INITIAL VALUES

EXPLANATORY VBL: ADVPNK
PREWHITENING PARAMETERS
  AR( 1)    0.8296E-01
  AR( 2)   -0.2587E+00
  AR( 3)   -0.8765E-01
  AR( 4)    0.3567E+00
```

```
TRANSFER FUNCTION PARAMETERS
NUM( 0)   0.4805E+00
NUM( 1)  -0.1887E+00

ERROR PARAMETERS
 AR( 1)   0.2771E+00
```

OBS	95% LOWER CONF LIMIT	FORECAST	95% UPPER CONF LIMIT	ACTUAL	ERROR (A-F)	%ERROR (A-F)/A

```
ORIGIN = 1946.

EXPLANATORY VBL: ADVPNK
1947.                1035.
1948.                1045.
1949.                1067.
1950.                1017.
1951.                1014.
1952.                1028.
1953.                1043.
1954.                1022.
1955.                1015.
1956.                1023.
1957.                1033.
1958.                1025.
1959.                1018.
1960.                1022.
```

OBS	95% LOWER CONF LIMIT	FORECAST	95% UPPER CONF LIMIT	ACTUAL	ERROR (A-F)	%ERROR (A-F)/A
1947.	1579.	2051.	2523.	1920.	-130.7	-6.809
1948.	1248.	2029.	2810.	1910.	-118.9	-6.227
1949.	1029.	2033.	3036.	1984.	-48.89	-2.464
1950.	830.8	2011.	3190.	1787.	-223.5	-12.51
1951.	651.3	1999.	3347.	1689.	-310.0	-18.36
1952.	495.7	2005.	3515.	1866.	-139.3	-7.467
1953.	364.4	2015.	3665.	1896.	-118.8	-6.268
1954.	234.5	2008.	3781.	1684.	-323.8	-19.23
1955.	108.3	2000.	3892.	1633.	-367.3	-22.49
1956.	-6.435	2003.	4012.	1657.	-345.9	-20.88
1957.	-109.7	2009.	4128.	1569.	-440.1	-28.05
1958.	-212.0	2007.	4226.	1390.	-617.1	-44.40
1959.	-313.3	2002.	4318.	1387.	-615.4	-44.37
1960.	-408.4	2003.	4414.	1289.	-714.0	-55.39

```
OPTION? ?
AVAILABLE OPTIONS:
    $  STOP
```

REFERENCES

Aaronson, D. E., Dienes, C. T., and M. C. Musheno (1978). "Changing the Public Drunkenness Laws: the Impact of Decriminalization." *Law and Society Review, 12,* 3 (Spring), 405–436.

Anderson, R. L. (1942). "Distribution of the Serial Correlation Coefficient." *The Annals of Mathematical Statistics, 13,* 1 (March), 1–13.

Anderson, T. W. (1978). "Repeated Measurements on Autoregressive Processes." *Journal of the American Statistical Association, 73,* 362 (June), 371–378.

Ansley, C. F. (1979). "An Algorithm for the Exact Likelihood of a Mixed Autoregressive-Moving Average Process." *Biometrika, 66,* 1 (April), 59–65.

Ashley, R., Granger, C. W. J., and R. Schmalensee (1980). "Advertising and Aggregate Consumption: An Analysis of Causality." *Econometrica, 48,* 5 (July), 1149–1167.

Bartlett, M. S. (1935). "Some Aspects of the Time-Correlation Problem in Regard to Tests of Significance." *Journal of the Royal Statistical Society, 98,* 3, 536–543.

Bartlett, M. S. (1946). "On the Theoretical Specification and Sampling Properties of Autocorrelated Time-Series." *Journal of the Royal Statistical Society, Series B, 8,* 1 (April), 27–41, 85–97.

Bartlett, M. S. (1966). *An Introduction to Stochastic Processes with Special Reference to Methods and Applications* (2nd Ed.). Cambridge, England: Cambridge University Press.

Box, G. E. P., and D. R. Cox (1964). "An Analysis of Transformation." *Journal of the Royal Statistical Society, Series B, 26,* 2, 211–252.

Box, G. E. P., and G. M. Jenkins (1976). *Time Series Analysis: Forecasting and Control* (Rev. Ed.). San Francisco: Holden–Day, Inc. (First Ed., 1970.)

Box, G. E. P., and P. Newbold (1971). "Some Comments on a Paper of Coen, Gomme, and

Kendall." *Journal of the Royal Statistical Society, Series A, 134,* 2, 229–240.

Box, G. E. P., and D. A. Pierce (1970). "Distribution of Residual Autocorrelations in Autoregressive–Integrated Moving Average Time Series Models." *Journal of the American Statistical Association, 65,* 332 (December), 1509–1526.

Box, G. E. P., and G. C. Tiao (1975). "Intervention Analysis with Applications to Economic and Environmental Problems." *Journal of the American Statistical Association, 70,* 349 (March), 70–79.

Brigham, E. O. (1974). *The Fast Fourier Transform.* Englewood Cliffs. NJ: Prentice-Hall, Inc.

Brubacher, S. R., and G. T. Wilson (1976). "Interpolating Time Series with Application to the Estimation of Holiday Effects on Electricity Demand." *The Journal of the Royal Statistical Society, Series C (Applied Statistics), 25,* 2, 107–116.

Caines, P. E., Sethi, S. P., and T. W. Brotherton (1977). "Impulse Response Identification and Causality Detection for the Lydia-Pinkham Data." *Annals of Economic and Social Measurement, 6,* 2 (Spring), 147–163.

Campbell, D. T., and H. L. Ross (1968). "The Connecticut Crackdown on Speeding: Time-Series Data in Quasi-experimental Analysis." *Law and Society Review, 3,* 1 (August), 33–53.

Chan, K. H., Hayya, J. C., and J. K. Ord (1977). "A Note on Trend Removal Methods: the Case of Polynomial Regression Versus Variate Differencing." *Econometrica, 45,* 3 (April), 737–744.

Chatfield, C., and D. L. Prothero (1973). "Box–Jenkins Seasonal Forecasting: Problems in a Case-study." *Journal of the Royal Statistical Society, Series A, 136,* 3, 295–352.

Chiang, A. C. (1967). *Fundamental Methods of Mathematical Economics.* New York: McGraw-Hill Book Company.

Clarke, D. G. (1976). "Econometric Measurement of the Duration of the Advertising Effect on Sales." *Journal of Marketing Research, 13,* (November), 345–357.

Cleveland, W. P., and G. C. Tiao (1976). "Decomposition of Seasonal Time Series: A Model for the Census X-11 Program." *Journal*

of the American Statistical Association, 71, 355 (September), 581–587.

Cleveland, W. S., and S. J. Devlin (1980). "Calendar Effects in Monthly Time Series: Detection by Spectrum Analysis and Graphical Methods." *Journal of the American Statistical Association, 75,* 371 (September), 487–496.

Cleveland, W. S., and S. J. Devlin (1982). "Calendar Effects in Monthly Time Series: Modeling and Adjustment." *Journal of the American Statistical Association, 77,* 379 (September), 520–528.

Cleveland, W. S., Dunn, D. M., and I. J. Terpenning (1978). "SABL: A Resistant Seasonal Adjustment Procedure with Graphical Methods for Interpretation and Diagnosis." *In:* Zellner, A. (Ed.) *Seasonal Analysis of Economic Time Series.* U.S. Department of Commerce, Economic Research Report (ER-1), 201–231.

Dagum, E. B. (1978). "Modelling, Forecasting, and Seasonally Adjusting Economic Time Series with the X-11 ARIMA Method." *The Statistician, 27,* 203–216.

Davies, N., and P. Newbold (1979). "Some Power Studies of a Portmanteau Test of Time Series Model Specification." *Biometrika, 66,* 1 (April), 153–155.

Davies, N., Triggs, C. M., and P. Newbold (1977). "Significance Levels of the Box–Pierce Portmanteau Statistic in Finite Samples." *Biometrika, 64,* 3 (December), 517–522.

DeGroot, M. H. (1973). "Doing What Comes Naturally: Interpreting a Tail Area as a Posterior Probability or as a Likelihood Ratio." *Journal of the American Statistical Association, 68,* 344 (December), 966–969.

DeGroot, M. H. (1975). *Probability and Statistics.* Reading, MA: Addison-Wesley.

Deutsch, S. J., and F. B. Alt (1977). "The Effect of Massachusetts' Gun Control Law on Gun-related Crimes in the City of Boston." *Evaluation Quarterly, 1,* 4 (November), 544–568.

Dickey, J. M. (1977). "Is the Tail Area Useful as an Approximate Bayes Factor?" *Journal of the American Statistical Association, 72,* 357 (March), 138–142.

Durbin, J. (1960). "The Fitting of Time-Series

Models." *Review of the International Institute of Statistics, 28,* 3, 233–244.

Fama, E. F. (1970). "Efficient Capital Markets: A Review of Theory and Empirical Work." *Journal of Finance, 25,* 2 (May), 383–417.

Fox, A. J. (1972). "Outliers in Time Series." *Journal of the Royal Statistical Society, Series B, 34,* 3, 350–363.

Geurts, M. D., and I. B. Ibrahim (1975). "Comparing the Box–Jenkins Approach with the Exponentially Smoothed Forecasting Model Application to Hawaii Tourists." *Journal of Marketing Research, 12,* 2 (May), 182–188.

Geweke, J. (1983). "Inference and Causality in Economic Time Series Models." *In:* Griliches, Z. and M. Intrilligator (Eds.) *Handbook of Econometrics.* Volume II. Amsterdam: North-Holland (in press).

Gibbons, J. D., and J. W. Pratt (1975). "*P*-Values: Interpretation and Methodology." *The American Statistician, 29,* 1 (February), 20–25.

Glass, G. V. (1968). "Analysis of Data on the Connecticut Speeding Crackdown as a Time-Series Quasi-experiment." *Law and Society Review, 3,* 1 (August), 55–76.

Glass, G. V., Wilson, V. L., and J. M. Gottman (1975). *Design and Analysis of Time-Series Experiments.* Colorado: Colorado University Press.

Granger, C. W. J. (1963). "The Effect of Varying Month-length on the Analysis of Economic Time-Series." *L'Industria, 1,* 3, 41–53.

Granger, C. W. J. (1977). "Comment on 'Relationships—and the Lack Thereof—Between Economic Time Series, with Special Reference to Money and Interest Rates' by D. A. Pierce." *Journal of the American Statistical Association, 72,* 357 (March), 22–23.

Granger, C. W. J. (1980). "Testing for Causality: A Personal Viewpoint." *Journal of Economic Dynamics and Control, 2,* 4 (November), 329–352.

Granger, C. W. J., and P. Newbold (1973). "Some Comments on the Evaluation of Economic Forecasts." *Applied Economics, 5,* 1 (March), 35–47.

Granger, C. W. J., and P. Newbold (1977). *Forecasting Economic Time Series.* New York: Academic Press.

Gregg, J. R., Hossell, C. V., and J. T. Richardson (1964). "Mathematical Trend Curves: An Aid to Forecasting." I.C.I. Monograph No. 1. Edinburgh, Scotland: Oliver and Boyd.

Griliches, Z. (1967). "Distributed Lags: A Survey." *Econometrica, 35,* 1 (January), 16–49.

Hanssens, D. (1977). "An Empirical Study of Time-Series Analysis in Marketing Model Building." Unpublished Ph.D. thesis, Purdue University, Krannert Graduate School of Industrial Administration.

Harrison, P. J., and S. F. Pearce (1972). "The Use of Trend Curves as an Aid to Market Forecasting." *Industrial Marketing Management, 1,* 2 (January), 149–170.

Haugh, L. D. (1976). "Checking the Independence of Two Covariance-Stationary Time Series: A Univariate Residual Cross-Correlation Approach." *Journal of The American Statistical Association, 71,* 354 (June), 378–385.

Helmer, R. M., and J. K. Johansson (1977). "An Exposition of the Box–Jenkins Transfer Function Analysis with an Application to the Advertising-Sales Relationship." *Journal of Marketing Research, 14,* 2 (May), 227–239.

Izenman, A. J., and S. L. Zabel (1981). "Babies and the Blackout: The Genesis of a Misconception." *Social Science Research, 10,* 3 (September), 282–299.

Jenkins, G. M. (1979). *Practical Experiences with Modelling and Forecasting Time Series.* Jersey, Channel Islands: A GJP Publication.

Jenkins, G. M., and D. G. Watts (1968). *Spectral Analysis and its Applications.* San Francisco: Holden-Day, Inc.

Jorgenson, D. W. (1966). "Rational Distributed Lag Functions." *Econometrica, 32,* 1 (January), 135–149.

Judge, G. G., Griffiths, W. E., Hill, R. C., and T-C. Lee (1980). *The Theory and Practice of Econometrics.* New York: John Wiley & Sons.

Kendall, M. G. (1976). *Time Series* (2nd Ed.). London: Charles Griffin and Company, Ltd. (First Ed., 1973)

Kleiner, B., Martin, R. D., and D. J. Thomson (1979). "Robust Estimation of Power Spectra." *Journal of the Royal Statistical Society, Series B, 41,* 3, 313–351.

Kyle, P. W. (1978). "Lydia Pinkham Revisited:

A Box–Jenkins Approach." *Journal of Advertising Research, 18,* 2 (April), 32–39.

Ledolter, J. (1976). "ARIMA Models and Their Use in Modelling Hydrologic Sequences." International Institute for Applied Systems Analysis Research Memorandum, RM-76-69 (September).

Leskinen, E., and T. Teräsvirta (1976). "Forecasting the Consumption of Alcoholic Beverages in Finland: A Box–Jenkins Approach." *European Economic Review, 8,* 5 (December), 349–369.

Leuthold, R. M., MacCormick, A. J. A., Schmitz, A., and D. G. Watts (1970). "Forecasting Daily Hog Prices and Quantities—A Study of Alternative Forecasting Techniques." *Journal of the American Statistical Association, 65,* 329 (March), 90–107.

Levenbach, H., and B. E. Reuter (1976). "Forecasting Trending Time Series with Relative Growth Rate Models." *Technometrics, 18,* 3 (August), 261–272.

Ljung, G. M., and G. E. P. Box (1978). "On a Measure of Lack of Fit in Time Series Models." *Biometrika, 65,* 2 (August), 297–303.

Maddala, G. S. (1977). *Econometrics.* New York: McGraw-Hill Book Co.

Marquardt, D. W. (1963). "An Algorithm for Least Squares Estimation of Nonlinear Parameters." *Journal of the Society of Industrial and Applied Mathematics, 11,* 2 (June), 431–441.

Martin, R. D. (1980). "Robust Methods for Time Series." Mimeographed. U. of Washington, Department of Statistics, Seattle, WA.

Martin, R. D., Samarov, A., and W. Vandaele (1983). "Robust Methods for ARIMA Models." In *Applied Times Series Analysis of Economic Data,* Proceedings of the Conference on Applied Time Series Analysis of Economic Data, October 13–15, 1981. U.S. Department of Commerce, Bureau of Census, Economic Research Report (ER-5).

McCleary, R., and R. A. Hay, Jr. (1980). *Applied Time Series Analysis for the Social Sciences.* Beverly Hills, CA: Sage Publications.

McSweeny, A. J. (1978). "Effects of Response Cost on the Behavior of a Million Persons: Charging for Directory Assistance in Cincin-

nati." *Journal of Applied Behavior Analysis, 11,* 1 (Spring), 47–51.

Montgomery, D. C., and L. A. Johnson (1976). *Forecasting and Time Series Analysis.* New York: McGraw-Hill Book Company.

Morris, M. J. (1977). "Forecasting the Sunspot Cycle." *Journal of the Royal Statistical Society, Series A, 140,* 4, 437–448.

Mosteller, F., and J. W. Tukey (1977). *Data Analysis and Regression: A Second Course in Statistics.* Reading, MA: Addison-Wesley.

Nelson, C. R. (1973). *Applied Time Series Analysis for Managerial Forecasting.* San Francisco: Holden–Day, Inc.

Nelson, C. R. (1976). "The Interpretation of R^2 in Autoregressive-Moving Average Time Series Models." *The American Statistician, 30,* 4 (November), 175–180.

Nelson, C. R., and H. Kang (1981). "Spurious Periodicity in Inappropriately Detrended Time Series." *Econometrica, 49,* 3 (May), 741–751.

Palda, K. (1964). *The Measurement of Cumulative Advertising Effects.* Englewood Cliffs, NJ: Prentice-Hall.

Palm, F., and A. Zellner (1980). "Large Sample Estimation and Testing Procedures for Dynamic Equation Systems." *Journal of Econometrics, 12,* 3 (April), 251–283.

Pierce, D. A. (1972). "Residual Correlations and Diagnostic Checking in Dynamic-Disturbance Time Series Models." *Journal of the American Statistical Association, 67,* 339 (September), 636–640.

Pierce, D. A. (1980). "A Survey of Recent Developments in Seasonal Adjustment." *The American Statistician, 34,* 3 (August), 125–134.

Pierce, D. A., and L. D. Haugh (1977). "Causality in Temporal Systems: Characterizations and a Survey." *Journal of Econometrics, 5,* 3 (May), 265–293.

Plosser, C. I., and G. Wm. Schwert (1977). "Estimation of a Non-invertible Moving Average Process. The Case of Overdifferencing." *Journal of Econometrics, 6,* 2 (September), 199–224.

Press, S. J. (1982). *Applied Multivariate Analysis: Using Basenian and Frequentist Methods of Inference* (2nd Ed.). Melbourne, FL: Robert E. Krieger Publishing Co.

Quenouille, M. H. (1949). "Approximate Tests of Correlation in Time-Series." *Journal of the Royal Statistical Society, Series B, 11,* 1, 68–84.

Quenouille, M. H. (1957). *Analysis of Multiple Time Series.* New York: Hafner Publishing Co.

Quenouille, M. H. (1958). "Discrete Autoregressive Schemes with Varying Time-intervals." *Metrika, 1,* 21–27.

Roberts, H. V. (1974). *Conversational Statistics.* Palo Alto, CA: The Scientific Press.

Ross, H. L., Campbell, D. T., and G. V. Glass (1970). "Determining the Effects of a Legal Reform: The British "Breathalyzer" Crackdown of 1967." *American Behavioral Scientist, 13,* 3 (January/February), 493–509.

Roy, R. (1977). "On the Asymptotic Behaviour of the Sample Autocovariance Function for an Integrated Moving Average Process." *Biometrika, 64,* 2 (August), 419–421.

Schlaifer, R. (1981). *User's Guide to the AQD Collection.* (8th Ed.) Boston, MA: Harvard Business School.

Shiskin, J., Young, A. H., and J. C. Musgrave (1967). "The X-11 Variant of the Census Method-II Seasonal Adjustment Program." Technical Paper No. 15. Washington, DC: U.S. Department of Commerce, Bureau of the Census.

Singleton, R. C. (1967). "On Computing the Fast Fourier Transform." *Communications of the ACM, 10,* 10 (October), 647–654.

Snedecor, G. W., and W. G. Cochran (1980). *Statistical Methods* (7th Ed.). Ames, IA: Iowa State University Press.

Thompson, H. E., and G. C. Tiao (1971). "Analysis of Telephone Data: A Case Study of Forecasting Seasonal Time Series." *The Bell Journal of Economics, 2,* 2 (Autumn), 515–541.

Tiao, G. C., Box, G. E. P., and W. J. Hamming (1975a). "Analysis of Los Angeles Photochemical Smog Data: A Statistical Overview." *Journal of Air Pollution Control Association, 25,* 3 (March), 260–268.

Tiao, G. C., Box, G. E. P., and W. J. Hamming (1975b). "A Statistical Analysis of the Los Angeles Ambient Carbon Monoxide Data 1955–1972." *Journal of Air Pollution Control Association, 25,* 11 (November), 1129–1136.

U.S. Department of Commerce. *1975 Business Statistics.* Bureau of Economic Analysis. The Biennial Supplement to the *Survey of Current Business,* and earlier supplements.

U.S. Department of Commerce. *Survey of Current Business.* Bureau of Economic Analysis.

Vatter, P. A., Bradley, S. P., Frey, S. C., Jr., and B. B. Jackson (1978). *Quantitative Methods in Management: Text and Cases.* Homewood, IL: Richard D. Irwin, Inc.

Walker, G. (1931). "On Periodicity in Series of Related Terms." *Proceedings of Royal Society London, A131,* 518–532.

Wallis, K. F. (1977). "Multiple Time Series Analysis and the Final Form of Econometric Models." *Econometrica, 45,* 6 (September), 1481–1498.

Wheelwright, S. C., and S. Makridakis (1980). *Forecasting Method for Management* (3rd Ed.). New York: John Wiley & Sons.

Wichern, D. W., and R. H. Jones (1977). "Assessing the Impact of Market Disturbances Using Intervention Analysis." *Management Science, 24,* 3 (November), 329–337.

Wold, H. O. (1938). *A Study in the Analysis of Stationary Time Series* (2nd Ed.: 1954). Uppsala Sweden: Almqvist and Wiksell Book Co.

Yule, G. U. (1926). "Why Do We Sometimes Get Nonsense-Correlations Between Time-Series?—A Study in Sampling and the Nature of Time-Series." *Journal of the Royal Statistical Society, 89,* 1, 1–64.

Yule, G. U. (1927). "On a Method for Investigating Periodicities in Distributed Series, With Special Reference to Wolfer's Sunspot Numbers." *Philosophical Transactions of the Royal Society London, Series A, 226,* (July), 267–298.

Zellner, A. (1979). "Causality and Econometrics." *Carnegie-Rochester Conference Series on Public Policy, 10,* 9–54.

Zellner, A., and F. Palm (1974). "Time Series Analysis and Simultaneous Equation Econometric Models." *Journal of Econometrics, 2,* 1 (May), 17–54.

Zellner, A., and F. Palm (1975). "Time Series and Structural Analysis of Monetary Models of the U.S. Economy." *Sankhyā, Series C, 37,* 2, 12–56.

Zimring, F. E. (1975). "Firearms and Federal Law: The Gun Control Act of 1968." *Journal of Legal Studies, 4,* 1 (January), 133–198.

NAME INDEX

SUBJECT INDEX

A   3

B   4

C   5

D   6

E   7

F   8